Cognitive Science Compendium, Volume 1

COGNITIVE SCIENCE COMPENDIUM, VOLUME 1

MIAO-KUN SUN
EDITOR

Nova Science Publishers, Inc.
New York

For permission to use material from this book please contact us:
Telephone 631-231-7269; Fax 631-231-8175
Web Site: http://www.novapublishers.com

LIBRARY OF CONGRESS CATALOGING-IN-PUBLICATION DATA

Cognitive Science Compendium/Miao-Kun Sun(editor).
p. cm.
ISBN 978-1-60456-597-3(v-1:hardcover)
1. Cognitive neuroscience.2.Congnitive science.I. Sun,Miao-Kun.
QP360.5.C647 2008
612.8'233—dc22
2008012912

Published by Nova Science Publishers, Inc. ≃ New York

CONTENTS

PREFACE: LEARNING COGNITIVE SCIENCES

At the center of the workings of the mammalian nervous system are the neural processes governing cognition: vital functions including perception, attention, memory, consciousness, language, judgment, reasoning, problem solving, and decision-making. Cognitive function is widely used to solve mathematics, economics, and spatial or mechanical problems. The ways through which problems are solved are also governed by the organisms' internal representation of the outside world. Perception, judgment, reason, problem solving, decision and overt response are impacted by expectations, or 'internal value', based on prior knowledge and experience. The obvious goals of cognitive sciences through investigation include: 1) a better understanding of the underlying mechanisms that define cognition, 2) a clearer picture of pathophysiology of cognitive deficits in a variety of disorders, and 3) more effective therapeutic treatments of cognitive diseases. Part of the cognitive science is also the development of artificial intelligence and computational theory of cognition.

We learn to understand neural/molecular devices of cognition through intensive investigation, both in humans and animal models. While important information about cognition has been obtained from studies in humans, both normal subjects and patients with cognitive deficits, the human brain is certainly not one observable organ. Fortunately, a majority of cognitive functions exists in other species. Animal models simulating most of the highest human cognitive functions with the exception of language are now available. In fact, much more information about cognitive sciences has been learned from detailed investigation, including a correlational approach, in a variety of animal models. For instance, cognition cannot exist without memory. Experience is stored in working or short-term memory and compared with knowledge retrieved from long-term memory. The memory processes, such as storage, search, consolidation/reconsolidation, retrieval, and extinction, can be grossly assessed using neuro-scientific tests, EEG measures, imaging, and clinical ratings in humans, and can also been defined in greater detail (synaptic plasticity, specific neural network and gene product involvements and mechanisms) in animal models. Many non-human species possess the general ability to reason, i.e., to infer, deduce, predict effects, and solve problems in a partially unknown situation. For example, cognitive mapping, a common type of cognitive test, is used to evaluate the ability of animals to combine information at the same time and/or in order to solve cognitive tasks.

We need to learn more about what and how cognitive functions are impaired or lost because of diseases or aging. Cognition is affected by a variety of factors, including development, integrity of neural/molecular network/cascades involved in cognition, stress, behavior, neurodegenerative disorders, state of consciousness, sleep, evolution of cognitive processes, and addiction and substance abuse. Cognitive disorders devastate millions of human lives. An interventional approach has been used successfully to determine the factors and to test whether the molecular/cellular changes recorded play a key role in the cognitive processes. The knowledge of precise molecular, cellular, and network basis of impaired cognition would have dramatic consequences on our ability to develop medications that are effective in the treatment of cognitive disorders.

Much more needs to be learned for medical and life scientists to restore normal cognitive functions. Better knowledge of the neural/molecular basis of cognition would make it possible to adjust more targeted therapeutic strategies. Cognitive neuropharmacology, the pharmacological study of cognitive processes, covers basic research and clinical research/needs and is becoming an active research field. Cognitive neuropharmacology consists of modifying cognitive abilities, together with the ability to dynamically process data that have already been acquired in order to infer new information, to find new behavioral or cognitive strategies, or to use acquired information in new contexts. Another line of cognitive neuropharmacology is the development of cognitive enhancers. Currently, cognitive neuropharmacology is in its infancy and its benefits will become apparent with time.

There has been a rapid increase in the interest of clinician, scientists, and the pharmacological industry in defining neural correlates and mechanisms that underlie cognition and cognitive impairments and in developing new classes of drugs designed especially for treatment of cognitive disorders including memory deficits. However, we still face two major challenges. First, cognitive disorders usually persist. What molecular and neural mechanisms might directly underlie the lasting cognitive changes? None of the neural and molecular events identified to date appear to persist nearly long enough to be responsible for the lasting cognitive changes. Second, how might such changes lead to the functional alterations in cognition?

Miao-Kun Sun, Ph.D.

In: Cognitive Science Compendium, Volume 1
Editor: Miao-Kun Sun
ISBN: 978-1-60456-597-3

Chapter 1

COGNITIVE DISRUPTION IN ANGELMAN SYNDROME: FROM MOUSE MODELS TO MOLECULAR MECHANISMS

Edwin J. Weeber[2], Jonathan M. Levenson[1] and J. David Sweatt[1,*]
[1] Department of Neuroscience, Baylor College of Medicine, One Baylor Plaza, Houston Texas, 77030
[2] Department of Molecular Physiology and Biophysics Kennedy Center for Research on Human Development Vanderbilt University Medical Center Nashville, TN 37232-0615

ABSTRACT

Gross morphological changes in the brain due to genetic abnormalities are often associated with severe mental retardation. This perhaps most obscures our ability to differentiate biochemical versus morphological alterations as the perpetrator behind the loss of cognitive ability in humans diagnosed with mental retardation. However, there are certain instances of human mental retardation syndromes in which morphological changes in the CNS are absent, but severe cognitive problems remain. One such identified form of mental retardation is Angelman Syndrome (AS). A unique facet of AS is that the *Ube3a* gene, responsible for AS, undergoes maternal imprinting in a brain-region specific manner. In AS, the product of maternal *Ube3a* gene is absent in the hippocampus, a brain region known to be involved in memory processing. Thus, AS results in a severe mental retardation state that arises from a delimited hippocampal lesion without any accompanying morpho-logical changes in the brain. Progress in understanding the molecular disruption in the hippocampus using a mouse model for AS has shed light on molecular components involved in human cognition and strengthened

* To whom correspondence should be addressed: J. David Sweatt, Ph.D., Baylor College of Medicine, Department of Neuroscience, One Baylor Plaza, S607, Houston, Texas 77030, jsweatt@bcm.tmc.edu

our view of the importance of the hippocampus in developing and maintaining cognitive ability in humans.

Keywords: Hippocampus, CaMKII, Protein Phosphatase, Autophosphorylation, Long-term Potentiation, Memory

INTRODUCTION

A simplified but useful way to think about the neurobiological basis of human cognition is to consider that the CNS comprises two basic components. The first is anatomical groups of neurons and the neural circuits they encompass. These are the information processing circuits mediating cognition and behavior. By-and-large these are laid down based on genetically encoded programs and in response to environmental signals early in the developmental process. The second component of cognition is the cellular and molecular processes that mediate behavior and behavioral change on a more dynamic basis. These processes predominate after the relevant circuits in which these cells and molecules reside have been established and solidified through development. Information processing in these domains occurs through the dynamic regulation of cellular response properties and via the intracellular signal transduction processes that allow computation at the molecular level. Thus, cognition is subserved by both the hard-wired circuitry that is the information processing machinery, and by the molecular and cellular processes that operate within this circuitry to control and modify its output on a moment-to-moment basis.

Mental retardation can result from either disruption of the neural circuitry underlying cognition or disruption of the cellular and molecular processes underlying cognition. Historically the anatomical and circuit-based disruptions have received the lion's share of attention by researchers in this field. This makes sense. The most readily identifiable mental retardation syndromes are embedded within a complex of recognizable and consistent morphological disruptions. In fact this is what defines a mental retardation syndrome – a cognitive deficiency that consistently presents in combination with other identifiable clinical features such as dysmorphic features. In addition, frank anatomical disruption of the CNS is often observable in these syndromes as well – direct evidence for a derangement of the information-processing circuitry of the CNS.

Molecular and cellular disruptions as a basis for mental retardation have received much less attention. In some cases mental retardation as a result of these mechanisms appears to occur in the complete absence of anatomical disruption, as in the case of certain "non-syndromic" diagnoses of mental retardation. However, even in the case of syndromic mental retardation where anatomical malformation is evident, it is important to keep in mind that disruption of cellular and molecular mechanisms can play an important role in cognitive decline.

Recent discoveries highlight the importance of this consideration in two particular instances. The first is Neurofibromatosis type 1 (NF1) - linked mental retardation. Recent work, largely from the laboratory of Alcino Silva, has demonstrated that disruption of a particular biochemical signaling cascade, the ERK/MAP Kinase cascade, plays a part in the cognitive defects in NF1 mental retardation (Costa et al., 2002; Costa et al., 2001). The second example will be the focus of this review – Angelman Syndrome (AS). Recent work from several laboratories has shown that AS is associated with disruption of an enzymatic signaling cascade that has been the focus of studies in the learning and memory field for many years – the calcium/calmodulin-dependent protein kinase II (CaMKII).

In this review we will first describe AS as it presents clinically. We then will describe a number of current findings concerning the molecular and cellular basis of cognitive dysfunction in AS based on studies of a genetically engineered mouse model of AS. These studies have implicated disruption of both CaMKII specifically and Long-term Potentiation (LTP) of synaptic transmission in general in the mental retardation associated with AS. Moreover, much of the disruption of these processes occurs in the absence of discernable disruption of neuroanatomical development, indicating that the locus of AS is at the cellular and molecular level. Therefore in the final part of the review we will describe the normal function of CaMKII in synaptic function, in order to explain how disruption of its normal functional roles may contribute to a loss of cognitive ability in AS.

UNDERSTANDING ANGELMAN SYNDROME

Angelman Syndrome (AS) occurs in one of every 15,000–20,000 births (Clayton-Smith, 1993) and is commonly accompanied by severe mental retardation, epilepsy, a puppet-like gait, dysmorphic facial features, a happy disposition with bouts of inappropriate laughter, hyperactivity, sleep disorders, and lack of speech (Buoni et al., 1999; Fryburg et al., 1991; Laan et al., 1999; Williams et al., 1995). Understanding the genetic aberration that defines Angelman syndrome (AS) begins in understanding how genes can be inherited in an active state from one parent and, alternatively, in an inactive state from the other parent. This phenomenon, referred to as genetic imprinting, is a poorly understood mechanism affecting a relatively few known autosomal mammalian genes (DeChiara et al., 1991; Guillemot et al., 1995; Rachmilewitz et al., 1992). During maternal imprinting, paternally allelic DNA methylation allows only maternal genes to be expressed in a certain subset of tissues (Glenn et al., 2000; Solter, 1998). Through the silencing of the paternal gene, the remaining functional genes in this locus are exclusively inherited maternally. The process of imprinting lends *Ube3a* to genetic and epigenetic mutations that are sufficient to confer a complete lack of the *Ube3a* protein in the imprinted brain regions. Nearly 70% of AS cases involve the de novo deletion (~4 Mb) on the maternal chromosome at 15q11-13 (Class I) (Kaplan et al., 1987; Magenis et al., 1990). These deletions contain a 2-3 Mb section of known imprinted genes of which *Ube3a* is located. 7-9% of AS cases are the result of imprinting mutations in which both the maternal and paternal copies undergo paternal-like methylation (Class III) (Buiting et al., 1995; Gabriel et al., 1999), and 2-3% show paternal unpaired disomy (UPD) (Class II) (Knoll et al., 1989; Malcolm et al., 1991), i.e., the inheritance of two copies of the paternal chromosome region with no maternal copies present. The remaining patients with clinically diagnosable AS do not fall into the aforementioned three groups but rather represent familial-derived mutations of one or more genes in the 15q11-13 critical region.

The *Ube3a* gene product was initially identified following viral studies of p53 tumor suppressor protein degradation by the oncogenic human papilloma virus (HPV). Researchers interested in HPV *E6* gene expression following infection found that the HPV E6 protein alone had no effect on p53, however in the presence of an associated protein p53 degradation was observed (Mietz et al., 1992; Waddell and Jenkins, 1998) (Harris, 1996; Park et al., 1997). This associated protein, referred to as the E6-associated protein or E6-AP, was found to be a 100 kDa protein that functioned as an ubiquitin ligase enzyme. Geneticists investigating the cause of AS eventually found that the AS gene was in fact *Ube3a*, which codes for E6-AP (Huibregtse et al., 1991; Huibregtse et al., 1993; Kishino et al., 1997; Matsuura et al., 1997). This was an incredibly fortuitous find since E6-AP ubiquitin ligase falls into the E3 category of ubiquitin ligases, which confers a high degree of substrate specificity on the targeted protein complex. Thus, it was believed that identifying the protein target of E6-AP would unlock the mystery surrounding the disorder. Once the gene for AS was discovered an intense search for the putative E6-AP protein targets involved in the disorder ensued. To date only 4 proteins have been identified as targets of E6-AP-dependent ubiquination. These include: the p53 tumor suppressor protein, the HHR23A protein, a human homologue to the yeast DNA repair protein Rad23 (Kumar et al., 1999); the multicopy maintenance protein (Mcm) 7 subunit involved in the initiation of DNA replication (Kuhne and Banks, 1998); and E6-AP which is a target for itself (Nuber et al., 1998). Unfortunately, no obvious connection can be made between these identified targets and the etiology of AS.

The AS Mouse Model

The lack of a clear connection between E6-AP function and AS, coupled with the scarce availability of AS post-mortem brain tissue, necessitated producing a mouse model for AS. Two mouse models were developed, each with a different strategy to disrupt *Ube3a*. The first mouse line, created by Gabriel and colleagues, utilized an Epstein-Barr virus latent membrane protein 2A (LMP2A) transgenic insertion, in effect deleting the entire equivalent AS critical region (Albrecht et al., 1997; Gabriel et al., 1999). This is essentially what is seen in humans that exhibit the most common genetic defect leading to AS; a ~4 mb deletion in the 15q11-13 region. Therefore, a comparable deletion to the homologous region in mouse chromosome 7, an area exhibiting a similar gene complement, was made. Alternatively, the mouse model developed by Jiang and colleagues utilized a null mutation in *Ube3a* via a single gene knockout (Jiang et al., 1998). While both of these mouse models effectively disrupt *Ube3a* there exists some controversy over whether *Ube3a* deficiency in and of itself is responsible for the manifestation of AS, or if the full phenotype is due to the combined affect of disrupting several genes in the human chromosome 15q11-13 critical region (Lee and Wevrick, 2000).

A phenotypic assessment in patients with the *de novo* deletion (class I) exhibit a profile more true to what is commonly considered to be the clinically classical AS. These patients show the full realm of AS characteristics including microcephaly and hypopigmentation (Saitoh et al., 1994; Smith et al., 1996). In contrast, patients where AS arises from specific *Ube3a* mutation (class II) or *Ube3a* mis-methylation (class III) results in a patient with fewer instances of hypopigmentation, microcephaly, and seizure (Burger et al., 1996; Gillessen-Kaesbach et al., 1999; Saitoh et al., 1999; Smith et al., 1997). This indicates that AS caused by Ube3A disruption is exacerbated by the disruption of genes located at the same locus. Among these genes is the $GABA_A$ receptor β3 subunit (GABRB3) (DeLorey et al., 1998; DeLorey and Olsen, 1999). There is recent evidence that the inhibitory neurotransmitter

GABA is involved in the suppression of seizures and may explain the reduced seizure in class II and III AS patients. Supporting this hypothesis, a lower instance of seizure activity in the Jiang et al., mouse model compared to the other mouse model was observed. Since nearly all available evidence points to *Ube3a* as the locus for AS, and given the inherent difficulties in interpreting the effects of multiple gene deletions in mice, the consensus was that the single *Ube3a* null mutation mouse would provide a better instrument to investigate the molecular mechanisms underlying the AS etiology. The following description is of experiments performed using the Jiang et al. AS mouse model.

It should be noted that the *Ube3a* maternal deficient (m-/p+) mouse is also valuable for studying the specific patterns and mechanisms of maternal *Ube3a* imprinting in the CNS of developing and adult animals. The utility of the *Ube3a* m-/p+ mutant, and other mouse models for human CNS disorders, becomes clear when considering the difficulty in a thorough determination of the imprinting regions in post-mortem AS patient brain tissue. The identification of imprinted regions through the use of immunohistochemical techniques in mice is commonly performed, and relatively easy to interpret. Importantly, both AS patients and *Ube3a* deficient mice exhibit normal brain histopathology, suggesting that maternal E6-AP deficiency does not effect neuronal development or organization. In adult mice the detection of *Ube3A* mRNA revealed high expression levels in areas CA1 and CA3 of the hippocampus, basal ganglia, cerebellum, cerebral cortex and the periglomerular, granual and mitral cells of the olfactory bulb. This was compared to maternal deficient mice that showed no detectable hippocampal expression, reduced cerebellar expression with no detectable expression in the purkinje cell layer, and an overall reduced expression in the olfactory region associated with a lack of detectable expression in the mitral cell layer (Albrecht et al., 1997; Jiang et al., 1998). Brain-specific maternal *Ube3a* expression within the mouse model was consistent with the observed phenotype of severe seizure, motor dysfunction, and associative learning and LTP deficits. These results taken together strengthen the general belief that the AS etiology is predominately due to the deficiency of expression from the maternal *Ube3a* allele. The studies on maternal imprinting in the mouse reemphasize the point that the patterns of imprinting are consistent with cognitive disruption primarily, if not exclusively, due to the lack of *Ube3a* expression in the hippocampus of AS patients.

The initial characterization of the maternal deficient *Ube3a* null mouse mutant (referred from here on as the *Ube3a* m-/p+ mutant) revealed striking similarities to the human phenotype (Jiang et al., 1998). These phenotypic similarities are more easily discernable when sub-divided into categories of physical and physiologic characteristics.

Physical Similarities

The brains of *Ube3a* m-/p+ mice showed an overall normal morphology, with no obvious abnormalities in any brain regions including regions of maternal imprinting. However, by 18 days of age *Ube3a* m-/p+ mice were observed to have smaller bodies and brains than wild type mice. This size difference was seen in both the cortex and cerebellum of the brain. Motor function and coordination was assessed using hind-paw footprint analysis, the bar crossing test and the accelerating rotorod test. Each of these tests confirmed a distinct and significant motor deficit in the *Ube3a* m-/p+ mice, which mimics the tremor, ataxia and motor incoordination described in human AS patients. In addition, seizure in AS patients is very prominent and found to affect greater than 90% of diagnosed AS patients. Likewise, seizures in *Ube3a* m-/p+ mice could easily be induced through audiogenic means by simply running an object vigorously over the metal grate lid of the mouse's home cage.

Of course physical changes are more easily quantifiable and tend to be a more convincing argument for the recapitulation of a human "mouse model". However, there are several ways to measure the behavioral correlates to human cognitive ability in the mouse by assessing its learning ability through behavioral testing, and function at a synaptic level through electrophysiologic techniques.

Cognitive Similarities

In humans the hippocampus is intimately involved in spatial memory processes and is linked to numerous cortical regions. Acquisition of explicit memory involves the hippocampus as part of a polymodal sensory integration scheme that processes visual, auditory and somatosensory inputs (Figure 1). A great deal of evidence now implicates the hippocampus as an important brain structure involved in cognitive processing in humans. The maternal imprinting effects on the hippocampus in AS thereby translate into one of the most devastating characteristics of AS in humans; severe cognitive deficits.

How can one test for cognitive deficits in mice? Although determining cognition in mice is problematic on a number of levels, we can test *Ube3a* m-/p+ mice by assessing their ability for associative learning, specifically using the fear conditioning learning paradigm. In this learning paradigm an aversive stimulus (unconditioned stimulus, US), consisting of a mild foot shock, is paired twice with an auditory component (conditioned stimulus, CS). The animals are then tested 24 hours following training through the reintroduction of the mouse into the conditioning chamber or placing the mouse in a novel context while repeating the CS. Learning and memory is assessed by observing freezing behavior, which is a learned behavioral fear response to either the association to the CS or context. Although both contextual and cued fear-conditioned learning is dependent on the proper function of the amygdala, only the contextual learning is hippocampus-dependent. As expected, the *Ube3a* m-/p+ mutant mice showed normal associative learning to the CS, but had a severe contextual fear conditioning deficit.

Physiologic Similarities

The lack of hippocampal *Ube3a* expression coupled with the severe hippocampus-dependent learning deficit made the characterization of hippocampal synaptic function in the AS model mice essential. The hippocampus is separated into three distinct synaptic pathways; the perforant path, the mossy fiber synapse and the Schaffer collateral synapses (Figure 2A). Of these three major synaptic connections the Schaffer collaterals in area CA1 of the hippocampus is by far the most studied. It is in this region that synaptic transmission and plasticity were characterized in the *Ube3a* m-/p+ mice. Jiang and collegues found that synaptic transmission was unaltered in the AS mouse. This indicated that overall connectivity

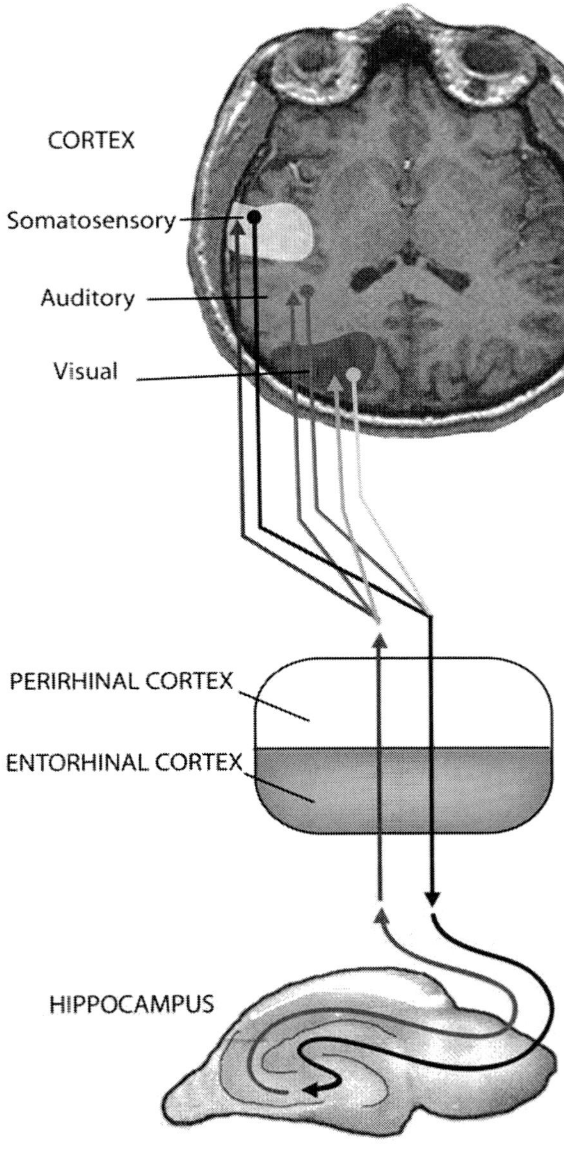

Figure 1. Diagram representing the multimodal nature of the hippocampus, which allows two-way processing to higher-order cortical areas, such as the somatosensory, auditory and visual cortices. (Brain fMRI of Edwin Weeber, courtesy of the Human Neuroimaging Laboratory, Baylor College of Medicine)

and single stimulus synaptic function were normal. Next, synaptic plasticity induced with high frequency stimulation (HFS) was tested. It was shown that *Ube3a* m-/p+ mice exhibited a severe deficit in LTP induction. Although controversial, parallels can be drawn between the mechanisms of synaptic strengthening which occur during hippocampal LTP and those postulated as occurring during hippocampus-dependent learning and memory events. Thus, a deficit in hippocampal LTP is consistent with a deficit in learning and memory. To reemphasize this important point, the uniqueness of a specific hippocampal lesion in the AS mouse makes the relevance and information from investigation of the hippocampus essential for its elucidation of possible mechanistic alteration caused by maternal *Ube3a* deficiency in humans. Also, the observation that AS mice with a specific hippocampus lesion exhibit learning and LTP deficits, supports the current dogma that the hippocampus is essential for normal cognitive function in humans.

AS Mouse Hippocampal Physiology 'Under the Microscope'

It took nearly three years for our group to identify a specific biochemical alteration in the AS mouse. Initially, the strategy employed identification of proteins that exhibited: 1) A lack of ubiquitination in the AS mouse compared to normal mouse brain protein complement 2) Increased levels due to lack of degradation through the ubiquitin proteosome pathway. Although sound in design, identifying short-lived ubiquinated proteins proved to be problematic, and no gross changes in protein levels were observed in AS mouse imprinted areas. Being grounded in investigating the biochemistry of learning and memory processes and the mechanisms involved in long-term potentiation forced us to think about the evidence in hand, and where the potential biochemical lesion might lie. We felt that the key was in understanding the deficit in hippocampal LTP. The initial studies by Jiang et al., were only a superficial examination of hippocampal LTP induction. Thus, with the groundwork laid by Jiang and his collegues, we pursued a more in-depth analysis of hippocampal synaptic plasticity that revealed new and interesting implications of *Ube3a* deficiency in the hippocampus of AS mice.

To better understand the basis of the CA1 LTP deficit in the AS mouse we varied the ways in which LTP was induced (Figure 2B). A standard LTP-inducing stimulation consisting of a 1 second, 100 Hz stimulation elicits a long-lasting increase in synaptic plasticity. Alternatively, increasing the number of high frequency trains of stimulation can produce a potentiation that is substantially higher in magnitude and longer lasting. Manipulating these variables in the LTP induction protocol is often used to determine the efficacy or potency of an applied drug or, as in this case, evaluate the severity or penetrance of the *Ube3a* deficiency. To test the possibility that the LTP deficits in m-/p+ mutants were due to an increase in the threshold of LTP induction, LTP was induced with a saturating amount of HFS designed to give the maximum potentiation in a given hippocampal slice (Figure 2B). We found that the saturating HFS could rescue the LTP deficit in the AS mouse. What was the mechanism by which the LTP deficit could be over come? The induction of LTP is highly dependent on upon an influx of postsynaptic Ca^{++}. Increasing the number of stimulations would increase the amount of Ca^{++} influx significantly and could explain the LTP rescue. The question then would be whether the derangement of LTP induction was up-stream (receptor activation) or down-stream of calcium influx. The major neurotransmitter-activated calcium channel at the CA1 hippocampal synapse in the N-methyl-D-aspartate receptor (NMDAR). To determine if the increase in LTP threshold observed in m-/p+ *Ube3a* mutants was due to insufficient postsynaptic NMDA receptor-dependent Ca^{++} influx, slices were stimulated with repeated very high frequency stimulation in the presence of the NMDA receptor antagonist DL-2-Amino-5-phosphonopentanoic acid (AP5). Under these conditions the LTP deficit in the *Ube3a* m-/p+ mice remained (Figure 2B). The inability of m-/p+ mutants to achieve NMDA receptor-independent LTP induction suggested that the LTP deficit, and importantly the potential site of hippocampal synaptic dysfunction, resided down-stream of calcium influx. The studies described above set the stage for the investigation into the molecular derangements that could be responsible for the cognitive loss in AS patients.

MOLECULAR CHANGES IN THE AS MOUSE

We hypothesized that the deficit in LTP and learning was biochemical in origin. The lack of any morphological changes in the CNS coupled with the ability to rescue LTP suggested that a disruption in a biochemical signal transduction pathway was the culprit. The rescue of LTP also suggested that the mechanism responsible for proper synaptic function were present, but the system needed to be pushed, (i.e. during repeated HFS) in order for the system to function at a level for LTP induction to occur. More importantly, the in-depth electrophysiology described above reduced the number of proteins to be examined from the greater than 10,000 present in the CNS to just a handful. The proteins chosen to be examined

A

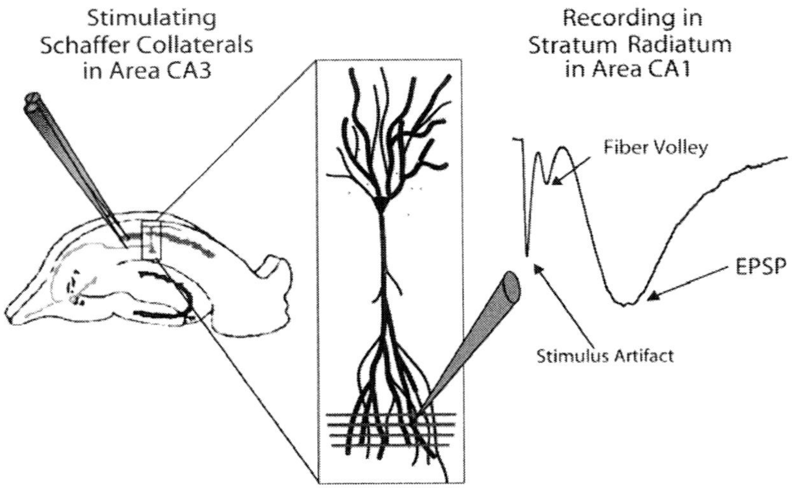

Figure 2. (A) Schematic drawing of the process of field recording at the Schaffer collateral synapse in area CA1 of the hippocampus, and the resulting waveform of the population excitatory postsynaptic potential (pEPSP). The presynaptic stimulation of area CA3 (fiber volley) results in the combined activation of a population of postsynaptic neurons in area CA1. This depolarization, measured in the dendritic field represents itself as the slope of the pEPSP. Determination of the slope of the pEPSP before and after high frequency stimulation provides a quantifiable measurement of the amount of potentiation incurred.

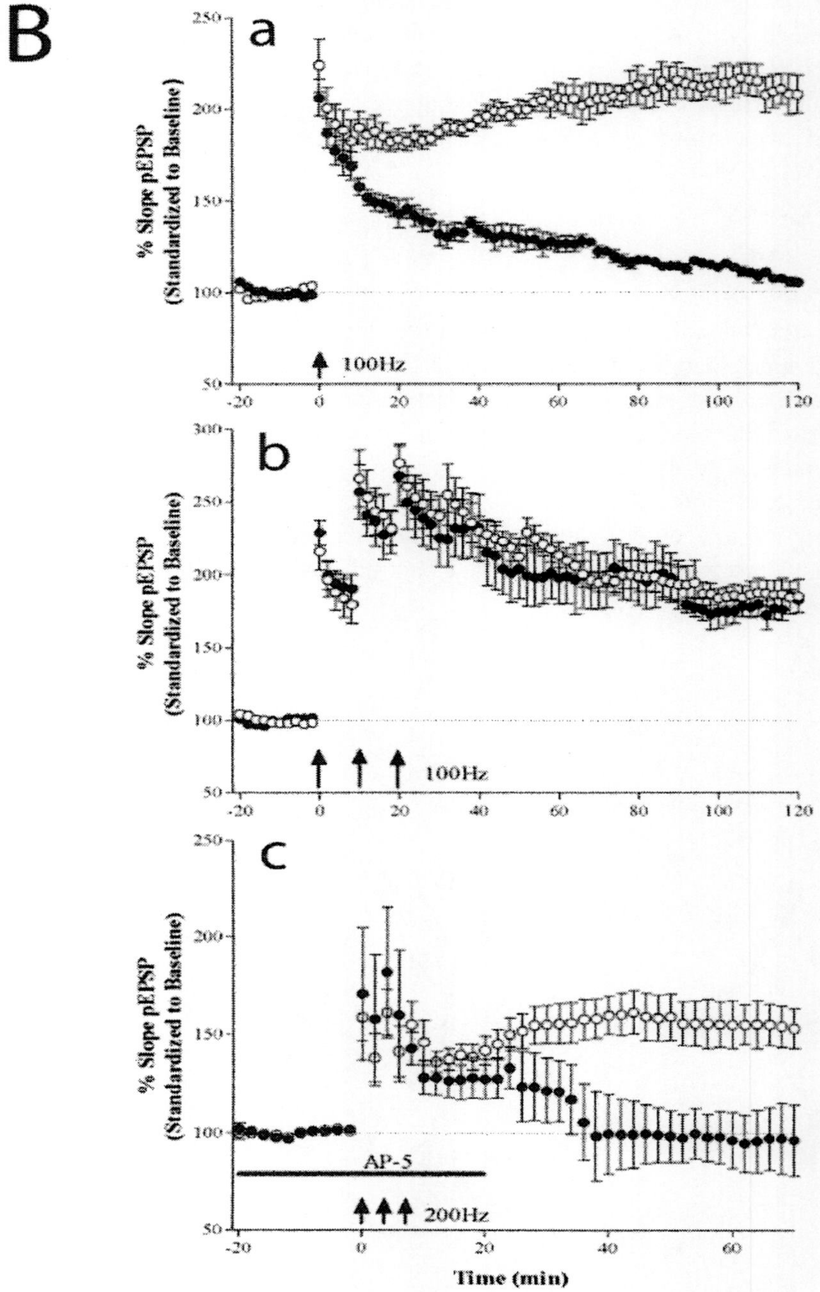

Figure 2. (B) Electrophysiology studies on *Ube3a* m-/p+ Angelman Syndrome mice (closed circles) compared to wild type mice (open circles). Different LTP inducing protocols are used including a) Two 1 second trains of 100 Hz stimulation (modest stimulation) b) LTP induced with saturating HFS (6 x 1 second, 100 Hz stimulation), and c) NMDA receptor-independent LTP induced with 3 trains of 200 Hz high frequency stimulation in the presence of AP-5. Adapted from (Weeber et al., 2003).

closer had to meet the following specific criterion: 1) Known to be involved in LTP induction processes. 2) Known to be involved in learning and memory processes. 3) Activated or modified in the presence of Ca^{++}. Again, our thinking was that *Ube3a*-dependent changes in protein levels of one or more of these proteins would be the key to unlocking the mystery surrounding *Ube3a* deficiency. After an arduous investigation of a number of protein kinases no alterations in protein levels were found. Using a similar reasoning as described above, but also with a mixture of frustration and desperation, we began to look at the phosphorylation states of these protein kinases. The important regulatory actions of phosphorylation in neuronal function, plasticity and learning are well established with synaptic protein kinases. Moreover, there are several examples of protein phosphorylation acting as a molecular switch to target a protein for ubiquitination and degradation through the proteosome pathway (Basu and Haldar, 2002; Zheng et al., 2002). To test this hypothesis, we examined the level of phosphorylated PKC, PKA, ERK and CaMKII in the AS mouse model. No changes were seen in the phosphorylation levels of PKC, PKA or ERK (p42) isolated from hippocampal homogenates of *Ube3a* m-/p+ mice. However, a significant increase was detected in phospho-αCaMKII at the autophosphorylation site Threonine 286 (Thr^{286}) (Weeber et al., 2003; Figure 3). This single observation represented the first identified biochemical alteration in the AS mouse model that could potentially explain the learning and LTP deficits in the mouse and the cognitive deficits in human AS. To help explain the implications of CaMKII misregulation, the next section will address what is currently known about CaMKII regulation and its importance in normal signaling in the mammalian CNS.

THE NATURE OF CAMKII

The plasma membrane presents a unique challenge in coordinating intracellular physiology with external stimuli. Most changes in the external environment cannot be directly transduced through the plasma membrane. In biological systems, external stimuli affect intracellular physiology by activating a variety of second messenger signaling systems. These second messengers come in a variety of forms, including molecules like cyclic adenosine monophosphate (cAMP), arachidonic acid (AA), reactive oxygen species (ROS) and perhaps the most ubiquitous, Ca^{++}. Second messengers modulate the activity of several intracellular enzymes, causing immediate changes in intracellular physiology. Most second messenger signaling pathways, if activated with the proper spatiotemporal patterns, can elicit long-lasting changes in cellular physiology, which are mediated by changes in gene transcription.

Formation of long-term memory requires the transduction of distinct patterns of synaptic activity into long-term changes in neuronal physiology (Berry and Thompson, 1978; Givens, 1996; Landfield et al., 1972; Larson and Lynch, 1986; Larson and Lynch, 1989; Larson et al.,, 1986; O'Keefe and Recce, 1993; Seager et al., 2002; Winson, 1978). Flux of Ca^{++} through the plasma membrane and into the cytoplasm is believed to mediate much of the early intracellular signaling events responsible for establishing long-term changes in synaptic efficacy that are responsible for establishing long-term memory (Selcher et al., 2002). Once elevated in the cytoplasm, Ca^{++} can impinge on a number of signaling pathways.

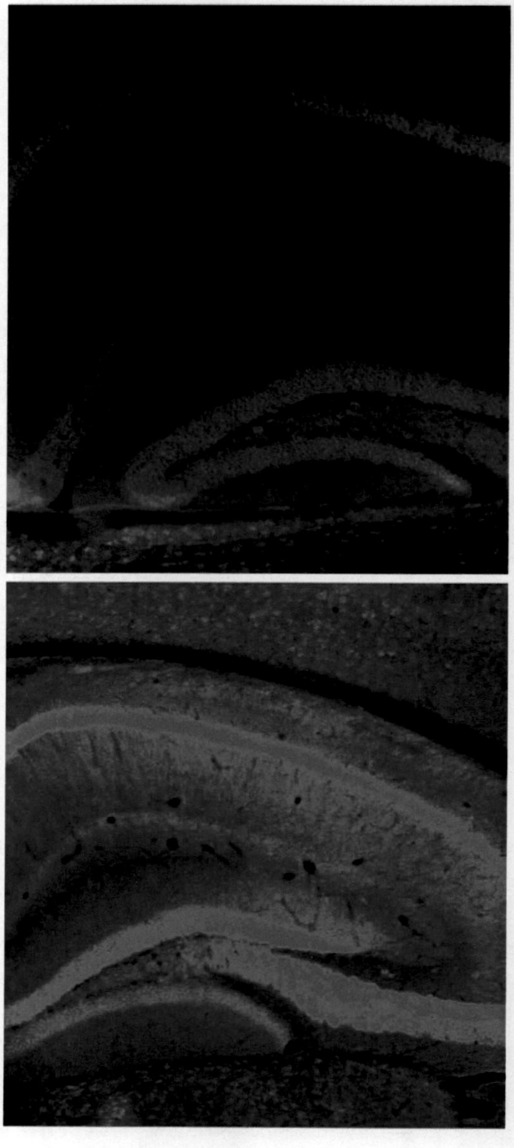

Figure 3. Increased immunoreactivity to Phosphorylated CaMKII at Thr^{286} in Angelman mouse hippocampus. Phosphorylated CaMKII at Thr^{286} was detected immunohistochemically in the hippocampi of wild-type (top panel) and Angelman (lower panel) mice. Increases in immunoreactivity are seen in stratum pyramidali, stratum oriens, and stratum radiatum of CA1 and CA3 as well as the molecular layer and granule cell layer of the dentate gyrus. 100 X magnification of the hippocampus. Adapted from (Weeber et al., 2003).

For Ca^{++} to be an effective regulator of cellular physiology, proteins sensitive to intracellular Ca^{++} concentration must exist in the cytoplasm at sites where Ca^{++} flux is most likely to occur. Some enzymes contain Ca^{++} binding sites and are directly regulated by changes in intracellular Ca^{++} concentration. Examples of these types of proteins are Ca^{++}-activated K^{+} ion channels and some forms of protein kinase C. Other Ca^{++}-binding proteins serve to couple changes in Ca^{++} concentration with modulation of enzymatic activity. An example of this second class of Ca^{++}-binding protein is calmodulin. Calmodulin is highly expressed in every tissue, and serves as the major intracellular receptor for Ca^{++} (Chin and Means, 2000). In addition, calmodulin has been shown to bind to and regulate a number of proteins in a wide variety of organisms (Chin and Means, 2000).

The initial stages of long-term memory formation are typified by the activation of several protein kinase cascades. One class of kinases whose activity is regulated by calmodulin are the Ca^{++}-calmodulin dependent protein kinases (CaMK). These protein kinases must be bound by Ca^{++}-activated calmodulin in order to function. Four CaMKs have been identified to date (Hudmon and Schulman, 2002). However, CaMKII is incredibly abundant in the CNS and accounts for 1% of the total protein in the forebrain and 2% in the hippocampus (Erondu and Kennedy, 1985), making it the most abundant CaMK in the brain.

CaMKII represents a family of serine / threonine protein kinases. There are 4 genes coding for CaMKII, designated α, β, δ, and γ. Each CaMKII message undergoes alternative splicing to produce several distinct isoforms (Bayer et al., 1998; Bayer et al., 1996; Bennett and Kennedy, 1987; Brocke et al., 1995; Bulleit et al., 1988; Hanley et al., 1987; Hoch et al., 1999; Karls et al., 1992; Kwiatkowski and McGill, 1995; Lin et al., 1987; Nghiem et al., 1993; Schworer et al., 1993; Srinivasan et al., 1994; Takeuchi and Fujisawa, 1998; Tobimatsu and Fujisawa, 1989; Tobimatsu et al., 1988; Tombes and Krystal, 1997; Urquidi and Ashcroft, 1995). The basic structure of the CaMKII enzyme includes a catalytic domain, a regulatory domain and an association domain (Figure 4). The catalytic domain is responsible for enzymatic activity of the CaMKII enzyme. The regulatory domain is involved in inhibition of the catalytic domain in the absence of Ca^{++}. The association domain is responsible for mediating protein-protein interactions. CaMKII does not exist in the cytoplasm as a monomer, but rather as a multimeric enzyme. A typical CaMKII holoenzyme consists of 8 – 12 α and β CaMKII molecules, depending on brain region (Dosemeci et al., 1999; Kanaseki et al., 1991; Kolodziej et al., 2000). The holoenzyme is arranged in a hexameric ring, with the catalytic, ATP-binding and Ca^{++}-calmodulin domains protruding out from the central core structure (Dosemeci et al., 1999; Kanaseki et al., 1991; Kolodziej et al., 2000). Assembling CaMKII into multimeric enzymes allows for more fine-tuning of Ca^{++} sensitivity (De Koninck and Schulman, 1998), provides a unique structure for protein-protein interactions (Walikonis et al., 2001), and provides a convenient mechanism for autophosphorylation to occur (Figure 4).

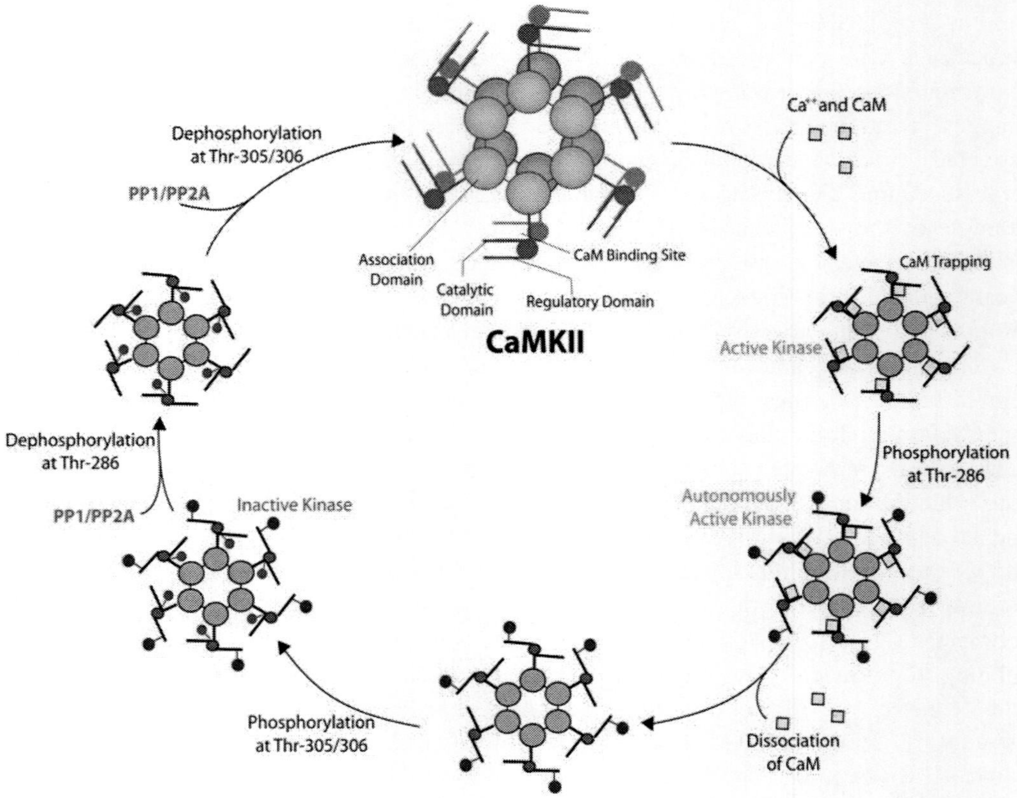

Figure 4. Regulation of CaMKII activity. CaMKII exists in the cytoplasm as stack of 2 hexmeric rings. The catalytic, ATP-binding and regulatory domains all protrude from the central ring. In its native state, CaMKII is inactive. CaMKII is activated by binding with Ca^{++}-calmodulin, which causes a conformational change in the molecule exposing the catalytic domain. After activation, CaMKII can undergo autophosphorylation at Thr^{286}. Autophosphorylation at Thr^{286} blocks association of the regulatory with the catalytic domain, resulting in an enzyme that is active in the absence of Ca^{++}-CaM. Upon dissociation of Ca^{++}-CaM, CaMKII undergoes another autophosphorylation at $Thr^{305/306}$, which renders the kinase inactive. Through the action of various phosphatases, CaMKII is dephosphorylated at Thr^{286} and $Thr^{305/306}$, resetting the kinase to its basal, inactive state.

CaMKII in Synaptic Plasticity

Initial studies of CaMKII's involvement in memory formation were performed using an *in vitro* model for memory formation: long-term potentiation (LTP) of the Schaffer-colateral synapses in area CA1 of the hippocampus. The Schaffer-Collateral synapses are excitatory, and use L-glutamate as their excitatory neurotransmitter (Bramham et al., 1990; Nadler et al., 1976). It has long been appreciated that induction of LTP requires high-frequency synaptic activity (Alger and Teyler, 1976; Bliss and Lomo, 1973) that results in activation of 2 types of glutamate receptors on postsynaptic CA1 pyramidal neurons: AMPA and NMDA receptors. AMPA receptors have an ion-channel that, when activated by glutamate, allows Na^{+} to enter the cell, resulting in depolarization of the postsynaptic membrane. NMDA receptors are also coupled to an ion channel, however under resting conditions this ion channel is blocked by a molecule of Mg^{++} (Mayer et al., 1984; Nowak et al., 1984). Once the membrane is depolarized, the Mg^{++} block is removed and Ca^{++} ions flow through the NMDA-receptor ion channel into the cytoplasm (Mayer et al., 1984; Nowak et al., 1984). The resulting increase in intracellular Ca^{++} mediated by the NMDA-R is the primary signal for induction of LTP (Harris et al., 1984; Lynch et al., 1983; Malenka et al., 1988; Morris et al., 1986). The increase in cytoplasmic Ca^{++} concentration mediated by NMDA-Rs is relatively short-lived, and therefore cannot account for the long-term increase in synaptic efficacy observed after induction of LTP. Therefore, it has been proposed that the initial increase in Ca^{++} leads to the activation of several downstream kinases, which modify the function of the various synaptic proteins and induce long-lasting changes in synaptic efficacy (Bading and Greenberg, 1991; Malinow et al., 1988; O'Dell et al., 1991). This intimate relationship between NMDA receptor mediated Ca^{++} influx and CaMKII activation brings us back to the LTP deficit seen in the AS mouse. Saturating amounts of high frequency stimulation, causing an optimum amount of Ca^{++} influx, was used overcome the LTP deficit. The signaling and autoregulatory aspects of Ca^{++} on CaMKII will become more apparent in the following section.

The regulation of CaMKII through autophosphorylation mechanisms is extraordinarily dynamic (Figure 4). In its native state the CaMKII catalytic domain is associated with the regulatory domain, rendering the enzyme inactive. The regulatory domain inhibits the catalytic domain primarily by acting as a pseudosubstrate (Yang and Schulman, 1999), however the regulatory domain also appears to interfere with ATP binding (Colbran et al., 1989; Kelly et al., 1988; Smith et al., 1992). Ca^{++}-calmodulin activates CaMKII by disrupting the association between the catalytic and inhibitory domains (Colbran et al., 1989). Once bound to Ca^{++}-calmodulin, CaMKII undergoes autophosphorylation at amino acid residue Thr^{286}, a residue located on the inhibitory domain (Colbran et al., 1989). Phosphorylation of Thr^{286} has 3 consequences for CaMKII function: the kinase becomes autonomously active (Saitoh and Schwartz, 1985), the affinity of Ca^{++}-calmodulin for CaMKII increases 1000-fold (Meyer et al., 1992), and a site becomes exposed allowing CaMKII to bind NMDA glutamate receptors (Bayer et al., 2001; Leonard et al., 1999; Strack and Colbran, 1998). Once the initial calcium signal has been sent and the CaMKII has completed its job of phosphorylating target proteins, it must be switched off and reset in order to be ready for the next wave of calcium influx. This is accomplished through another autophosphorylation mechanism at a different site on the regulatory domain of the enzyme. Thus, once Ca^{++}-calmodulin dissociates from an autonomously active CaMKII, another autophosphorylation at Threonine 305 and 306 ($Thr^{305/306}$) occurs on the alpha and beta subunits, respectively. All of the potential functional consequences of phosphorylation at $Thr^{305/306}$ are relatively poorly understood. However, it has been shown that $Thr^{305/306}$ phosphorylation can render the CaMKII enzyme inactive (Colbran, 1993a). In addition, the $Thr^{305/306}$ resides in the Ca^{++}-calmodulin binding site of CaMKII and inhibits subsequent binding of Ca^{++}-calmodulin to CaMKII (Hanson and Schulman, 1992b). Phosphorylation of $Thr^{305/306}$ can also occur in the absence of activation of CaMKII, and possibly preventing subsequent activation by Ca^{++}-calmodulin (Colbran, 1993b; Hanson and Schulman, 1992b). This management of activation, alterations in Ca^{++} and calmodulin sensitivity and autophosphorylation is highly orchestrated. Thus, even slight disruptions of this process potentially could have devastating affects on synaptic function and cognitive ability.

Using many different techniques, several labs have provided strong evidence for the involvement of CaMKII in the induction of synaptic plasticity. Initial studies using pharmacological inhibitors of CaMKII or calmodulin observed complete loss of LTP induction in the absence of CaMKII function (Malenka et al., 1989; Malinow et al., 1989). Later studies exploiting transgenic mouse models demonstrated that loss of CaMKII expression is associated with either a partial or complete loss of LTP induction (Hinds et al., 1998; Silva et al., 1992b). These studies, however, only give a cursory picture of the underlying necessity for CaMKII in synaptic plasticity. To truly appreciate this kinase an understanding of the process governing its regulation is needed.

Regulation of CaMKII Activity

How is autophosphoylation tied to the regulation of synaptic plasticity? Recent studies have examined the role of the Thr^{286} autophosphorylation site in CaMKII-dependent synaptic function. Genetically modifying the Thr^{286} site to an alanine (T286A), which prevents phosphorylation at the Thr^{286} site, results in complete loss of LTP induction (Giese et al., 1998). Genetically modifying the Thr^{286} site to an aspartate (T286D), which mimics phosphorylation, also leads to impairment of LTP induction (Bach et al., 1995; Rotenberg et al., 1996). More refined studies suggest that low levels of the autonomously active kinase facilitate, while high levels of autonomously active kinase inhibit, LTP induction (Bejar et al., 2002).

Much research indicates that mis-phosphorylation at Thr^{286} can potentially result in loss of synaptic plasticity; but what about the other autophosphorylation sites? As mentioned above, autophosphorylation can also occur at the $Thr^{305/306}$ site on the Alpha and Beta CaMKII subunit respectively. In fact, recent studies investigating the physiologic role of this phosphorylation site in LTP induction found that mutating the site from a threonine to an alanine (T305/306A), which blocks phosphorylation, increases the amount of CaMKII in the postsynaptic density (PSD) and enhances LTP induction (Elgersma et al., 2002). Furthermore, mutating the threonine to an aspartate (T305/306D), which mimics phosphorylation, decreases the amount of CaMKII in the PSD and blocks induction of LTP (Elgersma et al., 2002). Together, these results indicate that CaMKII is critical in the induction of LTP and must be active in a precise spatio-temporal pattern (see model in Figure 4). Were the T305/306 phosphorylation sites also altered in our AS mutant animals? Through a collaborative effort with Alcino Silva's laboratory at UCLA, we determined that the hippocampus of AS mice showed a significant increase in CaMKII phosphorylated at Thr^{305} and Thr^{306}. These results confirmed that the CaMKII enzyme in the maternal imprinted region of the hippocampus in our AS mouse model exhibited numerous alterations in the phosphorylation state and therefore in the activity, regulation and, as we will discuss in the next section, the cellular location of the enzyme.

CaMKII and the Post Synaptic Density

Derangements at $Thr^{305/306}$ can have effects other than changes in enzyme activity, such as influencing the subcellular association of the holoenzyme (Figure 5). Recently, a T305D CaMKII transgenic mouse was generated which mimics persistent inhibitory phosphorylation. This mutation is capable of decreasing the association of CaMKII with the structure known as the post-synaptic density (PSD). The PSD is a specialization of the cytoskeleton at the synaptic junction which allows for the assembly and clustering of proteins necessary for proper synaptic function. Alterations in Thr^{305} phosphorylation can detrimentally affect LTP and learning (Elgersma et al., 2002). In contrast, blocking inhibitory phosphorylation at $Thr^{305/306}$ increases CaMKII in the PSD. These remarkable results raise the interesting question of whether increased Thr^{305} phosphorylated αCaMKII in our *Ube3a* m-/p+ mice could result in decreased PSD-associated CaMKII. To test this hypothesis, we examined levels of total and autophosphorylated CaMKII from PSDs isolated from the hippocampus of *Ube3a* m-/p+ mice. We found that the total PSD-associated CaMKII was significantly reduced to nearly half of that found in wild-types. These results suggest that in the AS mice there are sufficient amounts of phosphorylated Thr^{305} αCaMKII to cause a considerable translocation of the CaMKII complement away from the PSD. In fact, when we examined the phosphorylated state of the remaining CaMKII in the PSD, we found a 2.5-fold increase P-Thr^{286} CaMKII, but no change in P- Thr^{305} CaMKII. This supports the model that Thr^{286} phosphorylation promotes PSD association by binding to the NMDAR and is in agreement with the observation that the association with the NMDAR effectively blocks inhibitory phosphorylation at Thr^{305} (Bayer et al., 2001). Furthermore our results indicate that Thr^{305} phosphorylation promotes PSD dissociation (Elgersma et al., 2002). The misphosphorylation of CaMKII at these sites does not preclude there being a population of unaffected CaMKII available for normal signaling and activation in the AS mouse. We hypothesize that there must be some complement of normal CaMKII since not all of the CaMKII is phosphorylated and we can rescue the LTP with saturating HFS. This again underscores the exquisitely orchestrated process underlying CaMKII regulation and the consequences of misregulation through which synaptic plasticity can be disrupted.

CaMKII in Long-Term Memory

Synaptic plasticity in the hippocampus is believed to be involved in the formation spatial memories. Disruption of CaMKII function has profound affects on hippocampal LTP, therefore one would expect disruption of CaMKII to have dramatic effects on long-term memory formation. Animals lacking the CaMKII gene show profound deficits in spatial learning tasks (Bach et al., 1995; Silva et al., 1992a). In addition, both sites of autophosphorylation appear to be involved in formation of long-term memory. Deficits in spatial learning tasks are observed in animals that have either the T286A or T286D mutation (Bach et al., 1995; Bejar et al., 2002; Giese et al., 1998). Moreover, animals possessing the T305/6A mutation exhibited enhanced spatial learning while animals with the T305D mutation had severe deficits in spatial learning (Elgersma et al., 2002). All of these observations indicate that CaMKII is critical for the proper formation of hippocampus-dependent spatial memory.

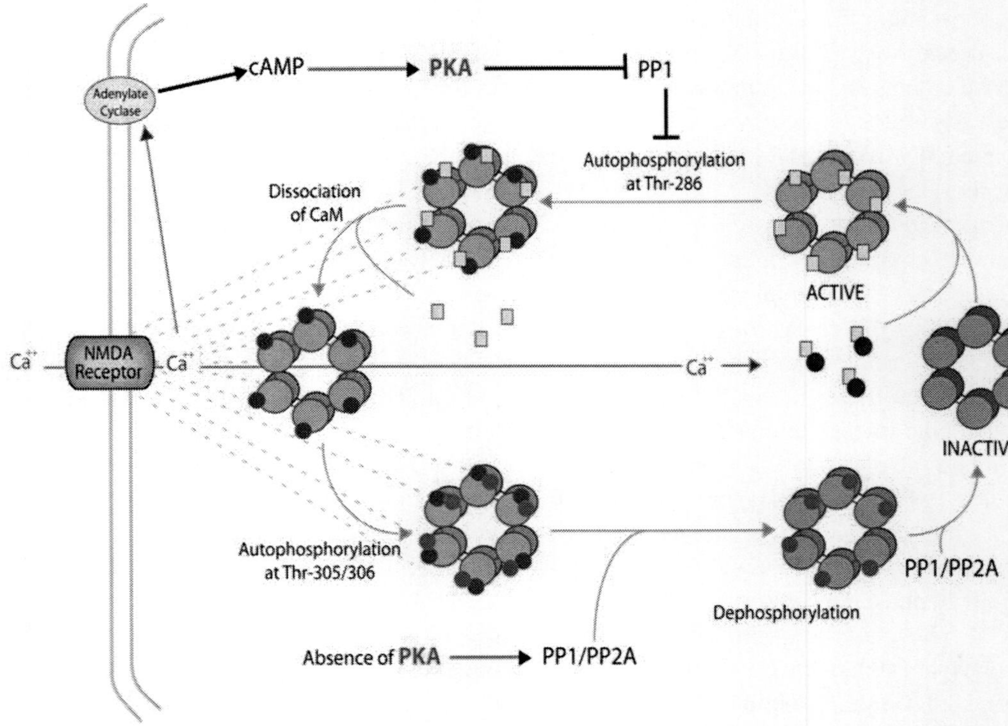

Figure 5. Intracellular Targeting of CaMKII. Inactive CaMKII is located in the cytoplasm. Once activated by Ca^{++}-CaM and phosphorylated at Thr^{286}, CaMKII translocates into the postsynaptic density, where it associates with NMDA-receptors. Autophosphorylation at $Thr^{305/306}$ decreases the affinity of CaMKII for NMDA-receptors, leading to the translocation of CaMKII back into the cytoplasm. Dephosphorylation of CaMKII by various phosphatases allows the cycle of CaMKII translocation to begin again.

Proposed Mechanisms Underlying CaMKII Misregulation

The identification of a derangement in CaMKII led to the next logical question: what underlies the misregulation of CaMKII phosphorylation? As discussed above, CaMKII cycles between phosphorylated and non-phosphorylated states. The cycling state of CaMKII is not unlike many other proteins involved in signal transduction processes that undergo endless cycles of phosphorylation and dephosphorylation. Nature has optimized these processes that provide an incredibly rapid and potent mechanism for the regulation of protein function. However, if left unchecked, phosphorylation of proteins could rapidly lead to derangement of cellular physiology. Soluble enzymes, known as phosphatases exist within the cytoplasm and remove phosphate groups from various proteins. CaMKII, like most other cytoplasmic proteins, is subject to regulation by both kinases and phosphatases (Figure 6).

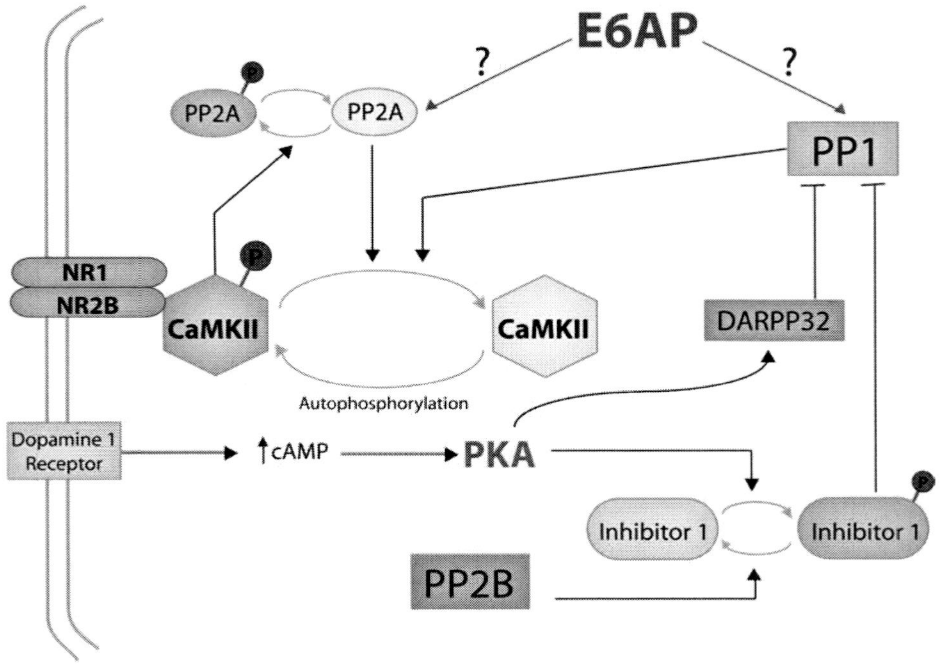

Figure 6. Schematic drawing of the proteins involved in regulating the three major phosphatases in the CNS; Protein Phosphatase-1 (PP1), Protein Phosphatase-2A (PP2A), and Protein Phosphatase-2B (PP2B).

There are three major protein phosphatases (PP) found in the brain: PP1, PP2A and PP2B (calcineurin). PP1 and PP2A appear to be primarily responsible for dephosphorylation of CaMKII at Thr^{286}, with PP2A regulating CaMKII phosphorylation in the cytoplasm and PP1 in the postsynaptic density (Bennecib et al., 2001; Shields et al., 1985; Strack et al., 1997). Through the actions of PP1 and PP2A, the activity of CaMKII is held in check and allows the enzyme to return to a state where Ca^{++} influx can again activate the proper signal transduction cascades involving CaMKII. The presence of PP1 and PP2A cause an apparent conundrum in the regulation of CaMKII. CaMKII is a Ca^{++}-dependent kinase, however under conditions conducive to induction of long-term synaptic plasticity or memory formation, CaMKII undergoes autophosphorylation and becomes a Ca^{++}-independent autonomous enzyme. How does this occur in the presence of PP1 and PP2A? Several proteins have been identified that act as inhibitors of PP activity. One example is the Inhibitor 1 (I1) protein (Huang and Glinsmann, 1976). When phosphorylated via protein kinase A, I1 can bind to and inhibit the activity of PP1 (Blitzer et al., 1998; Brown et al., 2000). Inhibition of PP1 via I1 facilitates the transition of CaMKII from a Ca^{++}-dependent to an autonomous state (Blitzer et al., 1998; Brown et al., 2000). The functional consequences of PPs on CaMKII activity are not solely negative. As noted above, CaMKII contains an inhibitory autophosphorylation site at $Thr^{305/306}$ (Patton et al., 1990). Phosphorylation at this site downregulates CaMKII activity and prevents binding by Ca^{++}-calmodulin. PP1 can dephosphorylate $Thr^{305/306}$, which then allows for reactivation of CaMKII by Ca^{++}-calmodulin (Patton et al., 1990).

Figure 6 shows the highly regulated action of protein phosphatases and the proteins that control their activity as well. Thus, the changes in the phosphorylation state of AS mouse hippocampal CaMKII could be due changes in protein levels in one or more of these proteins. Again we focused on the possibility that the lack of ubiquitin ligase results in an up-regulation of a specific protein, in this instance perhaps an inhibitor of PP1 or PP2A. Unfortunately, we found no changes in any of the proteins involved in the PP1 and PP2A pathways. However, we did find that PP1 and/or PP2A phosphatase activity was significantly reduced in the m-/p+ mutants by a prodigious 2.5 fold decrease (Weeber et al., 2003). This finding strongly suggests that the aberrant state of P-Thr^{286} and P-Thr^{305} αCaMKII phosphorylation is due to changes in the activity of one or both of these important phosphatases. This also suggests that the increase in Thr^{305} αCaMKII is due to both the increase P-Thr^{286} αCaMKII, since phosphorylation at Thr^{286} precedes phosphorylation at Thr^{305} (Colbran and Soderling, 1990; Hanson and Schulman, 1992a), and reduced PP1 and/or PP2A activity, which are known to dephosphorylate P-Thr^{305} αCaMKII *in vitro* (Patton et al., 1990).

This is the current state of our knowledge about the biochemistry of AS, and our 'detective' story pauses briefly here. In a relatively short period of time, as scientific research goes, workers in this area have gone from the discovery of the AS gene, to identifying specific areas of biochemical derangement associated with the disorder. As we momentarily linger at this point, we can begin to consider the next step of this research by hypothesizing concerning the potential molecular mechanisms underlying the derangement of CaMKII phosphorylation via down-regulation of PP1 and PP2A activity. The last section of this review will focus on the potential areas of future research and the conclusions that we have drawn from our data thus far.

SEARCHING FOR THE AS 'SMOKING GUN'

What may be the underlying mechanism of reduced PP1/PP2A activity? The observed alterations in P-Thr^{286} and P-Thr^{305} CaMKII may be due to the endogenous alterations associated with hippocampal E6-AP deficiency, specifically an increase in proteins usually targeted by a now-absent ubiquitin ligase. For example, p53 is a known target of E6-AP and accumulates in the hippocampus of m-/p+ mice as well as in humans with AS. An interesting possibility arises from the identification of a p53 associated protein, designated p53 binding protein 2 (p53BP2), which can bind p53 and PP1 in a mutually exclusive manner (Helps et al., 1995). Very little is known about p53BP2, and its physiological importance in the CNS is unclear. However, it is known that p53BP2 can inhibit the phosphatase activity of PP1 at nanomolar concentrations. It is hypothesized that p53BP2 may be involved in the dephosphorylation and regulation of p53, but its effect on CaMKII phosphorylation is completely unknown at present. This raises the interesting possibility that the chronic presence and considerable increase in p53 in the imprinted areas of AS mice and humans can directly affect PP1 activity through p53BP2. An alternative hypothesis is that phosphorylation of copious p53 protein can cause phosphatase inhibition. This hypothesis is supported by recent research showing that inhibition of PP1 can occur following p53 phosphorylation in rat cardiomyocytes (Long et al., 2002). The actions of p53BP2 or p53 phosphorylation in Thr^{286}-CaMKII dephosphorylation represent an exciting area of future research. An interesting side note to this story is that when we hit upon the possibility of a p53-CaMKII connection we pursued a genetic approach to test our hypothesis. This is relatively easy to accomplish by crossing p53-deficient mice with our AS mice and in an attempt to rescue the AS phenotype. Interestingly, both AS mice and p53 mice are viable and breed readily, however when these two mice are crossed the offspring die shortly after birth (Weeber and Sweatt, unpublished observations). This suggests that p53 and CaMKII may interact at a molecular level in an unknown matter.

Conclusion

There is accumulating evidence that points to the hippocampus as an important brain structure in cognitive processing, and memory formation. This evidence is strongly supported through the investigation of cognitive disorders such as AS. The production of the AS mouse model was paramount in the identification of the molecular players involved in AS-associated hippocampal dysfunction and, by inference, cognition. Although CaMKII is long-held as a protein involved in learning and memory processes, investigation of the AS mouse has supplied the first report that an alteration in CaMKII has the potential for causing a human disorder associated with a severe cognitive deficiency.

References

Albrecht, U.; Sutcliffe, J. S.; Cattanach, B. M.; Beechey, C. V.; Armstrong, D.; Eichele, G.; Beaudet, A. L. Imprinted expression of the murine Angelman syndrome gene, Ube3a, in hippocampal and Purkinje neurons. *Nat Genet* 17: 75-8, 1997.

Alger, B. E.; Teyler, T. J. Long-term and short-term plasticity in the CA1, CA3, and dentate regions of the rat hippocampal slice. *Brain Res* 110: 463-80, 1976.

Bach, M. E.; Hawkins, R. D.; Osman, M.; Kandel, E. R.; Mayford, M. Impairment of spatial but not contextual memory in CaMKII mutant mice with a selective loss of hippocampal LTP in the range of the theta frequency. *Cell* 81: 905-15, 1995.

Bading, H.; Greenberg, M. E. Stimulation of protein tyrosine phosphorylation by NMDA receptor activation. *Science* 253: 912-4, 1991.

Basu, A.; Haldar, S. Signal-induced site specific phosphorylation targets Bcl2 to the proteasome pathway. *Int J Oncol* 21: 597-601, 2002.

Bayer, K. U.; De Koninck, P.; Leonard, A. S.; Hell, J. W.; Schulman, H. Interaction with the NMDA receptor locks CaMKII in an active conformation. *Nature* 411: 801-5, 2001.

Bayer, K. U.; Harbers, K.; Schulman, H. alphaKAP is an anchoring protein for a novel CaM kinase II isoform in skeletal muscle. *Embo J* 17: 5598-605, 1998.

Bayer, K. U.; Lohler, J.; Harbers, K. An alternative, nonkinase product of the brain-specifically expressed Ca2+/calmodulin-dependent kinase II alpha isoform gene in skeletal muscle. *Mol Cell Biol* 16: 29-36, 1996.

Bejar, R.; Yasuda, R.; Krugers, H.; Hood, K.; Mayford, M. Transgenic calmodulin-dependent protein kinase II activation: dose-dependent effects on synaptic plasticity, learning, and memory. *J Neurosci* 22: 5719-26, 2002.

Bennecib, M.; Gong, C.; Grundke-Iqbal, I.; Iqbal, K. Inhibition of PP-2A upregulates CaMKII in rat forebrain and induces hyperphosphorylation of tau at Ser 262/356. *FEBS Lett* 490: 15-22., 2001.

Bennett, M. K.; Kennedy, M. B. Deduced primary structure of the beta subunit of brain type II Ca2+/calmodulin-dependent protein kinase determined by molecular cloning. *Proc Natl Acad Sci U S A* 84: 1794-8, 1987.

Berry, S. D.; Thompson, R. F. Prediction of learning rate from the hippocampal electroencephalogram. *Science* 200: 1298-300, 1978.

Bliss, T. V.; Lomo, T. Long-lasting potentiation of synaptic transmission in the dentate area of the anaesthetized rabbit following stimulation of the perforant path. *J Physiol* 232: 331-56, 1973.

Blitzer, R. D.; Connor, J. H.; Brown, G. P.; Wong, T.; Shenolikar, S.; Iyengar, R.; Landau, E. M. Gating of CaMKII by cAMP-regulated protein phosphatase activity during LTP. *Science* 280: 1940-2., 1998.

Bramham, C. R.; Torp, R.; Zhang, N.; Storm-Mathisen, J.; Ottersen, O. P. Distribution of glutamate-like immunoreactivity in excitatory hippocampal pathways: a semiquantitative electron microscopic study in rats. *Neuroscience* 39: 405-17, 1990.

Brocke, L.; Srinivasan, M.; Schulman, H. Developmental and regional expression of multifunctional Ca2+/calmodulin-dependent protein kinase isoforms in rat brain. *J Neurosci* 15: 6797-808, 1995.

Brown, G. P.; Blitzer, R. D.; Connor, J. H.; Wong, T.; Shenolikar, S.; Iyengar, R.; Landau, E. M. Long-term potentiation induced by theta frequency stimulation is regulated by a protein phosphatase-1-operated gate. *J Neurosci* 20: 7880-7., 2000.

Buiting, K.; Saitoh, S.; Gross, S.; Dittrich, B.; Schwartz, S.; Nicholls, R. D.; Horsthemke, B. Inherited microdeletions in the Angelman and Prader-Willi syndromes define an imprinting centre on human chromosome 15 [published erratum appears in Nat Genet 1995 Jun;10(2):249]. *Nat Genet* 9: 395-400, 1995.

Bulleit, R. F.; Bennett, M. K.; Molloy, S. S.; Hurley, J. B.; Kennedy, M. B. Conserved and variable regions in the subunits of brain type II Ca2+/calmodulin-dependent protein kinase. *Neuron* 1: 63-72, 1988.

Buoni, S.; Grosso, S.; Pucci, L.; Fois, A. Diagnosis of Angelman syndrome: clinical and EEG criteria. *Brain Dev* 21: 296-302, 1999.

Burger, J.; Kunze, J.; Sperling, K.; Reis, A. Phenotypic differences in Angelman syndrome patients: imprinting mutations show less frequently microcephaly and hypopigmentation than deletions. *Am J Med Genet* 66: 221-6, 1996.

Chin, D.; Means, A. R. Calmodulin: a prototypical calcium sensor. *Trends Cell Biol* 10: 322-8, 2000.

Clayton-Smith, J. Clinical research on Angelman syndrome in the United Kingdom: observations on 82 affected individuals. *Am J Med Genet* 46: 12-5, 1993.

Colbran, R. J. Inactivation of Ca2+/calmodulin-dependent protein kinase II by basal autophosphorylation. *J Biol Chem* 268: 7163-70., 1993a.

Colbran, R. J. Inactivation of Ca2+/calmodulin-dependent protein kinase II by basal autophosphorylation. *J Biol Chem* 268: 7163-70, 1993b.

Colbran, R. J.; Smith, M. K.; Schworer, C. M.; Fong, Y. L.; Soderling, T. R. Regulatory domain of calcium/calmodulin-dependent protein kinase II. Mechanism of inhibition and regulation by phosphorylation. *J Biol Chem* 264: 4800-4, 1989.

Colbran, R. J.; Soderling, T. R. Calcium/calmodulin-independent autophos-phorylation sites of calcium/calmodulin-dependent protein kinase II. Studies on the effect of phosphorylation of threonine 305/306 and serine 314 on calmodulin binding using synthetic peptides. *J Biol Chem* 265: 11213-9, 1990.

Costa, R. M.; Federov, N. B.; Kogan, J. H.; Murphy, G. G.; Stern, J.; Ohno, M.; Kucherlapati, R.; Jacks, T.; Silva, A. J. Mechanism for the learning deficits in a mouse model of neurofibromatosis type 1. *Nature* 415: 526-30, 2002.

Costa, R. M.; Yang, T.; Huynh, D. P.; Pulst, S. M.; Viskochil, D. H.; Silva, A. J.; Brannan, C. I. Learning deficits, but normal development and tumor predisposition, in mice lacking exon 23a of Nf1. *Nat Genet* 27: 399-405, 2001.

De Koninck, P.; Schulman, H. Sensitivity of CaM kinase II to the frequency of Ca2+ oscillations. *Science* 279: 227-30, 1998.

DeChiara, T. M.; Robertson, E. J.; Efstratiadis, A. Parental imprinting of the mouse insulin-like growth factor II gene. *Cell* 64: 849-59, 1991.

DeLorey, T. M.; Handforth, A.; Anagnostaras, S. G.; Homanics, G. E.; Minassian, B. A.; Asatourian, A.; Fanselow, M. S.; Delgado-Escueta, A.; Ellison, G. D.; Olsen, R. W. Mice lacking the beta3 subunit of the GABAA receptor have the epilepsy phenotype and many of the behavioral characteristics of Angelman syndrome. *J Neurosci* 18: 8505-14, 1998.

DeLorey, T. M.; Olsen, R. W. GABA and epileptogenesis: comparing gabrb3 gene-deficient mice with Angelman syndrome in man. *Epilepsy Res* 36: 123-32, 1999.

Dosemeci, A.; Reese, T. S.; Petersen, J. D.; Choi, C.; Beushausen, S. Localization of the linker domain of Ca2+/calmodulin-dependent protein kinase II. *Biochem Biophys Res Commun* 263: 657-62, 1999.

Elgersma, Y.; Fedorov, N. B.; Ikonen, S.; Choi, E. S.; Elgersma, M.; Carvalho, O. M.; Giese, K. P.; Silva, A. J. Inhibitory autophosphorylation of CaMKII controls PSD association, plasticity, and learning. *Neuron* 36: 493-505, 2002.

Erondu, N. E.; Kennedy, M. B. Regional distribution of type II Ca2+/calmodulin-dependent protein kinase in rat brain. *J Neurosci* 5: 3270-7, 1985.

Fryburg, J. S.; Breg, W. R.; Lindgren, V. Diagnosis of Angelman syndrome in infants. *Am J Med Genet* 38: 58-64, 1991.

Gabriel, J. M.; Merchant, M.; Ohta, T.; Ji, Y.; Caldwell, R. G.; Ramsey, M. J.; Tucker, J. D.; Longnecker, R.; Nicholls, R. D. A transgene insertion creating a heritable chromosome deletion mouse model of Prader-Willi and angelman syndromes. *Proc Natl Acad Sci U S A* 96: 9258-63, 1999.

Giese, K. P.; Fedorov, N. B.; Filipkowski, R. K.; Silva, A. J. Autophosphorylation at Thr286 of the alpha calcium-calmodulin kinase II in LTP and learning. *Science* 279: 870-3, 1998.

Gillessen-Kaesbach, G.; Demuth, S.; Thiele, H.; Theile, U.; Lich, C.; Horsthemke, B. A previously unrecognised phenotype characterised by obesity, muscular hypotonia, and ability to speak in patients with Angelman syndrome caused by an imprinting defect. *Eur J Hum Genet* 7: 638-44, 1999.

Givens, B. Stimulus-evoked resetting of the dentate theta rhythm: relation to working memory. *Neuroreport* 8: 159-63, 1996.

Glenn, C. C.; Deng, G.; Michaelis, R. C.; Tarleton, J.; Phelan, M. C.; Surh, L.; Yang, T. P.; Driscoll, D. J. DNA methylation analysis with respect to prenatal diagnosis of the Angelman and Prader-Willi syndromes and imprinting. *Prenat Diagn* 20: 300-306, 2000.

Guillemot, F.; Caspary, T.; Tilghman, S. M.; Copeland, N. G.; Gilbert, D. J.; Jenkins, N. A.; Anderson, D. J.; Joyner, A. L.; Rossant, J.; Nagy, A. Genomic imprinting of Mash2, a mouse gene required for trophoblast development. *Nat Genet* 9: 235-42, 1995.

Hanley, R. M.; Means, A. R.; Ono, T.; Kemp, B. E.; Burgin, K. E.; Waxham, N.; Kelly, P. T. Functional analysis of a complementary DNA for the 50-kilodalton subunit of calmodulin kinase II. *Science* 237: 293-7, 1987.

Hanson, P. I.; Schulman, H. Inhibitory autophosphorylation of multifunctional Ca2+/calmodulin-dependent protein kinase analyzed by site-directed mutagenesis. *J Biol Chem* 267: 17216-24., 1992a.

Hanson, P. I.; Schulman, H. Inhibitory autophosphorylation of multifunctional Ca2+/calmodulin-dependent protein kinase analyzed by site-directed mutagenesis. *J Biol Chem* 267: 17216-24, 1992b.

Harris, C. C. p53: at the crossroads of molecular carcinogenesis and molecular epidemiology. *J Investig Dermatol Symp Proc* 1: 115-8, 1996.

Harris, E. W.; Ganong, A. H.; Cotman, C. W. Long-term potentiation in the hippocampus involves activation of N-methyl-D-aspartate receptors. *Brain Res* 323: 132-7, 1984.

Helps, N. R.; Barker, H. M.; Elledge, S. J.; Cohen, P. T. Protein phosphatase 1 interacts with p53BP2, a protein which binds to the tumour suppressor p53. *FEBS Lett* 377: 295-300, 1995.

Hinds, H. L.; Tonegawa, S.; Malinow, R. CA1 long-term potentiation is diminished but present in hippocampal slices from alpha-CaMKII mutant mice. *Learn Mem* 5: 344-54, 1998.

Hoch, B.; Meyer, R.; Hetzer, R.; Krause, E. G.; Karczewski, P. Identification and expression of delta-isoforms of the multifunctional Ca2+/calmodulin-dependent protein kinase in failing and nonfailing human myocardium. *Circ Res* 84: 713-21, 1999.

Huang, F. L.; Glinsmann, W. H. Separation and characterization of two phosphorylase phosphatase inhibitors from rabbit skeletal muscle. *Eur J Biochem* 70: 419-26, 1976.

Hudmon, A.; Schulman, H. Neuronal CA2+/calmodulin-dependent protein kinase II: the role of structure and autoregulation in cellular function. *Annu Rev Biochem* 71: 473-510, 2002.

Huibregtse, J. M.; Scheffner, M.; Howley, P. M. A cellular protein mediates association of p53 with the E6 oncoprotein of human papillomavirus types 16 or 18. *Embo J* 10: 4129-35, 1991.

Huibregtse, J. M.; Scheffner, M.; Howley, P. M. Cloning and expression of the cDNA for E6-AP, a protein that mediates the interaction of the human papillomavirus E6 oncoprotein with p53. *Mol Cell Biol* 13: 775-84, 1993.

Jiang, Y. H.; Armstrong, D.; Albrecht, U.; Atkins, C. M.; Noebels, J. L.; Eichele, G.; Sweatt, J. D.; Beaudet, A. L. Mutation of the Angelman ubiquitin ligase in mice causes increased cytoplasmic p53 and deficits of contextual learning and long-term potentiation. *Neuron* 21: 799-811., 1998.

Kanaseki, T.; Ikeuchi, Y.; Sugiura, H.; Yamauchi, T. Structural features of Ca2+/calmodulin-dependent protein kinase II revealed by electron microscopy. *J Cell Biol* 115: 1049-60, 1991.

Kaplan, L. C.; Wharton, R.; Elias, E.; Mandell, F.; Donlon, T.; Latt, S. A. Clinical heterogeneity associated with deletions in the long arm of chromosome 15: report of 3 new cases and their possible genetic significance. *Am J Med Genet* 28: 45-53, 1987.

Karls, U.; Muller, U.; Gilbert, D. J.; Copeland, N. G.; Jenkins, N. A.; Harbers, K. Structure, expression, and chromosome location of the gene for the beta subunit of brain-specific Ca2+/calmodulin-dependent protein kinase II identified by transgene integration in an embryonic lethal mouse mutant. *Mol Cell Biol* 12: 3644-52, 1992.

Kelly, P. T.; Weinberger, R. P.; Waxham, M. N. Active site-directed inhibition of Ca2+/calmodulin-dependent protein kinase type II by a bifunctional calmodulin-binding peptide. *Proc Natl Acad Sci U S A* 85: 4991-5, 1988.

Kishino T, Lalande M, Wagstaff J (1997) UBE3A/E6-AP mutations cause Angelman syndrome. *Nat Genet* 15:70-73.

Knoll, J. H.; Nicholls, R. D.; Magenis, R. E.; Graham, J. M., Jr.; Lalande, M.; Latt, S. A. Angelman and Prader-Willi syndromes share a common chromosome 15 deletion but differ in parental origin of the deletion. *Am J Med Genet* 32: 285-90, 1989.

Kolodziej, S. J.; Hudmon, A.; Waxham, M. N.; Stoops, J. K. Three-dimensional reconstructions of calcium/calmodulin-dependent (CaM) kinase IIalpha and truncated CaM kinase IIalpha reveal a unique organization for its structural core and functional domains. *J Biol Chem* 275: 14354-9, 2000.

Kuhne, C.; Banks, L. E3-ubiquitin ligase/E6-AP links multicopy maintenance protein 7 to the ubiquitination pathway by a novel motif, the L2G box. *J Biol Chem* 273: 34302-9, 1998.

Kumar, S.; Talis, A. L.; Howley, P. M. Identification of HHR23A as a substrate for E6-associated protein- mediated ubiquitination. *J Biol Chem* 274: 18785-92, 1999.

Kwiatkowski, A. P.; McGill, J. M. Human biliary epithelial cell line Mz-ChA-1 expresses new isoforms of calmodulin-dependent protein kinase II. *Gastroenterology* 109: 1316-23, 1995.

Laan, L. A.; v Haeringen, A.; Brouwer, O. F. Angelman syndrome: a review of clinical and genetic aspects. *Clin Neurol Neurosurg* 101: 161-70, 1999.

Landfield, P. W.; McGaugh, J. L.; Tusa, R. J. Theta rhythm: a temporal correlate of memory storage processes in the rat. *Science* 175: 87-9, 1972.

Larson, J.; Lynch, G. Induction of synaptic potentiation in hippocampus by patterned stimulation involves two events. *Science* 232: 985-8, 1986.

Larson, J.; Lynch, G. Theta pattern stimulation and the induction of LTP: the sequence in which synapses are stimulated determines the degree to which they potentiate. *Brain Res* 489: 49-58, 1989.

Larson, J.; Wong, D.; Lynch, G. Patterned stimulation at the theta frequency is optimal for the induction of hippocampal long-term potentiation. *Brain Res* 368: 347-50, 1986.

Lee, S.; Wevrick, R. Identification of Novel Imprinted Transcripts in the Prader-Willi Syndrome and Angelman Syndrome Deletion Region: Further Evidence for Regional Imprinting Control. *Am J Hum Genet* 66: 848-858, 2000.

Leonard, A. S.; Lim, I. A.; Hemsworth, D. E.; Horne, M. C.; Hell, J. W. Calcium/calmodulin-dependent protein kinase II is associated with the N-methyl-D-aspartate receptor. *Proc Natl Acad Sci U S A* 96: 3239-44, 1999.

Lin, C. R.; Kapiloff, M. S.; Durgerian, S.; Tatemoto, K.; Russo, A. F.; Hanson, P.; Schulman, H.; Rosenfeld, M. G. Molecular cloning of a brain-specific calcium/calmodulin-dependent protein kinase. *Proc Natl Acad Sci U S A* 84: 5962-6, 1987.

Long, X.; Wu, G.; Gaa, S. T.; Rogers, T. B. Inhibition of protein phosphatase-1 is linked to phosphorylation of p53 and apoptosis. *Apoptosis* 7: 31-9, 2002.

Lynch, G.; Larson, J.; Kelso, S.; Barrionuevo, G.; Schottler, F. Intracellular injections of EGTA block induction of hippocampal long-term potentiation. *Nature* 305: 719-21, 1983.

Magenis, R. E.; Toth-Fejel, S.; Allen, L. J.; Black, M.; Brown, M. G.; Budden, S.; Cohen, R.; Friedman, J. M.; Kalousek, D.; Zonana, J.; et al., Comparison of the 15q deletions in Prader-Willi and Angelman syndromes: specific regions, extent of deletions, parental origin, and clinical consequences. *Am J Med Genet* 35: 333-49, 1990.

Malcolm, S.; Clayton-Smith, J.; Nichols, M.; Robb, S.; Webb, T.; Armour, J. A.; Jeffreys, A. J.; Pembrey, M. E. Uniparental paternal disomy in Angelman's syndrome. *Lancet* 337: 694-7, 1991.

Malenka, R. C.; Kauer, J. A.; Perkel, D. J.; Mauk, M. D.; Kelly, P. T.; Nicoll, R. A.; Waxham, M. N. An essential role for postsynaptic calmodulin and protein kinase activity in long-term potentiation. *Nature* 340: 554-7, 1989.

Malenka, R. C.; Kauer, J. A.; Zucker, R. S.; Nicoll, R. A. Postsynaptic calcium is sufficient for potentiation of hippocampal synaptic transmission. *Science* 242: 81-4, 1988.

Malinow, R.; Madison, D. V.; Tsien, R. W. Persistent protein kinase activity underlying long-term potentiation. *Nature* 335: 820-4, 1988.

Malinow, R.; Schulman, H.; Tsien, R. W. Inhibition of postsynaptic PKC or CaMKII blocks induction but not expression of LTP. *Science* 245: 862-6, 1989.

Matsuura T, Sutcliffe JS, Fang P, Galjaard RJ, JIang YH, Benton CS, Rommens JM, Beaudet AL (1997) De novo truncating mutations in E6-AP ubiquitin-protein ligase gene (UBE3A) in Angelman syndrome. *Nat Genet* 15:74-77.

Mayer, M. L.; Westbrook, G. L.; Guthrie, P. B. Voltage-dependent block by Mg2+ of NMDA responses in spinal cord neurones. *Nature* 309: 261-3, 1984.

Meyer, T.; Hanson, P. I.; Stryer, L.; Schulman, H. Calmodulin trapping by calcium-calmodulin-dependent protein kinase. *Science* 256: 1199-202, 1992.

Mietz, J. A.; Unger, T.; Huibregtse, J. M.; Howley, P. M. The transcriptional transactivation function of wild-type p53 is inhibited by SV40 large T-antigen and by HPV-16 E6 oncoprotein. *Embo J* 11: 5013-20, 1992.

Morris, R. G.; Anderson, E.; Lynch, G. S.; Baudry, M. Selective impairment of learning and blockade of long-term potentiation by an N-methyl-D-aspartate receptor antagonist, AP5. *Nature* 319: 774-6, 1986.

Nadler, J. V.; Vaca, K. W.; White, W. F.; Lynch, G. S.; Cotman, C. W. Aspartate and glutamate as possible transmitters of excitatory hippocampal afferents. *Nature* 260: 538-40, 1976.

Nghiem, P.; Saati, S. M.; Martens, C. L.; Gardner, P.; Schulman, H. Cloning and analysis of two new isoforms of multifunctional Ca2+/calmodulin-dependent protein kinase. Expression in multiple human tissues. *J Biol Chem* 268: 5471-9, 1993.

Nowak, L.; Bregestovski, P.; Ascher, P.; Herbet, A.; Prochiantz, A. Magnesium gates glutamate-activated channels in mouse central neurones. *Nature* 307: 462-5, 1984.

Nuber, U.; Schwarz, S. E.; Scheffner, M. The ubiquitin-protein ligase E6-associated protein (E6-AP) serves as its own substrate. *Eur J Biochem* 254: 643-9, 1998.

O'Dell, T. J.; Kandel, E. R.; Grant, S. G. Long-term potentiation in the hippocampus is blocked by tyrosine kinase inhibitors. *Nature* 353: 558-60, 1991.

O'Keefe, J.; Recce, M. L. Phase relationship between hippocampal place units and the EEG theta rhythm. *Hippocampus* 3: 317-30, 1993.

Park, J. S.; Hwang, E. S.; Park, S. N.; Ahn, H. K.; Um, S. J.; Kim, C. J.; Kim, S. J.; Namkoong, S. E. Physical status and expression of HPV genes in cervical cancers. *Gynecol Oncol* 65: 121-9, 1997.

Patton, B. L.; Miller, S. G.; Kennedy, M. B. Activation of type II calcium/calmodulin-dependent protein kinase by Ca2+/calmodulin is inhibited by autophosphorylation of threonine within the calmodulin-binding domain. *J Biol Chem* 265: 11204-12., 1990.

Rachmilewitz, J.; Goshen, R.; Ariel, I.; Schneider, T.; de Groot, N.; Hochberg, A. Parental imprinting of the human H19 gene. *FEBS Lett* 309: 25-8, 1992.

Rotenberg, A.; Mayford, M.; Hawkins, R. D.; Kandel, E. R.; Muller, R. U. Mice expressing activated CaMKII lack low frequency LTP and do not form stable place cells in the CA1 region of the hippocampus. *Cell* 87: 1351-61, 1996.

Saitoh, S.; Harada, N.; Jinno, Y.; Hashimoto, K.; Imaizumi, K.; Kuroki, Y.; Fukushima, Y.; Sugimoto, T.; Renedo, M.; Wagstaff, J.; et al., Molecular and clinical study of 61 Angelman syndrome patients. *Am J Med Genet* 52: 158-63, 1994.

Saitoh, S.; Wada, T.; Kuno, T.; Kim, K. C.; Ohashi, H.; Hashimoto, K.; Niikawa, N. Clinical characteristics of Angelman syndrome patients with a non-IC-deleted imprinting mutation. *Clin Genet* 55: 277-8, 1999.

Saitoh, T.; Schwartz, J. H. Phosphorylation-dependent subcellular translocation of a Ca2+/calmodulin-dependent protein kinase produces an autonomous enzyme in Aplysia neurons. *J Cell Biol* 100: 835-42, 1985.

Schworer, C. M.; Rothblum, L. I.; Thekkumkara, T. J.; Singer, H. A. Identification of novel isoforms of the delta subunit of Ca2+/calmodulin-dependent protein kinase II. Differential expression in rat brain and aorta. *J Biol Chem* 268: 14443-9, 1993.

Seager, M. A.; Johnson, L. D.; Chabot, E. S.; Asaka, Y.; Berry, S. D. Oscillatory brain states and learning: Impact of hippocampal theta-contingent training. *Proc Natl Acad Sci U S A* 99: 1616-20, 2002.

Selcher, J. C.; Weeber, E. J.; Varga, A. W.; Sweatt, J. D.; Swank, M. Protein kinase signal transduction cascades in mammalian associative conditioning. *Neuroscientist* 8: 122-31, 2002.

Shields, S. M.; Ingebritsen, T. S.; Kelly, P. T. Identification of protein phosphatase 1 in synaptic junctions: dephosphorylation of endogenous calmodulin-dependent kinase II and synapse-enriched phosphoproteins. *J Neurosci* 5: 3414-22, 1985.

Silva, A. J.; Paylor, R.; Wehner, J. M.; Tonegawa, S. Impaired spatial learning in alpha-calcium-calmodulin kinase II mutant mice. *Science* 257: 206-11, 1992a.

Silva, A. J.; Stevens, C. F.; Tonegawa, S.; Wang, Y. Deficient hippocampal long-term potentiation in alpha-calcium-calmodulin kinase II mutant mice. *Science* 257: 201-6, 1992b.

Smith, A.; Marks, R.; Haan, E.; Dixon, J.; Trent, R. J. Clinical features in four patients with Angelman syndrome resulting from paternal uniparental disomy. *J Med Genet* 34: 426-9, 1997.

Smith, A.; Wiles, C.; Haan, E.; McGill, J.; Wallace, G.; Dixon, J.; Selby, R.; Colley, A.; Marks, R.; Trent, R. J. Clinical features in 27 patients with Angelman syndrome resulting from DNA deletion. *J Med Genet* 33: 107-12, 1996.

Smith, M. K.; Colbran, R. J.; Brickey, D. A.; Soderling, T. R. Functional determinants in the autoinhibitory domain of calcium/calmodulin-dependent protein kinase II. Role of His282 and multiple basic residues. *J Biol Chem* 267: 1761-8, 1992.

Solter, D. Imprinting. *Int J Dev Biol* 42: 951-4, 1998.

Srinivasan, M.; Edman, C. F.; Schulman, H. Alternative splicing introduces a nuclear localization signal that targets multifunctional CaM kinase to the nucleus. *J Cell Biol* 126: 839-52, 1994.

Strack, S.; Barban, M. A.; Wadzinski, B. E.; Colbran, R. J. Differential inactivation of postsynaptic density-associated and soluble Ca2+/calmodulin-dependent protein kinase II by protein phosphatases 1 and 2A. *J Neurochem* 68: 2119-28, 1997.

Strack, S.; Colbran, R. J. Autophosphorylation-dependent targeting of calcium/ calmodulin-dependent protein kinase II by the NR2B subunit of the N-methyl- D-aspartate receptor. *J Biol Chem* 273: 20689-92, 1998.

Takeuchi, M.; Fujisawa, H. New alternatively spliced variants of calmodulin-dependent protein kinase II from rabbit liver. *Gene* 221: 107-15, 1998.

Tobimatsu, T.; Fujisawa, H. Tissue-specific expression of four types of rat calmodulin-dependent protein kinase II mRNAs. *J Biol Chem* 264: 17907-12, 1989.

Tobimatsu, T.; Kameshita, I.; Fujisawa, H. Molecular cloning of the cDNA encoding the third polypeptide (gamma) of brain calmodulin-dependent protein kinase II. *J Biol Chem* 263: 16082-6, 1988.

Tombes, R. M.; Krystal, G. W. Identification of novel human tumor cell-specific CaMK-II variants. *Biochim Biophys Acta* 1355: 281-92, 1997.

Urquidi, V.; Ashcroft, S. J. A novel pancreatic beta-cell isoform of calcium/calmodulin-dependent protein kinase II (beta 3 isoform) contains a proline-rich tandem repeat in the association domain. *FEBS Lett* 358: 23-6, 1995.

Waddell, S.; Jenkins, J. R. Defining the minimal requirements for papilloma viral E6-mediated inhibition of human p53 activity in fission yeast. *Oncogene* 16: 1759-65, 1998.

Walikonis, R. S.; Oguni, A.; Khorosheva, E. M.; Jeng, C. J.; Asuncion, F. J.; Kennedy, M. B. Densin-180 forms a ternary complex with the (alpha)-subunit of Ca2+/calmodulin-dependent protein kinase II and (alpha)-actinin. *J Neurosci* 21: 423-33, 2001.

Weeber, E. J.; Jiang, Y. H.; Elgersma, Y.; Varga, A. W.; Carrasquillo, Y.; Brown, S. E.; Christian, J. M.; Mirnikjoo, B.; Silva, A.; Beaudet, A. L.; Sweatt, J. D. Derangements of hippocampal calcium/calmodulin-dependent protein kinase II in a mouse model for Angelman mental retardation syndrome. *J Neurosci* 23: 2634-44, 2003.

Williams, C. A.; Angelman, H.; Clayton-Smith, J.; Driscoll, D. J.; Hendrickson, J. E.; Knoll, J. H.; Magenis, R. E.; Schinzel, A.; Wagstaff, J.; Whidden, E. M.; et al., Angelman syndrome: consensus for diagnostic criteria.Angelman Syndrome Foundation. *Am J Med Genet* 56: 237-8, 1995.

Winson, J. Loss of hippocampal theta rhythm results in spatial memory deficit in the rat. *Science* 201: 160-3, 1978.

Yang, E.; Schulman, H. Structural examination of autoregulation of multifunctional calcium/calmodulin-dependent protein kinase II. *J Biol Chem* 274: 26199-208, 1999.

Zheng, Y. J.; Furukawa, T.; Ogura, T.; Tajimi, K.; Inagaki, N. M-phase specific expression and phosphorylation-dependent ubiquitination of the ClC-2 channel. *J Biol Chem*, 2002.

In: Cognitive Science Compendium, Volume 1
Editor: Miao-Kun Sun
ISBN: 978-1-60456-597-3

Chapter 2

ROLE OF GENETIC AND NONGENETIC VASCULAR RISK FACTORS IN SPORADIC LATE-ONSET ALZHEIMER'S DISEASE

***Francesco Panza*[1,*], *Alessia D'Introno*[1], *Anna M. Colacicco*[1], *Cristiano Capurso*[1,2], *Vincenza Pellicani*[3], *Orlando Todarello*[3], *Antonio Capurso*[1], *Sabrina Capurso*[1] *and Vincenzo Solfrizzi*[1]**

[1] Department of Geriatrics - Center for Aging Brain - Memory Unit
University of Bari
Policlinico - Piazza Giulio Cesare, 11
70124 Bari, Italy
[2] Department of Geriatrics, University of Foggia, Foggia, Italy
[3] Department of Neurological and Psychiatric Sciences, University of Bari, Bari, Italy

ABSTRACT

In recent years, family and twin studies as well as family and sibling-pair linkage studies continue to argue that genetic factors may play a primary role in Alzheimer's disease (AD). The three familial AD genes (PSEN1, PSEN2, and APP) in which mutations are known to cause early onset autosomal-dominant familial AD account for only a very small proportion of cases of the disease. Most AD cases are represented by sporadic and late onset forms, where no obvious inheritance pattern is seen. Furthermore, based on a growing vascular hypothesis for the pathogenesis of late-onset AD (LOAD) and other dementias, there is increasing interest for environmental and genetic vascular factors. Polymorphisms in different susceptibility genes already implicated in vascular disease risk are now also being suggested as possible genetic markers for increased risk of developing LOAD; however, many of these studies have shown conflicting results. This points to other genetic or nongenetic factors having a role in the greater proportion of AD cases. While theories have been held as to the role of vascular risk factors in the

* Corresponding author: email: geriat.dot@geriatria.uniba.it

development of AD for a number of years, it is only in the last decade that active genetic and biological research has been conducted to try and unravel this more clearly. While vascular disease and AD seem to share various common genetic and nongenetic risk factors, the story relating to genes of mutual interest that have been investigated such as APOE, ACE1, MTHFR, NOS3, LRP1, PON1, and OLR1 is not entirely clear. However, what is apparent is that APOE remains largely the only risk factor to be accepted as having a role in both vascular disease and AD, while in the last five years ACE1 seems to be rapidly gaining support. Furthermore, many of the risk factors for cerebrovascular disease and vascular dementia, including serum total cholesterol, lipoprotein(a), diabetes mellitus, atrial fibrillation, hypertension, APOE levels, and atherosclerosis, have also been shown to increase the risk of AD. This review will focus on the current knowledge on genetic and nongenetic vascular factors likely to be involved in LOAD.

Introduction

Cognitive impairment and dementia are common disorders among elderly persons and influence the individual's ability to function independently. Due to prolonged human life expectancy and increase in the number of elderly people, the prevalence of cognitive impairment and dementia are expected to increase. Based on current nosology, in occidental countries, the two most common forms of dementia are Alzheimer's disease (AD) and vascular dementia (VaD), with respective frequencies of 70% and 15% for all dementias (Whitehouse et al., 1997). Therefore, AD is the most common dementia and primary neurodegenerative disorder in the elderly. This neurodegenerative disease gradually leads to a complete psychological and physical dependency and finally to death within one to two decades. It involves aberrant protein processing and is characterized by the presence of both intraneuronal protein clusters composed of paired helical filaments of hyperphosphorilated tau protein [neurofibrillary tangles (NFTs)], and extracellular protein aggregates [senile plaques (SPs)]. The SPs are the result of misprocessing of the amyloid precursor protein (*APP*) by β- and γ-secretases to form a toxic β-amyloid (*Aβ*) peptide that aggregates and initiates a pathogenic self-perpetuating cascade ultimately leading to neuronal loss and dementia. According to the "amyloid cascade hypothesis" (Hardy and Higgins, 1992), the development of SPs is thought to precede and precipitate the formation of NFTs as a result of the cellular changes invoked.

AD can be diagnosed with certainty at autopsy, but it can also be diagnosed with reliable clinical criteria established in 1984 by the National Institute for Neurological and Communicative Disorders and Stroke/Alzheimer's Disease and Related Disorders Association (NINCDS/ADRDA) (McKhann et al., 1984). On the other hand, VaD is more problematic as a nosologic entity. In fact, while criteria for AD have remained stable for about 20 years, various criteria for VaD have been proposed: the State of California Alzheimer's Disease Diagnostic and Treatment Centers (ADDTC) criteria (Chui et al., 1992); the National Institute for Neurological Disorders and Stroke-Association Internationale pour la Recherche et l'Enseignement en Neurosciences (NINDS-AIREN) criteria (Roman et al., 1993), and the DSM IV criteria (American Psychiatric Association, 1994). The definitive diagnosis for VaD also relies on post-mortem findings, and this diagnosis usually does not confirm the clinical suspect (Gold et al., 2002).

The clinical presentation of VaD varies greatly depending on the causes and location of cerebral damage (Roman, 2002). Large-vessel disease leads commonly to multiple cortical infarcts and a multifocal cortical dementia syndrome, whereas small-vessel disease, usually resulting from hypertension and diabetes, causes periventricular white matter ischemia and lacunar strokes characterized clinically by subcortical dementia with frontal lobe deficits, executive dysfunction, slow information processing, impaired memory, inattention, depressive mood changes, slowing of motor function, Parkinsonian features, small-step gait, urinary disturbances and pseudobulbar palsy (Roman and Royall, 1999).

It is still not known what causes AD, and specific risk factors for the disease are difficult to isolate. However, since rare mutations that occur in three genes: APP, presenilin1 (PSEN1) and presenilin2 (PSEN2) (Goate et al., 1991; Rogaev et al., 1995; Sherrington et al., 1995) cause familial autosomal-dominant AD with early-onset (less than 5% of all AD cases) and all result in increased production of Aβ, it is clear that this pathway is important. Furthermore, since the remaining 95% of AD cases, predominantly of sporadic and late-onset nature, are neuropathologically indistinguishable from familial forms, it is possible that the disease results from a combination of hereditary and environmental factors that somehow involve the APP pathway. Yet attempts to identify environmental and genetic risk factors associated with AD have not been conclusive. The risk factors identified so far are intriguing but not completely illuminating.

While one of the most popular theories for the pathogenesis of late-onset AD (LOAD) remains the "amyloid hypothesis", another exciting area is rapidly developing around vascular risk factors as a key part of AD pathology. This relationship between risk factors for vascular disease and AD appears at first glance contradictory, since vascular risk factors and the presence of cerebrovascular disease (CVD) have been considered as exclusion criteria for the clinical diagnosis of AD. Furthermore, for a number of years, epidemiological studies have considered that the co-occurrence of various forms of vascular disease and dementia has been coincidental and largely dependent on the fact they are both common disorders. However, more recently, these co-occurences, although unexplained as yet, are thought have a more pathological significance (Kehoe, 2003). In fact, recent studies suggest that microvascular disorder may contribute to AD pathogenesis and synergistically to cognitive decline related to AD pathology, but the role of cerebrovascular pathology in AD is a matter of controversy (Kalaria, 2000; Skoog, 2000; Farkas and Lutten, 2001, Panza et al., 2004a). Vascular pathology of the aging brain and AD includes cerebral amyloid angiopathy (CAA), causing lobar mass hemorrhages, small or recurrent bleeds and ischemic infarcts, microvascular degeneration, disorder of the blood-brain barrier, white matter lesions, microinfarctions, lacunes, and cerebral hemorrhages (Jellinger, 2002). Furthermore, many of the risk factors for CVD and VaD, including serum total cholesterol (TC), lipoprotein(a) [Lp(a)], diabetes mellitus, atrial fibrillation, hypertension, apolipoprotein E (*APOE*) levels, and atherosclerosis, have also been shown to increase the risk of AD. The degree to which these risk factors contribute to cognitive decline may be influenced by genetic factors, such as APOE, that have a role in both vascular disorders and AD (Wakutani et al., 2002).

This review will focus on the current knowledge on genetic vascular factors likely to be involved in LOAD, with special emphasis placed on the (APOE) and angiotensin I converting enzyme 1 (ACE1) genes. We will also comment on the role of the main nongenetic vascular factors and their reputed roles in AD.

APOLIPOPROTEIN E GENE AND AD

Numerous genes related to vascular disease have been shown to increase susceptibility for sporadic LOAD. Among these genetic risk factors, apolipoprotein E (APOE) which is located on chromosome 19 and occurs in three common alleles, ε2, ε3, and ε4, is the best-documented one. The *APOE* protein is one of the major constituents in very low density lipoproteins and plays a key role in the transport of cholesterol and other lipids among various cells of the body. The importance of *APOE* in the central nervous system (CNS) became evident with the association of the ε4 allele of APOE with familial and sporadic LOAD (Saunders et al., 1993). In the CNS and in the peripheral nervous system, *APOE* is thought to be implicated in mobilization and redistribution of cholesterol, in repair, growth, and maintenance of myelin and neuronal membranes during development and injury (Ignatius et al., 1987; Poirier, 1994; Cedazo-Minguez and Cowburn, 2001). Furthermore, autopsy-confirmed studies reported that the frequencies of APOE ε4 allele vary from 0.33 to 0.40 in AD patients, whereas they range from 0.05 to 0.14 in healthy subjects (Nalbantoglu et al., 1994). The mechanisms by which inheritance of the APOE ε4 allele increases the likelihood of developing AD are not known with certainty. In the brains of AD patients, *APOE* occurs in excessive amounts and is co-localized with the plaques (Strittmatter et al., 1993) while *in vitro* studies have shown that *APOE* has a strong stimulatory role in the polymerisation of *Aβ* into amyloid filaments (Ma et al., 1994). An explanation with respect to AD and APOE ε4 inheritance may therefore hold that the *E4* isoform of *APOE* binds *Aβ* with a higher affinity than does the *E3* isoform (Strittmatter et al., 1993). Furthermore, *APOE4* acts as a "pathological molecular chaperone" that binds to soluble *Aβ* and enhances β-pleated sheet formation and amyloid fibril stability, facilitating *Aβ* build-up in plaques, as supported by the fact that patients bearing one APOE ε4 allele have an increased number of SPs and *Aβ* deposition (Berr et al., 1994). Another explanation about the association of APOE with AD relies just on the vascular hypothesis. APOE could be implicated in the AD pathogenesis through its involvement in atherosclerosis and vascular disease. The APOE ε4 allele has been known to be associated with coronary artery disease (CAD), and the development of atherosclerosis well before its association with AD (Davignon et al., 1988). The association with CAD is presumably related to the higher levels of TC, low density lipoprotein cholesterol (LDL-C), and apolipoprotein B in subjects homozygous for the APOE ε4 genotype than in those ε2- or ε3-carriers.

In our recent study carried out in Southern Italy population (Panza et al., 1999), we showed that the frequency of APOE ε4 allele in both LOAD patients and middle aged controls is underexpressed in relation to that of other European populations. From our observations we provided a new finding that the APOE ε4 allele frequency decreases according to a geographic trend from Northern to Southern Europe in LOAD patients and normal, middle-aged adults. However, despite the low prevalence of the ε4 allele due to the geographic decreasing trend from Northern to Southern Europe, the strength of association of APOE ε4 with AD does not appear to be affected . In fact, in our subsequent study (Panza et al., 2000), we confirmed the association between AD and the presence of the APOE ε4 allele. However, after stratifying the whole sample into two subgroups to test for age of onset of AD symptoms, while an increased ε4 allele frequency was observed in LOAD patients compared to controls, the increase was not statistically significant. This finding therefore suggests that the decreasing trend of APOE ε4 allele from Northern to Southern Europe may influence the strength of association between the ε4 allele and sporadic AD with respect to age of onset of the disease. It also suggests that the observed geographic trend may be confounded by environmental variations, such as diet and lifestyle, that with APOE interaction may result in earlier onset of non-AD related illnesses thus selecting for decreased APOE ε4 allele frequencies in some aging populations.

A gene dose-dependent increase in risk of AD has also been reported, such that subjects with two copies of ε4 allele have an earlier age of onset than those with one or no copies of ε4 allele (Frisoni et al., 1995). In patients with selective memory deficits or with Mild Cognitive Impairment (the most widely used term to classify nondemented aged persons with a mild memory or cognitive impairment that cannot be accounted for any recognized medical or psychiatric condition) the presence of the ε4 allele appears to be a strong predictor of the subjects who will progress to AD in 2 or 3 years (Petersen et al., 1995). On the contrary, other studies showed that the rate of progression of AD as a whole is independent of the APOE genotype (Norrman et al., 1995; Growdon et al., 1996). Moreover, the effect of ε4 allele on annual decline in episodic memory is significantly stronger than its effect on decline in other cognitive systems (Wilson et al., 2002), suggesting that APOE genotype could influence cognitive 'phenotype' in patients with AD rather than the disease progression (Smith et al., 1998). However, the inheritance of one or two APOE ε4 alleles does not predict AD with certainty (National Institute on Aging/Alzheimer's Association Working Group, 1996) and so a person can have one or two APOE ε4 alleles and still not get the disease, or not have any APOE ε4 alleles and still develop the disease.

This fact that APOE is neither necessary nor sufficient to cause AD is the main reason why APOE is classified as a risk factor for a AD and not a causative one. Furthermore, APOE genotyping predictive value is poor and it is not very sensitive and specific, and this is the main case used to argue against proposals for the inclusion of APOE genotyping in clinical diagnoses of AD. Thus APOE testing can be performed only in patients suspected of having AD, but not in healthy people or relatives of AD patients as screening or predictive test (American College of Medical Genetics/American Society of Human Genetics Working Group on ApoE and Alzheimer disease, 1995). For a person with symptoms of dementia, APOE testing may offer additional support that any dementia may be due to AD, and thus potentially increase confidence to the clinical diagnosis (Mayeux et al., 1998). In AD Centers, clinical diagnosis is already deemed to be correct over 90% of the time without any APOE testing. When APOE genotyping is performed in combination with the clinical criteria for the disease, the specificity of the diagnosis is increased by an estimated additional 4%.

More recently, the issue of APOE involvement in AD has been broadened by reports that the quantitative expression of APOE alleles results from mutations in the promoter region of the gene (Laws et al., 2003). Two polymorphisms at sites –491 and –219 of the APOE gene promoter have been initially examined and found to produce variations in the transcriptional activity of the gene, most probably through differential binding of nuclear proteins (Artiga et al., 1998b). The effects of both these polymorphisms on AD risk have been analyzed in different case-control studies with inconsistent results (Artiga et al., 1998a; Lambert et al., 1998a; Song et al., 1998; Laws et al., 1999; Casadei et al., 1999; Chen et al., 1999; Rebeck et al., 1999; Zurutuza et al., 2000; Beyer et al., 2002; Lambert et al., 2002; Roks et al., 2002).

Further two polymorphisms have been identified in the APOE gene, respectively at –427 site of the promoter region (Artiga et al., 1998b) and at –186 base pairs (bp) of the gene TATA box, within a consensus sequence of a potential transcriptional binding site (Thl/E47cs) (Lambert et al., 1998b). The studies on the association among the –186 and –427 APOE polymorphisms and sporadic AD have also been contradictory (Artiga et al., 1998b; Lambert et al., 1998a; Lambert et al., 1998b; Wang et al., 2000; Zurutuza et al., 2000; Zill et al., 2001; Beyer et al., 2002), whereas a linkage between the alleles of all the APOE regulatory region polymorphisms and the APOE coding region alleles have been generally found. These findings confirm that, while remaining imperfect, the APOE ε4 status is still a good predictor of risk for AD. However, since it has been estimated that APOE accounts for only 45-60% of the genetic risk for sporadic LOAD (Rubinsztein and Easton, 1999), and that APOE represents only 7-9% of the total variance in age of onset of familial AD (Warwick Daw et al., 2000), it turns out that additional, yet to be identified risk factors, are involved in the development of LOAD.

Angiotensin I Converting Enzyme 1 Gene and AD

Building on the improving understanding of the genetics of AD, research has been continuing to look for clues as to which genetic loci related to vascular disease are associated with AD. Among these genetic candidate links between AD and vascular risk, the renin angiotensin system (RAS) has been considered attractive target. One of the key enzyme of the RAS is the angiotensin I converting enzyme 1 (*ACE*), a dipeptidyl carboxypeptidase that plays an important role in blood pressure regulation and electrolyte balance by hydrolyzing angiotensin I into angiotensin II, and inactivating bradykinin, a potent vasodilator. Besides being present as a membrane-bound enzyme on the surface of vascular endothelial cells, *ACE* also circulates in plasma. Circulating *ACE* levels show extensive interindividual variability and are highly genetically determined (Cambien et al., 1992). Furthermore, an increased *ACE* activity in tissue homogenates from brain regions specifically involved in acetylcholine production was found in AD cases, as well as reduced *ACE* levels in cerebrospinal fluid (CSF) of AD and other dementia types compared with neurologically normal controls, and selective increase of *ACE* density in AD temporal cortex compared with controls (Kehoe, 2003).

The *ACE* gene (ACE1) is located on chromosome 17, and contains an insertion/deletion (I/D) polymorphism, due to the presence or absence of a 287bp alu-type sequence, in its 16th intron. Variation in ACE1 has been shown to cosegregate with serum *ACE* levels and tissue *ACE* activities (Rigat et al., 1992). The physiological importance of the I/D polymorphism relies on the fact that subjects bearing one of two copies of the ACE1*D allele have higher circulating and tissue *ACE* levels than those with two copies of the ACE1*I allele (Tiret et al., 1992). Of interest is that homozygosity of the ACE1*D allele has been linked to an increased risk for myocardial infarction, left ventricular hypertrophy (Schunkert, 1997), and in-stent restenosis after angioplasty (Amant et al., 1995). More recently this polymorphism has been associated with an increased risk for development of sporadic LOAD (Kehoe et al., 1999a) and to date over 20 studies have been conducted to try and replicate these findings (Kehoe, 2003). Of these, a number of studies have supported the original findings that inheritance of the ACE1*I allele is a risk factor for AD (Hu et al., 1999; Narain et al., 2000; Cheng et al., 2002; Yang et al., 2002). Another study in a large French population found an association between ACE*D allele and a dementia cohort involving various different diagnoses (Richard et al., 2001), while other studies found either a slight but significantly increased frequency of ACE1*I in AD patients (Alvarez et al., 1996) or did not detect any effect of ACE1 polymorphism (Scacchi et al., 1998; Palumbo et al., 1999; Myllykangas et al., 2000; Zuliani et al., 2001; Buss et al., 2002). Furthermore, one US study suggested a gender-specific effect of ACE1*I allele on AD (Crawford et al., 2000a) and a different study provided evidence of an interaction between ACE1*I and APOE ε4 alleles in increasing the risk of AD (Mattila et al., 2000). Only one study on Canadian and Russian populations has demonstrated an association between ACE1*D allele inheritance and AD but only amongst persons 66-70 years old (Farrer et al., 2000). Recently, APOE and ACE1 genes were evaluated in 150 consecutive patients 3 months after suffering acute stroke; no association was found between patients who progress to post-stroke dementia (PSD) and these vascular genes, suggesting that additional factors may contribute to the pathogenesis of PSD (Arpa et al., 2003). Thus, it is becoming apparent that the possible involvement of the ACE1 polymorphism in increasing AD risk is complex. In a very recent study (Panza et al., 2003b), we investigated the association between the ACE1 polymorphism and increased risk of AD in Southern Italy population. We did not find any evidence to support the suggested role of ACE1*I allele in AD, but we found a non-significant relationship between lower age of onset of AD and patients with ACE1*I/I genotype compared to those with the ACE1*D/D genotype. This finding, although not statistically significant, suggests that the presence of ACE1*I allele might provide a basis for anticipating the onset of the disease. Our findings in a population from Southern Italy confirm the results of a recent study performed in Sicily, in which there were no significant differences in ACE1 genotype or allele frequencies between 149 cases and 149 age and sex matched controls, even after stratification for APO ε4 carrier status (Monastero et al., 2002).

What is noteworthy from all these findings is that, as with our previous observations on APOE, we have identified that ACE1*I allele frequency decreases according to a geographic trend from Northern to Southern Europe in AD patients and age-matched controls, possibly explaining to some extent the different and conflicting patterns of ACE1 and AD association reported by many groups. The geographic trends may be markers for differing environmental factors likely to be interacting with ACE1 and introducing different allele selection patterns at different ages in various countries. Very recently, Kehoe and colleagues (Kehoe et al., 2003) performed the largest meta-analysis to-date of previously published studies on ACE1 polymorphism, and found significant association between ACE1*I allele and risk for AD. In the same study the authors genotyped additional single nucleotide polymorphisms (SNPs) in the ACE1 gene and conducted haplotype analysis in several cohorts of AD and control subjects and used the genotype and haplotype data to test for genetic impact on CSF levels of the longer toxic form of *Aβ* (Aβ42). Kehoe and colleagues found that haplotypes associated with lower *ACE* enzyme levels and which would normally contain the ACE1*I allele were not only strongly associated with increased AD risk but were also linked with detrimental levels of CSF-Aβ42 levels. These *in vivo* findings on CSF are important to follow and support earlier *in vitro* findings that *ACE* can degrade synthetic forms of *Aβ* thus precluding the formation of *Aβ* aggregates (Hu et al., 2001). Thus it is apparent that, with biological interactions between *ACE* and *Aβ*, levels of enzyme that are linked to genetic inheritance do have a new relevance with respect to risk of developing AD [for a more detailed discussion of this and other interesting links between *ACE* and AD see the review by Kehoe (Kehoe, 2003)]. There are therefore strong suggestions that pathways linking the effects of *APOE*, *ACE*, lipid, and *Aβ* metabolism or modification may be implicated in AD aetiology; it remains to investigate if any additional vascular risk factors genes are involved with AD.

Current Search for Other Vascular Susceptibility Genes Related to LOAD

The methyltetrahydrofolatereductase (*MTHFR*) gene (MTHFR) has been considered another new plausible susceptibility candidate gene for AD, as it is involved in the metabolism of homocysteine, and high homocysteine plasma levels (hyperhomocystinemia) have been considered an independent risk factor for CAD and stroke (Riggs et al., 1996). A common mutation in the gene (C→T) has been identified that results in thermolability and mildly reduced activity of the encoded enzyme, and causes hyperhomocystinemia (Frosst et al., 1995). The question whether carrying the C→T mutation is a risk factor for VaD and AD has been addressed in different studies (Chapman et al., 1998; Gussekloo et al., 1999; Nishiyama et al., 2000; Pollak et al., 2000; Brunelli et al., 2001; Prince et al., 2001; Zuliani et al., 2001). However, just one study reported an association with MTHFR and senile dementia (Nishiyama et al., 2000), though several data supported the connection between hyperhomocystinemia and cognitive decline and AD. The overall lack of findings implicating MTHFR with AD could support the suggestion that the increased levels of homocysteine observed in AD patients represent an epiphenomenon of dementia rather than a cause of it.

However, in a recent study examining the relation of the total homocysteine plasma level measured at baseline and eight years earlier with the risk of newly diagnosed dementia on follow-up on a sample of 1,092 subjects without dementia from the Framingham Study, the risk of AD nearly doubled with homocysteine plasma level greater than 14 micromol per liter. Thus increased homocysteine plasma levels seem to be a strong, independent risk factor for the development of dementia and AD, suggesting that the elevated homocysteine levels precede the onset of dementia and do not result from dementia-related nutritional and vitamin deficiencies (Seshadri et al., 2002; Solfrizzi et al., 2003).

The overlap between vascular disease and AD has put forward other genetic vascular risk factors likely involved in AD, such as nitric oxide synthase 3 (*NOS3*), paraoxonase 1 (*PON1*), low density lipoprotein related-receptor protein (*LRP1*), oxidized LDL-receptor 1 (*OLR1*) genes (Table 1). The *NOS3* gene (NOS3), located on chromosome 7, encodes endothelial *NOS* which produces the majority of circulating nitric oxide, a factor with potent vasodilatatory properties that has been implicated in various disorders including hypertension and myocardial infarction (Hibi et al., 1998; Uwabo et al., 1998). NOS3 has a common structural polymorphism (G→A) in its 7th exon which has been reported to be associated with LOAD, independent of APOE gene effects (Dahiyat et al., 1999). However, subsequent studies in different populations failed to replicate this original finding (Alvarez et al., 1999; Crawford et al., 2000; Higuchi et al., 2000; Sanchez-Guerra et al., 2001; Singleton et al., 2001; Liou et al., 2002; Tedde et al., 2002; Monastero et al., 2003), so that, even if not wholly excluded, this genetic polymorphism is unlikely to be a significant susceptibility marker for AD.

Table 1. Vascular susceptibility genes for sporadic late-onset AD.

Genes (polymorphism)	Role in vascular risk
APOE (chromosome 19; exon 4 112 and 158 polymorphisms) APOE (promoter polymorphisms)	Effect on levels of total cholesterol, LDL-cholesterol, serum *APOE*, and apolipoprotein B Regulation of quantitative expression of *APOE*
ACE1 (chromosome 17; intron 16 I/D polymorphism)	Effect on plasma *ACE* levels
MTHFR (chromosome 1; 677C/T polymorphism)	Effect on homocysteine plasma levels
NOS3 (chromosome 7; exon 7 G/A polymorphism)	Strong effect on vasodilatation
PON1 (chromosome 7; 192 A/G polymorphism)	Participation in the prevention of LDL oxidation
LRP1 (chromosome 12; exon 3 C/T and tetranucleotide repeat polymorphisms)	Regulation of proteinase and lipoprotein levels
OLR1 (chromosome 12; intron 4 A/G, intron 5 G/T, and 3' UTR C/T polymorphisms)	Binding to atherogenic oxidized LDL and involvement in oxidized LDL-induced apoptosis

AD = Alzheimer's disease
APOE = apolipoprotein E
ACE1 = angiotensin I converting enzyme 1
MTHFR = methyltetrahydrofolatereductase
NOS3 = nitric oxide synhase 3
PON1 = paraoxonase 1
LRP1 = low-density lipoprotein receptor-related protein 1
OLR1 = oxidized LDL-receptor 1

Two polymorphisms in the gene encoding *PON1* [nucleotide position 55 (M or L) and position 192 (R or Q)] have been identified and associated with CAD (Serrato et al., 1995; Mackness et al., 1998a). *PON1* is an A-esterase with peroxidase-like activity present on the surface of high density lipoproteins (HDL) and participates in the prevention of LDL oxidation. Serum levels of *PON1* are genetically determined and strongly influenced by the polymorphism on the position 192 of the gene (Mackness et al., 1998b). A significant decrease in *PON1* activity has been reported in both AD and VaD patients compared to the healthy control subjects, suggesting that defect in this enzyme might play a role in the pathogenesis of both AD and VaD (Paragh et al., 2002). In another study involving AD, VaD, and mild cognitive impairment patients, compared with healthy control subjects, serum *PON1* activity has been shown to be a discriminating factor between AD and VaD (Dantoine et al., 2002). The PON1 192 polymorphism has been also reported as a possible reliable marker to distinguish patients with AD from patients with VaD and from healthy subjects, indicating that differences in 192 polymorphism distribution between AD and VaD might partially explain the significant difference in serum *PON1* found in these two types of dementia (Dantoine et al., 2002). In contrast, other association studies, if not numerous, did not find any link between the PON1 192 polymorphism and AD or VaD (Sodeyama et al., 1999; Zuliani et al., 2001; Pola et al., 2003). Some more studies are needed to further investigate and clarify the hypothetical role of this gene and polymorphisms in various forms of dementia.

The *LRP1* gene (LRP1), noted to be involved in vascular disorders, is a large endocytic and multifactorial receptor that regulates proteinase and lipoprotein levels by their catabolism. LRP1 is an attractive candidate gene for AD for several reasons. Beyond its involvement in lipid metabolism, *LRP1* is the main *APOE* receptor in the brain, and binds *APP* and alpha-2 macroglobulin (Rebeck et al., 1995, 2001), all of which are proteins involved in AD pathogenesis. Further, LRP1 is expressed in immunoreactive SPs, dystrophic neurites and reactive astrocytes (Tooyama et al., 1995; Arelin et al., 2002), and is located on chromosome 12 within the AD linked region (Pericak-Vance et al., 1997; Rogaeva et al., 1998; Kehoe et al., 1999b; Scott et al., 1999). A silent C→T polymorphism in the exon 3 of LRP1 gene has been identified and associated with LOAD in some studies (Kang et al., 1997; Baum et al., 1998; Hollenbach et al., 1998; Lambert et al., 1998c; Kamboh et al., 1998), while contradictory results have been found in others (Woodward et al., 1998; Beffert et al., 1999; Hu et al., 2000; McIlroy et al., 2001; Sanchez-Guerra et al., 2001; Perry et al., 2001). In addition, a few studies have reported an LRP1 association with AD that is modulated by APOE ε4 allele and gender (Bullido et al., 2000; Hatanaka et al., 2000), and a meta-analysis of published studies has supported an overall involvement of the LRP1 gene in AD risk (Sanchez-Guerra et al., 2001b). This suggests therefore that the role of this polymorphism in AD risk might be explained by differing levels of linkage disequilibrium with other causative LRP1 or non LRP1 polymorphisms across different populations. Support of this hypothesis comes from the fact that a common tetranucleotide repeat polymorphism has also been reported in the 5' region of the LRP1 gene and that a few reports have shown evidence of differing strengths of association between different alleles in this polymorphism and AD (Lendon et al., 1997; Lambert et al., 1999; Wavrant-DeVrieze et al., 1997). Again, the majority of the studies investigating the tetranucleotide polymorphism did not observe any link with increased risk for the disease (Clatworthy et al., 1997; Fallin et al., 1997; Scott et al., 1998; Kamboh et al., 1998; McIlroy et al., 2001; Bullido et al., 2000; Hatanaka et al., 2000). We recently examined the relation between the polymorphism in the exon 3 of LRP1 gene and AD in a cohort of 166 sporadic AD patients and 225 sex- and age-matched nondemented controls from Southern Italy. No statistically significant differences were found in LRP1 genotype and allele frequencies between the whole AD sample and controls, nor in early- and late-onset subsets of AD patients, nor between LRP1 alleles and AD among APOE allele, age, or gender strata. However, comparing our results with the findings from other European populations, the LRP1 C allele frequency showed a statistically significant decreasing trend from Northern to Southern regions of Europe, with a concomitant increase in LRP1 T allele frequency, but only in AD patients. Finally, in AD sample, a decreasing geographical trend from North to South of Europe was found for LRP1 CC genotype, and an inverse trend for LRP1 CT genotype frequency. We suggest that these regional variations in LRP1 genotype and allele frequencies in AD could be related to the different patterns of association between this polymorphism and the disease in various European studies (Panza et al., 2004b), as for APOE and ACE1 genes (Panza et al., 2000; Panza et al., 2004b). Thus, despite the strong biological evidence implicating *LRP1* and AD pathogenesis, the idea that a genetic mutation in LRP1 could influence the risk for AD remains a hypothesis.

OLR1 is a type II membrane protein, which can act as an endocytosis receptor for atherogenic oxidized LDL and is expressed in vascular endothelium, as well as in smooth muscle cells, macrophages and platelets. *OLR1* also mediates oxidized LDL-induced apoptosis (Kataoka et al., 2001), suggesting its involvement in both lipid metabolism and apoptosis. *OLR1* expression has been found upregulated in hypertension, diabetes mellitus and atherosclerosis (Mehta 2002). An important role of *OLR1* in thrombosis and myocardial infarction has also been implied (Chen et al., 2001; Tatsuguchi et., 2003). Recently, the *OLR1* gene (OLR1), which is also located on chromosome 12 within the linkage region to AD (Pericak-Vance et al., 1997; Rogaeva et al., 1998; Kehoe et al., 1999b; Scott et al., 1999), has been screened and three different polymorphisms, in intron 4 (A→G), intron 5 (G→T) and 3' UTR of the non-coding exon 6 (C→T), have been found and associated with LOAD (Luedecking-Zimmer et al., 2002). The association with AD was significant after stratification for APOE alleles for all three polymorphisms. The strongest effect was observed for the 3' UTR polymorphism which has been also found to affect the binding of a transcription factor in an allele-dependent manner. Until other studies will investigate OLR1 polymorphisms for association with AD, it is too premature to decide whether observed OLR1 and APOE ε4 interactions will be implicated in AD.

Possible Role of Nongenetic Vascular Factors in AD

Various nongenetic factors known to increase the risk of vascular disease, including TC, Lp(a), diabetes mellitus, atrial fibrillation, hypertension, *APOE* levels, atherosclerosis, have been evaluated as risk factors for LOAD.

High serum TC during middle age or early old age seems to confer an increased risk of AD in old age (Notkola et al., 1998). Epidemiological studies showed that the onset of AD occurs earlier in APOE ε4-carriers with high serum TC (Jarvik et al., 1995). On the contrary, a cross sectional study reported an association between AD and lower TC (Kuusisto et al., 1997) which has been supported by a work by our group (Solfrizzi et al., 2002). Two Finnish population studies reported that individuals with AD had high TC levels 15-30 years before disease onset (Notkola, 1998) and cholesterol located in the brain cortex is increased in AD (Sparks, 1997). However, even if high TC often clusters with hypertension, these two entities seem to be independent risk factors of AD (Kivipelto, 2002).

Statins are used to lower TC in humans. Several experimental studies suggested that lowering TC level with statins may be important in AD. Simvastatin and lovastatin reduced intracellular and extracellular levels of Aβ-42 and Aβ-40 in primary cultures of hippocampal neurons and mixed cortical neurons (Fassbender, 2001). One explanation may come from a report that lowering cellular cholesterol with lovastatin reduced the conversion of APP to Aβ in cell culture (Simons, 1998).

In support of the hypothesis that statins may prevent AD, two case-based studies reported a reduced frequency of dementia and AD in statin-treated individuals (Wolozin, 2000). The recent longitudinal observation from the Canadian Study of Health and Aging (CSHA) (Rockwood, 2002), that incidence of AD was reduced in individuals using statins further supports the hypothesis that TC may be important in AD. A small hospital-based case-control study suggested that AD patients had higher mean plasma TC levels. A Finnish cohort study of 444 men found that subjects who developed AD had higher TC levels recorded 20-30 years earlier than those who did not develop the disease (OR 3.1; 95% CI 1.2-8.5) (Notkola, 1998).

Lipid levels, however were lower in AD than in non-demented subjects of a similar age in the Rotterdam Study and in a cross-sectional study (Kuusisto, 1997).

The Hisayama study found no association between serum TC and AD over a 7-year follow-up (Leibson, 1997). These findings were confirmed by another study that found no difference in serum TC or hyperlipidaemia between subjects with AD, VaD, or nondemented subjects (Boston, 1999). On the contrary, a recent population based prospective study with a 21- year follow-up in Finland found that high TC levels in middle age were associated with increased risk of AD in later life (Kivipelto, 2001). The Rotterdam study showed a reduced risk of AD associated with fish consumption (Kalmijn, 1997). Finally, two observational studies recently found that the prevalence of AD among patients taking statin drug was 60-73 % lower than in patients not on statins (Wolozin, 2000).

The effect of APOE genotype might depend upon serum TC levels (Yarvik, 1995). Some studies suggest that high TC levels are associated with an increased risk of AD. High caloric intake and fats in individuals with the APOE ε4 allele may also increase the risk for developing AD (Luchsinger, 2002). High serum lipoprotein levels, especially among people with cardiovascular and cerebrovascular disorders, may be a common and potentially modifiable risk factor for cognitive deterioration. High levels of LDL-C tend to be associated with an increased likelihood of cognitive impairment, and the reduction of LDL-C levels with statins seems to lower the likelihood of cognitive impairment, and the reduction of LDL-C levels with statins seems to lower the likelihood of cognitive decline (Yaffe, 2002).

Some epidemiological studies also indicate that the prevalence of AD might be decreased in patients treated with statins (Crisby, 2002). These substances show pleiotropic effects including regulation of eNOS activity and NO production, modulation and inflammatory processes, antioxidant activity, angiogenesis, and immunomodulation, as well as reduction of free TC and inhibition of cholesterol ester accumulation in macrophages by inhibiting LDL endocytosis and reducing mevalonate byproducts essential for cholesterol esterification. Other potential effects of statins include reduction of atherosclerotic plaque formation, endothelial protection, and reduction of oxidized LDL (Crisby, 2002).

Some authors have proposed the administration of statins as a preventive strategy in AD suggesting as probable actions for prevention of the following: stimulation of α-secretase activity; modulation of APP metabolism and Aβ production; immunomodulation leading to attenuation of inflammatory markers; reduction of lipoprotein oxidation and free radical injury; increase in cerebral endothelial NOS and cerebral circulation; activation of signal transcription through depletion of downstream isoprenoids. However, though the hypothesis looks attractive, the levels of TC and LDL-C in dementia are within a normal range and this cluster of patients does not fulfil criteria for receiving statins at conventional doses. Different players in lipid trafficking and metabolism are APOE and LRP1 genes-derived products (Cacabelos, 2002). The LRP1 is a multifunctional neuronal receptor for both APOE and α2-macroglobulin that mediates the clearance of APOE/Aβ and α2M/Aβ complexes. Isoforms of APP are also ligands for LRP1 gene and a neurovascular excess of LDL may interfere with APP processing and Aβ clearance contributing to increased Aβ accumulation in SPs and in microvascular angiopathy deposits thus incrementing neurotoxicity and cerebrovascular dysfunction, as well (Hyman et al., 2000).

Increased density of SPs but not of NFTs has been observed in brains of nondemented individuals with CAD, often associated with APOE ε4 allele and increased Aβ production has been found by increased cholesterol in both culture studies and transgenic mouse models of

AD (Sparks et al., 1996,1997). Furthermore, in our recent study we analyzed the relationship between sporadic AD and Lp(a), a LDL-like particle with apolipoprotein(a), that is believed to have atherogenic and thrombotic properties (Eaton et al., 1987) and has been associated with CVD as well as VaD (Zenker et al., 1986; Urakami et al., 2001). We found that Lp(a) serum levels were significantly associated, according to a nonlinear relationship, with an increased risk for AD, independently of APOE genotypes, and dependent on age (Solfrizzi et al., 2002).

Epidemiological and cross sectional studies have shown a link between hypertension and AD (Skoog et al., 1995; Kuusisto et al., 1997). Neuropathological evidence has also linked AD and hypertension, describing higher prevalence of SPs and NFTs in hypertensive patients without dementia than in normotensives (Sparks et al., 1995). The association between hypertension and dementia has been examined in the Goteborg study, a longitudinal population based examination of a cohort of 382 subjects without dementia at age 70 that found that the 18 partecipants who developed dementia at ages between 79 and 85 had had significantly higher systolic blood pressure (SBP) at 70 years and higher SBP at 70 and 75 years than those who did not develop dementia (Skoog, 1996). In addition, higher diastolic blood pressure (DBP) at 70 and higher DBP at 70 and 75 than those who did not develop dementia. Finally, higher DBP at age 70 was also associated with development of AD specifically in 10 of these subjects. This finding was verified in the Honolulu-Asia Aging Study where 3703 Japanese-American men were examined during 1965-71, and re-examined for dementia in 1991. In this study a high SBP (above 160 mmHg) or DPB (above 90 mmHg) predicted the diagnosis of AD and VaD 20-26 years later (Launer, 2000). The association was strongest among those who were never treated for hypertension.

The latest follow up in the Rotterdam study (Ruitemberg, 2000) reported that high blood pressure precedes AD approximately of nine years. In a recent report from the 6-year follow-up of the Kungsholmen Study, high SBP (above 180 mmHg) was related to an increased incidence of AD (Qui, 2003a). Also low DBP (below 140 mmHg), was related to an increased incidence of AD, indicating that pulse pressure may be important (Qui, 2003b). Others report that blood pressure declines during the course of AD and with increasing dementia severity. Finally a longitudinal study reported that low blood pressure was related to low Mini-Mental State Examination (MMSE) score at baseline, but not to incident AD when controlling for baseline MMSE (Guo, 1999). It's likely that the low blood pressure observed during pre-clinical and manifest dementia is secondary to the brain lesions in AD. A 20-year follow-up Swedish study of hypertensive men showed a link between high blood pressure and cognitive impairment (Kilander, 1998).

A neuropathological study described higher counts of NFTs and SPs in hypertensive individuals than in normotensives (Schmidt, 2000). A recent prospective population based study in Finland with a 21-year follow-up found that increased systolic blood pressure increased the risk of AD in later life (Kivipelto, 2001). Hypertension is a strong risk factor for cortical and lacunar infarction and for periventricular white matter rarefaction; it can cause deterioration in cognitive function or it may influence neurotransmission or regional cerebral blood flow. Pathologically, white matter hypodensities are more frequent in the brains of AD patients than in cognitively normal people.

Long-standing or untreated hypertension may cause lipohyalinosis and thickening of the vessel walls, which in turn narrow the lumen of the small perforating arterioles of the brain. The rigidity of cerebral vessels in response to decrease in cerebral perfusion pressure may cause lowered cerebral blood flow (DeJong, 1997; Farkas and Luiten, 2001). Hypertension is also known to cause breakdown of the blood-brain barrier (BBB) integrity which may consequently cause cerebral edema and induce systemic elements into the brain (Skoog, 2000). Low blood pressure, has also been suggested to be a risk factor for AD (Passant, 1997; Guo, 1997).

In the Rotterdam study, atherosclerosis was associated with both VaD and AD (Ott, 1998). The indicators of atherosclerosis used were carotid artery thickness and plaques measured by ultrasonography and ankle to brachial systolic blood pressure. Moreover, the degree of atherosclerosis was seen to correlate with the increased prevalence of AD. The OR for AD in those with severe atherosclerosis was 3.0 (95% CI 1.5-6) compared with those without any manifest atherosclerosis (Stewart, 1999). Neuropathologically, VaD may be viewed as a generalised vascular disorder which involves both large and small vessels, while AD is characterised by a slowly progressive capillary dysfunction in the absence of widespread focal infarctions (Ott, 1998).

Positive association between diabetes mellitus and AD has been reported from different longitudinal studies. Diabetes mellitus is known to be positively associated with vascular disease, and in previous studies on AD subjects with diabetes mellitus were often excluded. A retrospective cohort case note study of 1455 diabetic subjects from Rochester, New york reported an increased risk of AD in diabetics (RR 1.7; 95% CI 1.3-2.1) particularly in men (Leibson, 1997). A report from the Honolulu-Asia Aging Study, including 216 subjects who underwent autopsy, found that type 2 diabetes was associated with incidence of total dementia, AD and VaD (Peila, 2002). The association between diabetes and AD was particularly strong among carriers of the APOE ε4 allele. Participants with type 2 diabetes and the ε4 allele also had an increased number NFTs in the cortex and hippocampus, and they had a higher risk of CAA (Peila, 2002). There are several potential biological mechanism underlying relationship between diabetes mellitus and AD, including dysfunction of cerebral microvessels (Mooradian, 1997) and oxidative stress (Singh, 2001). CNS dysfunction in diabetes may also be mediated by other diabetes-related conditions in addition to hypertension, such as obesity, decreased peripheral circulation, ischemia, uremia, peripheral and autonomic neuropathy, the use of multiple drugs, other neuroendocrine, or neurochemical changes, and the consequences of hyper- or hypoglycaemia, hyperinsulinemia, and ketoacidosis (Luscher, 1994).

The Hisayama study, a seven year prospective population based study in elderly Japanese diabetics, showed a more than twofold increased risk of AD (RR 2.2; 95% CI 1.0-4.9) (Yoshitake, 1995). A large prospective study in Rotterdam among more than 6000 subjects followed over two years showed a relative risk of AD of 1.9 (95% CI 1.2-3.1). The risk of AD was highest among subjects who received insulin treatment (RR 4.3; CI 1.7-10.5) (Ott, 1998). The exact mechanism by which diabetes mellitus increases the risk of dementia is not known. Vascular risk factors associated with diabetes mellitus are important but there may be a direct effect of glucose on cerebral vasculature. One hypothesis is that in AD the hyperglycaemia may potentiate the deleterious effect of other pathological processes taking place, such as amyloid deposition (Messier, 1996). Advanced glycation end products have been demonstrated in NFTs and can cause cellular death associated with AD (Sasaki, 1998). A cross-sectional population based showed that not only diabetes but impaired glucose tolerance and hyperinsulinaemia in non-diabetics was associated with cognitive impairment (Kalmijin, 1995).

Hyperinsulinaemia associated with insulin resistance has been reported in AD patients in the Kuopio population study (Kuusisto, 1997). Insulin IGF-1 have been implicated in regulation of tau protein phosphorilation. Other pathogenic mechanism implicated in AD pathogenesis include BBB dysfunction, free radical damage, inflammatory mechanism and oxidative damage (Stewart, 1999).

The mechanism through which diabetes mellitus increases the risk of CVD is not entirely clear. Evidently diabetes is associated with micro- and macrovascular changes. However the risk of stroke is increased independently from other vascular risk factors, and it has been suggested that there might also be a direct effect of glucose on the cerebral vasculature. Intensive treatment of diabetes and good metabolic control reduce the risk of stroke (Kuusisto, 1994). It has been suggested that in AD the hyperglycemia has a direct effect on cerebral functioning, and the resulting damage may accumulate with the number of hypoglycaemic periods (Messier, 1996). In the Rotterdam study, atrial fibrillation was significantly more frequent in subjects with dementia (odds ratio 2.3; 95% CI 1.4-3.7) than in those without. There was a 1.8-fold excess risk of AD in patients with atrial fibrillation. This association was statistically significant when adjusted for age and sex. The strongest association occurred in AD patients with clinical or MRI evidence of CVD. The strongest associations was found with AD, not with VaD (Ott, 1997).

Both CVD and AD are common in the elderly. Current trends suggest they coincide more than by chance alone (Pasquier et al., 1998). This findings is in accord with the number of vascular variant of AD. Patients with CVD who progress to develop VaD may acquire AD changes in terms of amyloid deposits and CAA (Ballard et al., 2000; Esiri et al., 1997). It is plausible that ischemic events induce the amyloid precursor protein and other Alzheimer pathology associated factors, which may interact with glial cells to generate the lesions in hypoperfused areas of the brain (Kalaria et al., 1993).

Although an increased risk for stroke episodes or transient ischemic attacks have been implicated to induce AD-type degenerative changes, the converse can also occur. Individuals with AD may be at increased risk for stroke and cerebral infarction, the main diagnostic features in VaD. Two studies (Ferrucci et al., 1996; Gale et al., 1996) reported that elderly individuals without previous stroke and very low cognitive ability were subsequently at increased risk for later development of stroke. This observation has been corroborated by circumstantial evidence suggesting that rapid development of cerebral infarctions in some AD patients occurs because of a strong interaction between severity of amyloid infiltration of the media and adventitia of cerebral vessels, and hypertension (Olichney et al., 1995). After a stroke, about 25% of patients are found to be suffering from dementia, only a few from multi-infarct dementia; most show clinical features of AD (Tatemichi, 1994). Autopsy studies show that these AD patients have significant AD-type changes (Ballard, 2000). In patients cognitively intact a few months after stroke, the risk of developing delayed dementia was up to six times in hospital based studies (Tatemichi, 1994) and nine times in population-based studies (Kokman, 1996) compared with general population. Stroke may participate, aggravate or accelerate a pre-existing process or predisposition to AD (Snowdon, 1997). Characteristics, such as male, increased age, low socioeconomic status, low cognitive status before stroke, left-sided lesions, and presence of other vascular risk factors have been associated with a higher risk of dementia post stroke, while infarct volume seems less important (Kokman, 1996).

Various indicators of atherosclerosis have been related to AD. The prevalence of the disease appears to increase with the degree of atherosclerosis. An interaction between atherosclerosis and *APOE* has also been reported such that subjects bearing at least one APOE ε4 allele and severe atherosclerosis had the highest risk of AD (Hofman et al., 1997). APOE is the product of APOE gene on chromosome 19. It plays a role in triglyceride-rich lipoprotein metabolism and cholesterol homeostasis (Curtiss, 2000). The presence of the APOE ε4 allele has been associated with CAD in middle-aged men and with the development of atherosclerosis (Wilson, 1994). APOE has an important role in central nervous system repair. It influences recovery after a variety of neurological insults such as head trauma and stroke, and it may have neurotrophic, immunomodulatory and antioxidant effects (Laskowitz, 1998). The presence of the ε4 allele has been associated with poor repair after neuronal damage that results from amyloid β peptide deposition, oxidative stress and excitotoxic aminoacids (Mahley, 1997). The increased density of SPs seen in hypertensive patients seems to correlate with the APOE ε4 allele dose-frequency (Sparks, 1996). APOE ε4 allele is a genetic risk factor for dementia with stroke, including AD and VaD (Hofman, 1997). The presence of APOE ε4 was associated with poor performance on cognitive testing in the Cardiovascular Health Study (Kuller, 1998). In the Zutphen elderly study cohort (Kalmijn, 1996), CVD and the presence of APOE ε4 had a synergistic effect on cognitive decline. This finding was corroborated by a study that pooled data from two population-based studies (Rotterdam and New York). Patients with dementia and stroke had a higher frequency of APO ε4 allele than controls (Slooter, 1997).

Atherosclerosis was a risk factor for dementia and its major subtypes, AD and VaD, in patients enrolled in the Rotterdam study (Hofman, 1997), particularly in APOE ε4 carriers. Elevated TC concentrations in middle age may be an intermediate factor through which the APO ε4 allele is associated with the risk of AD (Notkola, 1998). However the risk of AD and VaD in the presence of APOE ε4 allele is independent of the presence of atherosclerosis (Slooter,1999). Patients older than 85 years with white matter lesions have an increased risk of dementia only if they also have the APOE ε4 allele. Possession of the APOE ε4 allele is associated with an increased risk of dementia only in subjects who have white matter lesions (Skoog, 1998). APOE may neither be sufficient nor necessary to produce dementia in old age. Dementia in the very old results from the interaction of several pathological processes (Skoog, 1998).

Serum *APOE* concentrations have been shown to modulate lipid metabolism (Bohnet et al., 1996), suggesting that *APOE* concentration might be a risk factor for CAD and CVD (Siest et al., 1995). A role of serum *APOE* concentration has been also suggested in AD, although the data on the association among serum *APOE* levels and AD are controversial (Taddei et al., 1997; Kuo, 1998; Scacchi et al., 1999). In a recent study performed by our group on AD patients, and age-matched healthy controls, centenarians, and young healthy adults, we found that the presence of the APOE ε4 allele was associated with lower serum *APOE* levels, whereas the ε2 allele with higher levels (Panza et al., 2003a). The difference in serum *APOE* concentration among AD cases and age-matched controls was not statistically significant, suggesting not only that serum *APOE* levels are unrelated to AD, but that AD patients are somewhat protected from age-associated increases in serum *APOE* levels. We also stressed the important role of the APOE locus in regulating serum *APOE* levels, and excluded the use of serum *APOE* levels as a biochemical marker for AD (Panza et al., 2003a).

In conclusion, ongoing family and twin studies as well as family and sibling-pair linkage studies continue to argue that genetic factors may play a primary role in AD. The three familial AD genes (PSEN1, PSEN2, and APP) in which mutations are known to cause early onset autosomal-dominant familial AD account for only a very small proportion of cases of the disease. Most AD cases are represented by sporadic and late onset forms, where no obvious inheritance pattern is seen but yet they closely resemble familial AD forms both clinically and neuropathologically. This points to other genetic or nongenetic factors having a role in the greater proportion of AD cases. While theories have been held as to the role of vascular risk factors in the development of AD for a number of years, it is only in the last decade that active genetic and biological research has been conducted to try and unravel this more clearly. While vascular disease and AD seem to share various common genetic and nongenetic risk factors, the story relating to genes of mutual interest that have been investigated such as APOE, ACE1, MTHFR, NOS3, LRP1, PON1, and OLR1 is not entirely clear. However, what is apparent is that APOE remains largely the only risk factor to be accepted as having a role in both vascular disease and AD, while in the last 5 years ACE1 seems to be rapidly gaining support. Furthermore, many of the risk factors for CVD and VaD, including TC, Lp(a), diabetes mellitus, atrial fibrillation, hypertension, APOE levels, and atherosclerosis, have also been shown to increase the risk of AD. The degree to which these risk factors contribute to cognitive decline may be influenced by genetic factors, such as APOE, that have a role in both vascular disorders and AD. Thus, if real relationships between vascular disease and AD exist, as has been discussed, it is evident that any genetic components will be underlain by complex interactions of genetic susceptibility factors, environmental conditions, ethnic, and/or regional/geographic differences.

Acknowledgments

This group is supported by Italian Longitudinal Study on Ageing (ILSA) (Italian National Research Council - CNR-Targeted Project on Ageing - Grants 9400419PF40 and 95973PF40), by AFORIGE ("Associazione per la Formazione e la Ricerca in Geriatria"). The authors wish to thank Ms. Damiana Calamita, Ms. Nicoletta Lobascio, Dr. Giovanni Castellaneta, and Dr. Amalia Terralavoro, Pfizer Inc., for skillful assistance.

References

Alvarez, R.; Alvarez, V.; Lahoz, C.H.; Martinez, C.; Pena, J.; Sanchez, J.M.; Guisasola, L.M.; Salas-Puig, J.; Moris, G.; Vidal, J.A.; Ribacoba, R.; Menes, B.B.; Uria, D.; Coto, E. Angiotensin converting enzyme and endothelial nitric oxide synthase DNA polymorphisms and late onset Alzheimer's disease. *J Neurol Neurosurg Psychiatry* 67: 733-736, 1999.

Amant, C.; Bauters, C.; Bodart, J.C.; Lablanche, J.M.; Grollier, G.; Danchin, N.; Hamon, M.; Richard, F.; Helbecque, N.; McFadden, E.P.; Amouyel, P.; Bertrand, M.E. D allele of the angiotensin I-converting enzyme is a major risk factor for restenosis after coronary stenting. *Circulation* 96: 56-60, 1997.

American College of Medical Genetics/American Society of Human Genetics Working Group on ApoE and Alzheimer disease. Statement on use of apolipoprotein E testing for Alzheimer disease. *JAMA* 274: 1627-1629, 1995.

Arelin, K.; Kinoshita, A.; Whelan, C.M.; Irizarry, M.C.; Rebeck, G.W.; Strickland, D.K.; Hyman, B.T. LRP and senile plaques in Alzheimer's disease: colocalization with apolipoprotein E and with activated astrocytes. *Brain Res Mol Brain Res* 104: 38-46, 2002.

Arpa, A.; del Ser, T.; Goda, G.; Barba, R.; Bornstein, B. Apolipoprotein E; angiotensin-converting enzyme and alpha-1-antichymotrypsin genotypes are not associated with post-stroke dementia. *J Neurol Sci* 210: 77-82, 2003

Artiga, M.J.; Bullido, M.J.; Frank, A.; Sastre, I.; Recuero, M.; Garcia, M.A.; Lendon, C.L.; Han, S.W.; Morris, J.C.; Vazquez, J.; Goate, A.; Valdivieso, F. Risk for Alzheimer's disease correlates with transcriptional activity of the APOE gene. *Hum Mol Genet* 7: 1887-1892, 1998a.

Artiga, M.J.; Bullido, M.J.; Sastre, I.; Recuero, M.; Garcia, M.A.; Aldudo, J.; Vazquez, J.; Valdivieso, F. Allelic polymorphisms in the transcriptional regulatory region of apolipoprotein E gene. *FEBS Lett* 421: 105-108, 1998b.

Ballard, C.; McKeith, I.; O'Brien, J.; Kalaria, R.; Jaros, E.; Ince, P.; Perry, R. Neuropathological substrates of dementia and depression in vascular dementia; with a particular focus on cases with small infarct volumes. *Dement Geriatr Cogn Disord* 11: 59-65, 2000.

Baum, L.; Chen, L.; Ng, H.K.; Chan, Y.S.; Mak, Y.T.; Woo, J.; Chiu, H.F.; Pang, C.P. Low density lipoprotein receptor related protein gene exon 3 polymorphism association with Alzheimer's disease in Chinese. *Neurosci Lett* 247: 33-36, 1998.

Beffert, U.; Arguin, C.; Poirier, J. The polymorphism in exon 3 of the low density lipoprotein receptor-related protein gene is weakly associated with Alzheimer's disease. *Neurosci Lett* 259: 29-32, 1999.

Berr, C.; Hauw, J.J.; Delaere, P.; Duyckaerts, C.; Amouyel, P. Apolipoprotein E allele epsilon 4 is linked to increased deposition of the amyloid beta-peptide (A-beta) in cases with or without Alzheimer's disease. *Neurosci Lett* 178: 221-224, 1994.

Beyer, K.; Lao, J.I.; Gomez, M.; Riutort, N.; Latorre, P.; Mate, J.L.; Ariza, A. Identification of a protective allele against Alzheimer disease in the APOE gene promoter. *Neuroreport* 13: 1403-1405, 2002.

Bohnet, K.; Pillot, T.; Visvikis, S.; Sabolovic, N.; Sies,t G. Apolipoprotein (apo) E genotype and apoE concentration determine binding of normal very low density lipoproteins to HepG2 cell surface receptors. *J Lipid Res* 37: 1316-1324, 1996.

Boston, P.; Dennis, M.S.; Jagger, C. Factors associated with vascular dementia in an elderly community population. *Int J Psychiatry* 14: 761-766, 1999.

Brunelli, T.; Bagnoli, S.; Giusti, B.; Nacmias, B; Pepe, G.; Sorbi, S.; Abbate, R. The C677T methylenetetrahydrofolate reductase mutation is not associated with Alzheimer's disease. *Neurosci Lett* 315: 103-105, 2001.

Bullido, M.J.; Guallar-Castillon, P.; Artiga, M.J.; Ramos, M.C.; Sastre, I.; Aldudo, J.; Frank, A.; Coria, F.; Rodriguez-Artalejo, F.; Valdivieso, F. Alzheimer's risk associated with human apolipoprotein E; alpha-2 macroglobulin and lipoprotein receptor related protein polymorphisms: absence of genetic interactions; and modulation by gender. *Neurosci Lett* 289: 213-216, 2000.

Buss, S.; Muller-Thomsen, T.; Hock, C.; Alberici, A.; Binetti, G.; Nitsch, R.M.; Gal, A.; Finckh, U. No association between DCP1 genotype and late-onset Alzheimer disease. *Am J Med Genet* 114: 440-445, 2002.

Cacabelos R. The histamine-cytokine network in Alzheimer disease: Etiopathogenetic and pharmacogenomic implications. Mizuno, Y.; Fisher, A.; Hanin, I. (Eds). Mapping progress of Alzheimer's and Parkinson's diseases. Advances in Behavioral Biology, vol 51, pp 59-64, New York, Pergamon Press, 2002.

Cambien, F.; Poirier, O.; Lecerf, L.; Evans, A.; Cambou, J.P.; Arveiler, D.; Luc, G.; Bard, J.M.; Bara, L.; Ricard, S.; Tiret, L.; Amouyel, P.; Alhenc-Gelas, F. ; Soubrier, F. Deletion polymorphism in the gene of angiotensin converting-enzyme is a potent risk factor for myocardial infarction. *Nature* 359: 641-644, 1992.

Casadei, V.M.; Ferri, C.; Veglia, F.; Gavazzi, A.; Salani, G.; Cattaneo, M.; Sorbi, S.; Annoni, G.; Licastro, F.; Mariani, C.; Franceschi, M.; Grimaldi, L.M. APOE-491 promoter polymorphism is a risk factor for late-onset Alzheimer's disease. *Neurology* 53: 1888-1889, 1999.

Cedazo-Minguez, A.; Cowburn, R.F. Apolipoprotein E: a major piece in the Alzheimer's disease puzzle. *J Cell Mol Med* 5: 254-266, 2001.

Chapman, J.; Wang, N.; Treves, T.A.; Korczyn, A.D.; Bornstein, N.M. ACE; MTHFR; Factor V Leiden and APOE polymorphisms in patients with vascular and Alzheimer's dementia. *Stroke* 29: 1401-1404, 1998.

Chen, L.; Baum, L.; Ng, H.K.; Chan, L.Y.; Sastre, I; Artiga MJ; Valdivieso F; Bullido MJ; Chiu HF; Pang CP. Apolipoprotein E promoter and alpha2-macroglobulin polymorphisms are not genetically associated with Chinese late onset Alzheimer's disease. *Neurosci Lett* 269: 173-177, 1999.

Chen, M.; Kakutani, M.; Naruko, T.; Ueda, M.; Narumiya, S.; Masaki, T.; Sawamura, T. Activation-dependent surface expression of LOX-1 in human platelets. *Biochem Biophys Res Commun* 282: 153-158, 2001.

Cheng, C.Y.; Hong, C.J.; Liu, H.C.; Liu, T.Y.; Tsai, S.J. Study of the association between Alzheimer's disease and angiotensin-converting enzyme gene polymorphism using DNA from lymphocytes. *Eur Neurol* 47: 26-29, 2002.

Chui, H.C.; Victoroff, J.I.; Margolin, D.; Jagust, W.; Shankle, R.; Katzman, R. Criteria for the diagnosis of ischemic vascular dementia proposed by the state of california Alzheimer's disease diagnostic and treatment centers. *Neurology* 42: 473-480, 1992.

Clatworthy, A.E.; Gomez-Isla, T.; Rebeck, G.W.; Wallace, R.B.; Hyman B.T. Lack of association of a polymorphism in the low-density lipoprotein receptor-related protein gene with Alzheimer disease. *Arch Neurol* 54: 1289-1292, 1997.

Crawford, F.; Abdullah, L.; Schinka, J.; Suo, Z.; Gold, M.; Duara, R.; Mullan, M. Gender-specific association of the angiotensin converting enzyme gene with Alzheimer's disease. *Neurosci Lett* 25;280: 215-219, 2002a.

Crawford, F.; Freeman, M.; Abdullah, L.; Schinka, J.; Gold M,; Duara, R.; Mullan, M. No association between the NOS3 codon 298 polymorphism and Alzheimer's disease in a sample from the United States. *Ann Neu*rol 47: 687, 2000b.

Crisby, M.; Carlson, L.A.; Winblad, B. Statins in the prevention and treatment of Alzheimer disease. *Alzheimer Dis Assoc Dis* 16: 131-136, 2002.

Curtiis, L.K. ; Boisvert, W.A. Apolipoprotein E and atherosclerosis. Curr Opin Lipidol 11: 243-251, 2001.

Dahiyat, M.; Cumming, A.; Harrington, C.; Wischik, C.; Xuereb, J.; Corrigan, F.; Breen, G.; Shaw, D.; St Clair, D. Association between Alzheimer's disease and the NOS3 gene. *Ann Neurol* 46: 664-667, 1999.

Dantoine, T.F.; Debord, J.; Merle, L.; Lacroix-Ramiandrisoa, H.; Bourzeix, L.; Charmes, J.P. Paraoxonase 1 activity: a new vascular marker of dementia? *Ann N Y Acad Sci* 977: 96-101, 2002a.

Dantoine, T.F.; Drouet, M.; Debord, J.; Merle, L.; Cogne, M.; Charmes, J.P. Paraoxonase 1 192/55 gene polymorphisms in Alzheimer's disease. *Ann N Y Acad Sci* 977: 239-244, 2002b.

Davignon, J.; Gregg, R.E.; Sing, C.F. Apolipoprotein E polymorphism and atherosclerosis. *Arteriosclerosis* 8: 1-21, 1988.

De Jong, G.I.; De Vos, R.A.; Steur, E.N.; Luiten, P.G. Cerebrovascular hypoprfusion: a risk factor for Alzheimer disease ? Animal model and post-mortem human studies. *Ann NY Acad Sci* 826: 56-74, 1997.

Eaton, D.L.; Fless, G.M.; Kohr, W.J.; McLean, J.W.; Xu, Q.T.; Miller, C.G.; Lawn, R.M.; Scanu, A.M. Partial amino acids sequence of apolipoprotein(a) shows that it is homologous to plasminogen. *Proc Natl Acad Sci USA* 84: 3224-3228, 1987.

Esiri, M.M.; Wilcock, G.K.; Morris, J.H. Neuropathological assessment of the lesions of significance in vascular dementia. *J Neurol Neurosurg Psychiatry* 63: 749-753, 1997.

Fallin, D.; Kundtz, A.; Town, T.; Gauntlett, A.C.; Duara, R.; Barker, W.; Crawford, F.; Mullan, M. No association between the low density lipoprotein receptor-related protein (LRP) gene and late-onset Alzheimer's disease in a community-based sample. *Neurosci Lett* 233: 145-147, 1997.

Farkas, E.; Luiten, P.G. Cerebral microvascular pathology in aging and Alzheimer's disease. *Prog Neurobiol* 64: 575-611, 2001.

Farrer, L.A.; Sherbatich, T.; Keryanov, S.A.; Korovaitseva, G.I.; Rogaeva, E.A.; Petruk, S.; Premkumar, S.; Moliaka, Y.; Song, Y.Q.; Pei, Y.; Sato, C.; Selezneva, N.D.; Voskresenskaya, S.; Golimbet, V.; Sorbi, S.; Duara, R.; Gavrilova, S.; St George-Hyslop, P.H.; Rogaev, E.I. Association between angiotensin-converting enzyme and Alzheimer disease. *Arch Neurol* 57: 210-214, 2000.

Fassbender, K.; Simons, M.; Bergmann, C.; Stroick, M.; Lutjhoan, D.; Keller, P.; Runz, H.; Kühl, S.; Bertsch, T.; von Bergmann, K.; Hennerici, M.; Beyreuther, K.; Hartmann, T. Simvastatin strongly reduces levels of Alzheimer's disease beta-amyloid peptides Abeta 42 and Abeta 40 *in vitro* and *in vivo*. *Proc Natl Acad Sci USA* 98, 5856-5861, 2001.

Ferrucci, L.; Guralnik, J.M.; Salive, M.E.; Pahor, M.; Corti, M.C.; Baroni, A.; Havlik, R.J. Cognitive impairment and risk of stroke in the older population. *J Am Geriatr Soc* 44: 237-241, 1996.

Frisoni, G.B.; Govoni, S.; Geroldi, C.; Bianchetti, A.; Calabresi, L.; Franceschini, G.; Trabucchi, M. Gene dose of the epsilon 4 allele of apolipoprotein E and disease progression in sporadic late-onset Alzheimer's disease. *Ann Neurol* 37: 596-604, 1995.

Frosst, P.; Blom, H.J.; Milos, R.; Goyette, P.; Sheppard, C.A.; Matthews, R.; Boers, G.H.J.; Heijer, M.D.; Kluijtmans, L.A.J.; van den Heuvel, L.P.; Rozen, R. A candidate genetic risk factor for vascular disease: a common mutation in methylenetetrahydrofolate reductase. *Nat Genet* 10: 111-113, 1995.

Gale, C.R.; Martyn, C.N.; Cooper, C. Cognitive impairment and mortality in a cohort of elderly people. *BMJ* 312: 608-611,1996.

Goate, A.; Chartier-Harlin, M.C.; Mullan, M.; Brown, J.; Crawford, F.; Fidani, L.; Giuffra, L.; Haynes, A.; Irving, N.; James, L.; Mant, R.; Newton, P.; Rooke, K.; Roques, P.; Talbot, C.; Pericak-Vance, M.; Roses, A.; Williamson, R.; Rossor, M.; Owen, M.; Hardy, J. Segregation of a missense mutation in the amyloid precursor protein gene with familial Alzheimer's disease. *Nature* 349: 704-706, 1991.

Gold, G.; Bouras, C.; Canuto, A.; Bergallo, M.F.; Herrmann, F.R.; Hof, P.R.; Mayor, P.A.; Michel, J.P.; Giannakopoulos, P. Clinicopathological Validation Study of Four Sets of Clinical Criteria for Vascular Dementia. *Am J Psychiatry* 159: 82-87, 2002.

Growdon, J.H.; Locascio, J.J.; Corkin, S.; Gomez-Isla, T.; Hyman, B.T. Apolipoprotein E genotype does not influence rates of cognitive decline in Alzheimer's disease. *Neurology* 47: 444-448, 1996.

Guo, Z.; Viitanen, M.; Winblad, B. Low blood pressure and five year mortality in a Stockolm cohort of the very old: possible confounding by cognitive impairment and other factors. *Am J Public Health* 87: 623-628, 1997.

Guo, Z.; Viitanen, M.; Winblad, B.; Fratiglioni, L. Low blood pressure and incidence of dementia in a very old sample: dependent of initial cognition. *J Am Geriatr Soc* 47: 723-726, 1999.

Gussekloo, J.; Heijmans, B.T.; Slagboom, P.E.; Lagaay, A.M.; Knook, D.L.; Westendorp, R.G. Thermolabile methylenetetrahydrofolate reductase gene and the risk of cognitive impairment in those over 85. *J Neurol Neurosurg Psychiatry* 67: 535-538, 1999.

Hardy, J.A.; Higgins, G.A. Alzheimer's disease: The amyloid cascade hypothesis. *Science* 256: 184-185, 1992.

Hatanaka, Y.; Kamino, K.; Fukuo, K.; Mitsuda, N.; Nishiwaki-Ueda, Y.; Sato, N.; Satoh, T.; Yamamoto, H.; Yoneda, H.; Imagawa, M.; Miki, T.; Ohta, S.; Ogihara, T. Low density lipoprotein receptor-related protein gene polymorphisms and risk for late-onset Alzheimer's disease in a Japanese population. *Clin Genet* 58: 319-323, 2000.

Hibi, K.; Ishigami, T.; Tamura, K.; Mizushima, S.; Nyui, N.; Fujita, T.; Ochiai, H.; Kosuge, M.; Watanabe, Y.; Yoshii, Y.; Kihara, M.; Kimura, K.; Ishii, M.; Umemura, S. Endothelial nitric oxide synthase gene polymorphism and acute myocardial infarction. *Hypertension* 32: 521-526, 1998

Higuchi, S.; Ohta, S.; Matsushita, S.; Matsui, T.; Yuzuriha, T.; Urakami, K.; Arai, H. NOS3 polymorphism not associated with Alzheimer's disease in Japanese. *Ann Neurol* 48: 685, 2000.

Hofman, A.; Ott, A.; Breteler, M.M.; Bots, M.L.; Slooter, A.J.; van Harskamp, F.; van Duijn, C.N.; Van Broeckhoven, C.; Grobbee, D.E. Atherosclerosis; apolipoprotein E; and prevalence of dementia and Alzheimer's disease in the Rotterdam Study. *Lancet* 349: 151-154, 1997.

Hollenbach, E.; Ackermann, S.; Hyman, B.T.; Rebeck, G.W. Confirmation of an association between a polymorphism in exon 3 of the low-density lipoprotein receptor-related protein gene and Alzheimer's disease. *Neurology* 50: 1905-1907, 1998.

Hu, C.J.; Sung, S.M.; Liu, H.C.; Hsu, W.C.; Lee, L.S.; Lee, C.C.; Tsai, C.H.; Chang, J.G. Genetic risk factors of sporadic Alzheimer's disease among Chinese in Taiwan. *J Neurol Sci* 181:127-131, 2000.

Hu, J.; Miyatake, F.; Aizu, Y.; Nakagawa, H.; Nakamura, S.; Tamaoka, A.; Takahash, R.; Urakami, K.; Shoji, M. Angiotensin-converting enzyme genotype is associated with Alzheimer disease in the Japanese population. *Neurosci Lett* 277: 65-67, 1999.

Hu, J.; Igarashi, A.; Kamata, M.; Nakagawa, H. Angiotensin-converting enzyme degrades Alzheimer amyloid β-peptide (Aβ); retards Aβ aggregation; deposition; fibril formation; and inhibits cytotoxicity. *J Biol Chem* 276; 51: 47863- 47868, 2001.

Hyman, B.T.; Strickland, D.; Rebeck, G.W. Role of the low-density lipoprotein receptor-related protein in beta-amyloid metabolism and Alzheimer disease. *Arch Neurol* 57: 646-650, 2000.

Ignatius, M.J.; Gebicke-Haerter, P.J.; Pitas, R,E; Shooter, E.M. Apolipoprotein E in nerve injury and repair. *Prog Brain Res* 71: 177-184, 1987.

Jarvik, G.P.; Wijsman, E.M.; Kukull, W.A.; Schellenberg GD; Yu C; Larson, E.B. Interactions of apolipoprotein E genotype; total cholesterol level; age; and sex in prediction of Alzheimer's disease: a case-control study. *Neurology* 45: 1092-1096, 1995.

Jellinger, K.A. Alzheimer disease and cerebrovascular pathology: an update. *J Neural Transm* 109: 813-836, 2002.

Kalaria, R.N.; Bhatti, S.U.; Palatinsky, E.A.; Pennington, D.H.; Shelton, E.R.; Chan, H.W.; Perry, G.; Lust, W.D. Accumulation of the β amyloid precursor protein at sites of ischemic injury in rat brain. *Neuroreport* 4: 211-214, 1993.

Kalmijn, S.; Feskens, E.J.; Launer, L.J.; Stijnen, T.; Kromhout, D. Glucose intolerance; hyperinsulinemia and cognitive function in a general population of elderly men. *Diabetologica* 38: 1096-1102, 1995.

Kalmijn, S.; Lanner, L.J.; Ott, A. Dietary fat intake and the risk of incident dementia in the Rotterdam Study. *Ann Neurol* 42: 776-782, 1997.

Kamboh MI; Ferrell RE; DeKosky ST. Genetic association studies between Alzheimer's disease and two polymorphisms in the low density lipoprotein receptor-related protein gene. *Neurosci Lett* 244: 65-68, 1998.

Kalaria, R.N. The role of cerebral ischemia in Alzheimer's disease. *Neurobiol Aging* 21: 321-330, 2000.

Kalmijn, S.; Feskens, E.J,M.; Launer, L.J.; Kromhout, D. Cerebrovascular disease; the apolipoprotein ε4 allele; and cognitive decline in a community-based study of elderly men. *Stroke* 27: 2230-2235, 1996.

Kataoka, H.; Kume, N.; Miyamoto, S.; Minami, M.; Morimoto, M.; Hayashida, K.; Hashimoto, N.; Kita, T. Oxidized LDL modulates Bax/Bcl-2 through the lectinlike Ox-LDL receptor-1 in vascular smooth muscle cells. *Arterioscler Thromb Vasc Biol* 21: 955-960, 2001.

Kehoe, P.G. The renin-angiotensin-aldosterone system and Alzheimer's disease? *J Renin Angiotensin Aldosterone Syst* 4: 80-93, 2003.

Kehoe, P.G.; Katzov, H.; Feuk, L.; Bennet, A.M.; Johansson, B.; Wiman, B.; de Faire, U.; Cairns, N.J.; Wilcock, G.K.; Brookes, A.J.; Blennow, K.; Prince, J.A. Haplotypes extending across ACE are associated with Alzheimer's disease. *Hum Mol Genet* 12: 859-867, 2003.

Kehoe, P.G.; Russ, C.; McIlory, S.; Williams, H.; Holmans, P.; Holmes, C.; Liolitsa, D.; Vahidassr, D.; Powell, J.; McGleenon, B.; Liddell, M.; Plomin, R.; Dynan, K.; Williams, N.; Neal, J.; Cairns, N.J.; Wilcock, G.; Passmore, P.; Lovestone, S.; Williams, J.; Owen, M.J. Variation in DCP1; encoding ACE; is associated with susceptibility to Alzheimer disease. *Nat Genet* 21: 71-72, 1999a.

Kehoe, P.; Wavrant-De Vrieze, F.; Crook, R.; Wu, W.S.; Holmans, P.; Fenton, I.; Spurlock, G.; Norton, N.; Williams, H.; Williams, N.; Lovestone, S.; Perez-Tur, J.; Hutton, M.;

Chartier-Harlin, M.C.; Shears, S,; Roehl, K,; Booth, J.; Van Voorst, W.; Ramic, D.; Williams, J.; Goate, A.; Hardy, J.; Owen, M.J. A full genome scan for late onset Alzheimer's disease. *Hum Mol Genet* 8: 237-245,1999b

Kilander, L.; Nyman, H.; Boberg, M.; Hansson, L.; Lithell, H. Hypertension is related to cognitive impairment. *Hypertension* 31: 780-786, 1998.

Kivipelto, M.; Helkala, E.L.; Laakso, M.P.; Hanninen, T.; Hallikainen, M.; Alhainen, K.; Iivonen, S.; Mannermaa, A.; Tuomilehto, J.; Nissinen, A.; Soininen, H. Apolipoprotein E epsilon4 allele; elevated midlife total cholesterol level; and high midlife systolic blood pressure are independent risk factor for late-life Alzheimer disease. *Ann Intern Med* 137: 149-155, 2002.

Kivipelto, M.; Helkala, E.L.; Laakso, M.P.; Hanninen, T., Hallikainen, M., Alhainen, K.; Soininen, H.; Tuomilehto, J.; Nissinen, A. Middlife vascular risk factors and Alzheimer's disease in later life: longitudinal population based study. *BMJ* 322: 1447-1451, 2001.

Kokmen, E.; Whisnant, J.P.; O'Fallon, W.M.; Chu, C.P.; Beard, C.M. Dementia after ischemic stroke: a population-based study in Rochester, Minnesota (1960-1984). *Neurology* 19: 154-159, 1996.

Kuo, Y.M.; Emmerling, M.R.; Bisgaier, C.L.; Essenburg, A.D.; Lampert, H.C.; Drumm, D.; Roher, A.E. Elevated low-density lipoprotein in Alzheimer's disease correlates with brain Abeta 1-42 levels. *Biochem Biophys Res Commun* 252: 711-715, 1998.

Kuller, L.H.; Shemanski, L.; Manolio, T.; Haan, M.; Fried, L.; Bryan, N.; Burke, G.L.; Tracy, R.; Bhadelia R. Relationship between ApoE, MRI findings, and cognitive function in the Cardiovascular Health Study. *Stroke* 29: 388-398, 1998.

Kuusisto, J; Koivisto K; Mykkanen L; Helkala EL; Vanhanen M; Hanninen T; Kervinen K; Kesaniemi YA; Riekkinen PJ; Laakso M. Association between features of the insulin resistance syndrome and Alzheimer's disease independently of apolipoprotein E4 phenotype: cross sectional population based study. *BMJ* 315: 1045-1049, 1997.

Kuusisto, J.; Mykkanen, L.; Pyorala, K.; Laakso M. Non insulin dependent diabetes and its metabolic control are important predictors of stroke in elderly subjects. *Stroke* 25: 1157-1164, 1994.

Lambert, J.C.; Berr, C.; Pasquier, F.; Delacourte, A.; Frigard, B.; Cottel, D.; Perez-Tur, J.; Mouroux, V.; Mohr, M.; Cecyre, D.; Galasko, D.; Lendon, C.; Poirier, J.; Hardy, J.; Mann, D.; Amouyel, P.; Chartier-Harlin, M.C. Pronounced impact of Th1/E47cs mutation compared with -491 AT mutation on neural APOE gene expression and risk of developing Alzheimer's disease. *Hum Mol Genet* 7: 1511-1516, 1998a

Lambert, J.C.; Pasquier, F.; Cottel, D.; Frigard, B.; Amouyel, P.; Chartier-Harlin, M.C. A new polymorphism in the APOE promoter associated with risk of developing Alzheimer's disease. *Hum Mol Gene*t 7: 533-540, 1998b.

Lambert, J.C.; Wavrant-De Vrieze, F.; Amouyel, P.; Chartier-Harlin, M.C. Association at LRP gene locus with sporadic late-onset Alzheimer's disease. *Lancet* 351: 1787-1788, 1998c.

Lambert, J.C.; Chartier-Harlin, M.C.; Cottel, D.; Richard, F.; Neuman, E.; Guez, D.; Legrain, S.; Berr, C.; Amouyel, P.; Helbecque, N. Is the LDL receptor-related protein involved in Alzheimer's disease? *Neurogenetics*. 2: 109-113, 1999.

Lambert, J.C.; Araria-Goumidi, L.; Myllykangas, L.; Ellis, C.; Wang, J.C.; Bullido, M.J.; Harris, J.M.; Artiga, M.J.; Hernandez, D.; Kwon, J.M.; Frigard, B.; Petersen, R.C.; Cumming, A.M.; Pasquier, F.; Sastre, I.; Tienari, P.J.; Frank, A.; Sulkava, R.; Morris,

J.C.; St Clair, D.; Mann, D.M.; Wavrant-DeVrieze, F.; Ezquerra-Trabalon, M.; Amouyel, P.; Hardy, J.; Haltia, M.; Valdivieso, F.; Goate, A.M.; Perez-Tur, J.; Lendon, C.L.; Chartier-Harlin, M.C. Contribution of APOE promoter polymorphisms to Alzheimer's disease risk. *Neurology* 59: 59-66, 2002.

Launer, L.J.; Ross, G.W.; Petrovitch, H.; Masaki, K.; Foley, D.; White, L.R.; Havlik, R.J. Midlife blood pressure and dementia: the Honolulu- Asia Aging Study. *Neurobiol Aging* 21: 49-55, 2000.

Laskowitz, D.T.; Horsburgh, K., Roses, A.D. Apolipoprotein E and the CNS response to injury. *J Cereb Blood Flow Metab* 18: 465-471,1998.

Laws, S.M.; Taddei, K.; Martins, G.; Paton, A.; Fisher, C.; Clarnette, R.; Hallmayer, J.; Brooks, W.S.; Gandy, S.E.; Martins, R.N. The -491AA polymorphism in the APOE gene is associated with increased plasma apoE levels in Alzheimer's disease. *Neuroreport* 10: 879-882,1999.

Laws, S.M.; Hone, E.; Gandy, S.; Martins, R.N. Expanding the association between the APOE gene and the risk of Alzheimer's disease: possible roles for APOE promoter polymorphisms and alterations in APOE transcription. *J Neurochem* 84: 1215-1236, 2003.

Leibson, C.; Rocca, W.; Hanson, V.; Cha, R.; Kokmen, E.; O'Brien, P.; Palumbo, P.J. Risk of dementia among persons with diabetes mellitus: a population based cohort study. *Am J Epidemiol*; 145: 301-308,1997.

Lendon, C.L.; Talbot, C.J.; Craddock, N.J.; Han, S.W.; Wragg, M.; Morris, J.C.; Goate, A.M. Genetic association studies between dementia of the Alzheimer's type and three receptors for apolipoprotein E in a Caucasian population. *Neurosci Lett* 222: 187-190,1997.

Liou, Y.J.; Hong, C.J.; Liu, H.C.; Liu, C.Y.; Liu, T.Y.; Chen, I.C.; Tsai, S.J. No association between the neuronal nitric oxide synthase gene polymorphism and Alzheimer Disease. *Am J Med Genet* 114: 687-688, 2002.

Luchsinger, J.A.; Tang, M-X.; Shea, S.; Mayeux, R. Caloric intake and the risk of Alzheimer disease. *Arch Neurol*; 59: 1258-1263, 2002.

Luedecking-Zimmer, E.; DeKosky, S.T.; Chen, Q.; Barmada, M.M.; Kamboh, M.I. Investigation of oxidized LDL-receptor 1 (OLR1) as the candidate gene for Alzheimer's disease on chromosome 12. *Hum Genet* 111: 443-451, 2002.

Luscher, T. Local relaxant and constricting factors in the vessel wall. In: Swales J; ed. Textbook of Hypertension; Oxford: Blackwell Scientific Publications, 1994.

Ma, J.; Yee, A.; Bryan Brewer, Jr H.; Das, S.; Potter, H. Amyloid-associated proteins α1-antichymotrypsin and Apolipoprotein E promote assembly of Alzheimer β-protein into filaments. *Nature* 372: 92-94, 1994.

Mackness, B.; Durrington, P.N.; Mackness, M.I. Human serum paraoxonase. *Gen Pharmacol* 31: 329-336, 1998a.

Mackness, B.; Mackness, M.I.; Arrol, S.; Turkie, W.; Durrington, P.N. Effect of the human serum paraoxonase 55 and 192 genetic polymorphisms on the protection by high density lipoprotein against low density lipoprotein oxidative modification. *FEBS Lett* 423: 57-60, 1998b.

Mahley, R.W.; Huang, Y. Apolipoprotein E : from atherosclerosis to Alzheimer disease and beyond. *Current Opinion in Lipidology*; 10: 207-217,1997.

Mattila, K.M.; Rinne, J.O.; Roytta, M.; Laippala, P.; Pietila, T.; Kalimo, H.; Koivula, T.; Frey H; Lehtimaki, T. Dipeptidil carboxypeptidase 1 (DCP1) and butyrilcholinesterase (BCHE) gene interactions with Alzheimer's disease and Parkinson's disease with coexisting Alzheimer pathology. *J Med Genet* 37: 766-770, 2000.

Mayeux, R.; Saunders, A.M.; Shea, S.; Mirra, S.; Evans, D.; Roses, A.D.; Hyman, B.T.; Crain, B.; Tang, M.X.; Phelps, C.H. Utility of the apolipoprotein E genotype in the diagnosis of Alzheimer's disease. Alzheimer's Disease Centers Consortium on Apolipoprotein E and Alzheimer's Disease. *N Engl J Med* 338: 506-511,1998.

McIlroy, S.P.; Dynan, K.B.; Vahidassr, D.J.; Lawson, J.T.; Patterson, C.C.; Passmore, P. Common polymorphisms in LRP and A2M do not affect genetic risk for Alzheimer disease in Northern Ireland. *Am J Med Genet* 105: 502-506, 2001.

McKhann, G.; Drachman, D.; Folstein, M.; Katzman, R.; Price, D.; Stadlan, E.M. Clinical diagnosis of Alzheimer's disease: Report of the NINCDS-ADRDA Work Group under the auspices of Department of Health and Human Services Task Force on Alzheimer's Disease. *Neurology* 34: 939-944, 1984.

Mehta, J.L.; Li, D. Identification, regulation and function of a novel lectin-like oxidized low-density lipoprotein receptor. *J Am Coll Cardiol* 39: 1429-1435, 2002.

Messier, C.; Gagnon, M. Glucose regulation and cognitive function : relation to Alzheimer's disease and diabetes mellitus. *Behav Brain Res*; 75: 1-11, 1996.

Monastero, R.; Caldarella, R.; Mannino, M.; Cefalu, A.B.; Lopez, G.; Noto, D.; Camarda, C.; Camarda, L.K.; Notarbartolo, A.; Averna, M.R.; Camarda, R. Lack of association between angiotensin converting enzyme polymorphism and sporadic Alzheimer's disease. *Neurosci Lett* 335: 147-149, 2002.

Monastero, R.; Cefalu, A.B.; Camarda, C.; Buglino, C.M.; Mannino, M.; Barbagallo, C.M.; Lopez, G.; Camarda, L.K.; Travali, S.; Camarda, R.; Averna, M.R. No association between Glu298Asp endothelial nitric oxide synthase polymorphism and Italian sporadic Alzheimer's disease. *Neurosci Lett* 341: 229-232, 2003.

Mooradian, A.D.; Chung, H.C.; Shah, G.N. Glu-1 expression in the cerebra of patients with Alzheimer disease. *Neurobiol Aging* 18: 469-474,1997.

Myllykangas, L.; Polvikoski, T.; Sulkava, R.; Verkkoniemi, A.; Tienari, P.; Niinisto, L.; Kontula, K.; Hardy, J.; Haltia, M.; Perez-Tur, J. Cardiovascular risk factors and Alzheimer's disease: a genetic association study in a population aged 85 or over. *Neurosci Lett* 292: 195-198, 2000.

Nalbantoglu, J.; Gilfix, B.M.; Bertrand, P.; Robitaille, Y.; Gauthier, S.; Rosenblatt, D.S.; Poirier, J. Predictive value of apolipoprotein E genotyping in Alzheimer's disease: results of an autopsy series and an analysis of several combined studies. *Ann Neurol* 36: 889-895,1994.

Narain, Y.; Yip, A.; Murphy, T.; Brayne, C.; Easton, D.; Evans, J.G.; Xuereb, J.; Cairns, N.; Esiri, M.M.; Furlong, R.A.; Rubinsztein, D.C. The ACE gene and Alzheimer's disease susceptibility. *J Med Genet* 37: 695-697, 2000.

National Institute on Aging/Alzheimer's Association Working Group Apolipoprotein E genotyping in Alzheimer's disease. *Lancet* 347: 1091-1095, 1996.

Nishiyama, M.; Kato, Y.; Hashimoto, M.; Yukawa, S.; Omori, K. Apolipoprotein E; methylenetetrahydrofolate reductase (MTHFR) mutation and the risk of senile dementia--an epidemiological study using the polymerase chain reaction (PCR) method. *J Epidemiol* 10: 163-172, 2000.

Norrman, J.; Brookes, A.J.; Yates, C.; St Clair, D. Apolipoprotein E genotype and its effect on duration and severity of early and late onset Alzheimer's disease. *Br J Psychiatry* 167: 533-536, 1995.

Notkola, I.L.; Sulkava, R.; Pekkanen, J.; Erkinjuntti, T.; Ehnholm, C.; Kivinen, P.; Tuomilehto, J.; Nissinen, A. Serum total cholesterol; apolipoprotein E epsilon 4 allele; and Alzheimer's disease *Neuroepidemiology* 17: 14-20, 1998.

Olichney, J.M.; Hansen, L.A.; Hofstetter, C.R.; Grundman,M.; Katzman, R.; Thal, L.J. Cerebral infarction in Alzheimer disease is associated with sever amyloid; angiopathy and hypertension. *Arch Neurol*; 52: 702-708, 1995.

Ott, A.; Breteler, M.M.; de Bruyne, M.C.; van Harskamp, F.; Grobbee, D.E.; Hofman, A. Atrial fibrillation and dementia in a population-based study. The Rotterdam Study. *Stroke* 28: 316-321,1997.

Ott, A.; Stolk, R.P.; van Harskamp, F.; Pols, H.A.; Hofman, A.; Breteler, M.M. Diabetes mellitus and the risk of dementia: The Rotterdam Study. *Neurology* 53: 1937-1942, 1999.

Ott, A.; Breteler, M.M.B.; Van Harskamp, F.; Stijnen, T, Hofman, A. Incidence and risk of dementia. The Rotterdam Study. *Am J Epidemiol*; 147: 574-580, 1998.

Palumbo, B.; Cadini, D.; Nocentini, G.; Filipponi, E.; Fravolini, M.L.; Senin, U. Angiotensin converting enzyme deletion allele in different kinds of dementia disorders. *Neurosci Lett* 267: 97-100, 1999.

Panza, F.; Solfrizzi, V.; Torres, F.; Mastroianni F; Del Parigi, A.; Colacicco, A.M.; Basile, A.M.; Capurso, C.; Noya, R.; Capurso, A. Decreased frequency of apolipoprotein E epsilon4 allele from Northern to Southern Europe in Alzheimer's disease patients and centenarians. *Neurosci Lett* 277: 53-56, 1999.

Panza, F.; Solfrizzi, V.; Torres, F.; Mastroianni, F.; Colacicco, A.M.; Basile, A.M.; Capurso, C.; D'Introno, A.; Del Parigi, A.; Capurso, A. Apolipoprotein E in Southern Italy: protective effect of e2 allele in early- and late-onset sporadic Alzheimer's disease. *Neurosci Lett* 292: 79-82, 2000.

Panza, F.; Solfrizzi, V.; Colacicco, A.M.; Basile, A.M.; D'Introno, A.; Capurso, C.; Sabba, M.; Capurso, S.; Capurso, A. Apolipoprotein E (APOE) polymorphism influences serum APOE levels in Alzheimer's disease patients and centenarians. *Neuroreport* 14: 605-608, 2003a.

Panza, F.; Solfrizzi, V.; D'Introno, A.; Colacicco, A.M.; Capurso, C.; Kehoe, P.G.; Capurso, A. Shifts in angiotensin I converting enzyme insertion allele frequency across Europe: implications for Alzheimer's disease risk. *J Neurol Neurosurg Psychiatry*; 74: 1159-1161, 2003b.

Panza, F.; D'Introno, A.; Colacicco, A.M.; Basile, A.M.; Capurso, C.; Kehoe, P.G.; Capurso, A.; Solfrizzi, V. Vascular risk and genetics of sporadic late-onset Alzheimer's disease. *J Neural Transm* 111: 69-89, 2004a.

Panza F.; D'Introno, A.; Colacicco, A.M.; Capurso, C.; Basile, A.M.; Capurso, S.; Capurso, A; Solfrizzi, V. Regional differences in allele and genotype frequencies of LDL receptor-related protein polymorphism in Alzheimer's disease in Europe. *Am J Med Genet* 126B: 69-73, 2004b.

Paragh, G.; Balla, P.; Katona, E.; Seres, I.; Egerhazi, A.; Degrell, I. Serum paraoxonase activity changes in patients with Alzheimer's disease and vascular dementia. *Eur Arch Psychiatry Clin Neurosci* 252: 63-67, 2002.

Pasquier, F.; Leys, D.; Scheltens, P. The influence of coincidental vascular pathology on symptomatology and course of Alzheimer's disease. *J Neural Transm Suppl*; 54: 117-127,1998.

Passant, U.; Warkentin, S.; Gustafson, L. Orthostatic hypotension and low blood pressure in organic dementia: a study of prevalence and related clinical characteristics. *Int J Geriatr Psychiatry,* 12: 395-403, 1997.

Peila, R.; Rodriguez, B.L.; Launer, L.J. Type 2 diabetes; APOE gene; and the risk for dementia and related pathologies: The Honolulu Asia Aging Study. *Diabetes*; 51: 1256-1262, 2002.

Pericak-Vance, M.A.; Bass, M.P.; Yamaoka, L.H.; Gaskell, P.C.; Scott, W.K.; Terwedow, H.A.; Menold, M.M.; Conneally, P.M.; Small, G.W.; Vance, J.M.; Saunders, A.M.; Roses, A.D.; Haines, J.L. Complete Genomic Screen in Late-Onset Familial Alzheimer Disease: Evidence for a New Locus on Chromosome 12. *JAMA* 278: 1237-1241, 1997.

Perry, R.T.; Collins, J.S.; Harrell, L.E.; Acton, R.T.; Go, R.C. Investigation of association of 13 polymorphisms in eight genes in southeastern African American Alzheimer disease patients as compared to age-matched controls. *Am J Med Genet* 105: 332-342, 2001.

Petersen, R.C.; Smith, G.E.; Ivnik, R.J.; Tangalos, E.G.; Schaid, D.J.; Thibodeau, S.N.; Kokmen, E.; Waring, S.C.; Kurland, L.T. Apolipoprotein E status as a predictor of the development of Alzheimer's disease in memory-impaired individuals. *JAMA* 273: 1274-1278, 1995.

Poirier, J. Apolipoprotein E in animal models of CNS injury and in Alzheimer's disease. *Trends Neurosci* 17: 525-530, 1994.

Pola, R.; Gaetani, E.; Flex, A.; Gerardino, L.; Aloi, F.; Flore, R.; Serricchio, M.; Pola, P.; Bernabei, R. Lack of Association between Alzheimer's Disease and Gln-Arg 192 Q/R Polymorphism of the PON-1 Gene in an Italian Population. *Dement Geriatr Cogn Disord* 15: 88-91, 2003.

Pollak, R.D.; Pollak, A.; Idelson, M.; Bejarano-Achache, I.; Doron, D.; Blumenfeld, A. The C677T mutation in the methylenetetrahydrofolate reductase (MTHFR) gene and vascular dementia. *J Am Geriatr Soc* 48: 664-668, 2000.

Prince, J.A.; Feuk, L.; Sawyer, S.L.; Gottfries, J.; Ricksten, A.; Nagga, K.; Bogdanovic, N.; Blennow, K.; Brookes, A.J. Lack of replication of association findings in complex disease: an analysis of 15 polymorphisms in prior candidate genes for sporadic Alzheimer's disease. *Eur J Hum Genet* 9: 437-444, 2001.

Qiu, C.; von Strauss, E.; Fastbom, J.; Winblad, B.; Fratiglioni, L. Low blood pressure and risk of dementia in the Kungsholmen project: a 6-year follow-up study. *Arch Neurol* 60: 223-228, 2003.

Qiu, C.; Winblad, B.; Viitanen, M.; Fratiglioni, L. Pulse pressure and risk of Alzheimer disease in persons aged 75 years and older: A community based; longitudinal study. *Stroke* 34: 594-599, 2003.

Rebeck, G.W.; Harr, S.D.; Strickland, D.K.; Hyman, B.T. Multiple; diverse senile plaque-associated proteins are ligands of an apolipoprotein E receptor; the alpha 2-macroglobulin receptor/low-density-lipoprotein receptor-related protein. *Ann Neurol* 37: 211-217, 1995.

Rebeck, G.W.; Cheung, B.S.; Growdon, W.B.; Deng, A.; Akuthota, P.; Locascio, J.; Greenberg, S.M.; Hyman, B.T. Lack of independent associations of apolipoprotein E promoter and intron 1 polymorphisms with Alzheimer's disease. *Neurosci Lett* 272: 155-158, 1999.

Rebeck, G.W.; Moir, R.D.; Mui, S.; Strickland, D.K.; Tanzi, R.E.; Hyman, B.T. Association of membrane-bound amyloid precursor protein APP with the apolipoprotein E receptor LRP. *Brain Res Mol Brain Res* 87: 238-245, 2001.

Relkin, N.R.; Tanzi, R.; Breitner, J.; Farrer, L.; Gandy, S.; Haines, J.; Hyman, B.; Mullan, M.; Poirer, J.; Strittmatter, W.; Folstein, M.; Mayeux, R.; Petersen, R.; Roses, A.; Schenk, D.; Small, G.; Vangool, W.; Cook-Deegan, R.; Fleck, L.; Kapp, M.; Karlinsky, H.; Pericak-Vance, M.; Post, S.; Wolpert, C. Apolipoprotein E genotyping in Alzheimer's disease. National Institute on Aging/Alzheimer's Association Working Group. *Lancet* 347: 1091-1095, 1996.

Richard, F.; Fromentin-David, I.; Ricolfi, F.; Ducimetiere, P.; Di Menza, C.; Amouyel, P.; Helbecque, N. The angiotensin I converting enzyme gene as a susceptibility factor for dementia. *Neurology* 56: 1593-1595, 2001.

Rigat, B.; Hubert, C.; Corvol, P.; Soubrier, F. PCR detection of the insertion/deletion polymorphism of the human angiotensin converting enzyme gene (DCP1) (dipeptidyl carboxypeptidase 1). *Nucleic Acids Res* 20: 1433, 1992.

Riggs, K.M.; Spiro, A. 3rd; Tucker, K.; Rush, D. Relations of vitamin B-12; vitamin B-6; folate; and homocysteine to cognitive performance in the Normative Aging Study. *Am J Clin Nutr* 63: 306-314, 1996.

Rockwood, K.; Kirkland, S.; Hogan, DB.; Macknight, C.; Merry, H.; Verreault, R.; Wolfson, C.; McDowell, I. Use of lipid-lowering agents; indication bias; and the risk of dementia in community-dwelling elderly people. *Arch Neurol* 59: 223-227, 2002.

Rogaev, E.I.; Sherrington, R.; Rogaeva, E.A.; Levesque, G.; Ikeda, M.; Liang, Y.; Chi, H.; Lin, C.; Holman, K.; Tsuda, T.; Mar, L.; Sorbi, S.; Nacmias, B.; Piacentini, S.; Chumakov, I.; Cohen, D.; Lannfelt, L.; Fraser, P.E.; Rommens, J.M.; St. George-Hyslop, P.H. Familial Alzheimer's disease in kindreds with missense mutations in a gene on chromosome 1 related to the Alzheimer's disease type 3 gene. *Nature* 376: 775-778, 1995.

Rogaeva, E.; Premkumar, S.; Song, Y.; Sorbi, S.; Brindle, N.; Paterson, A.; Duara, R.; Levesque, G.; Yu, G.; Nishimura, M.; Ikeda, M.; O'Toole, C.; Kawarai, T.; Jorge, R.; Vilarino, D.; Bruni, A.C.; Farrer, L.A.; St George-Hyslop, P.H. Evidence for an Alzheimer disease susceptibility locus on chromosome 12 and for further locus heterogeneity. *JAMA* 280: 614-618,1998.

Roks, G.; Cruts, M.; Houwing-Duistermaat, J.J.; Dermaut, B.; Serneels, S.; Havekes, L.M.; Hofman, A.; Breteler, M.M.; Van Broeckhoven, C.; van Duijn, C.M. Effect of the APOE-491A/T promoter polymorphism on apolipoprotein E levels and risk of Alzheimer disease: The Rotterdam Study. *Am J Med Genet* 114: 570-573, 2002.

Roman, G.C.; Tatemichi, T.K.; Erkinjuntti, T.; Cummings, J.L.; Masdeu, J.C.; Garcia, J.H.; Amaducci, L.; Brun, A.; Hofman, A.; Moody, D.M.; O'Brien, M.D.; Yamaguchi, T.; Grafman, J.; Drayer, B.P.; Bennett, D.A.; Fisher, M.; Ogata, J.; Kokmen, E.; Bermejo, F.; Wolf, P.A.; Gorelick, P.B.; Bick, K.L.; Pajeau, A.K.; Bell, M.A.; DeCarli, C.; Culebras, A.; Korczyn, A.D.; Bogousslavsky, J.; Hartmann, A.; Scheinberg, P. Vascular dementia: diagnostic criteria for research studies. Report of the NINDS-AIREN International Workshop. *Neurology* 43: 250-260, 1993.

Roman, G.C.; Royall, D.R. Executive control function: a rational basis for the diagnosis of vascular dementia. *Alzheimer Dis Assoc Disord* 13(Suppl. 3): S69-80,1999.

Roman, G.C. Defining dementia: clinical criteria for the diagnosis of vascular dementia. *Acta Neurol Scand Suppl* 178: 6-9, 2002.

Rubinsztein, D.C.; Easton D.F. Apolipoprotein E genetic variation and Alzheimer's disease. a meta-analysis. *Dement Geriatr Cogn Disord* 10: 199-209, 1999.

Ruitenberg, A.; Skoog, I.; Ott, A.; Aevarsson, O.; Witteman, J.C.; Lernfelt B; van Harskamp, F.; Hofman, A.; Breteler, M.M. Blood pressure and risk of dementia: results from the Rotterdam Study and the Gothenburg H-70 Study. Dement Geriatr Cogn Disord 12: 33-39, 2001.

Sanchez-Guerra, M.; Combarros, O.; Alvarez-Arcaya, A.; Mateo, I.; Berciano, J.; Gonzalez-Garcia, J.; Llorca, J. The Glu298Asp polymorphism in the NOS3 gene is not associated with sporadic Alzheimer's disease. *J Neurol Neurosurg Psychiatry* 70: 566-567, 2001a.

Sanchez-Guerra, M.; Combarros, O.; Infante, J.; Llorca, J.; Berciano, J.; Fontalba, A.; Fernandez-Luna, J.L.; Pena, N.; Fernandez-Viadero, C. Case-control study and meta-analysis of low density lipoprotein receptor-related protein gene exon 3 polymorphism in Alzheimer's disease. *Neurosci Lett* 316: 17-20, 2001b.

Sasaki, N.; Fukatsu, R.; Tsuzuki, K.; Hayashi, Y.; Yoshida, T.; Fujii, N.; Koike, T.; Wakayama, I.; Yanagihara, R.; Garruto, R.; Amano, N.; Makita, Z. Advanced glycation end products in Alzheimer's disease and other neurodegenerative diseases. *Am J Pathol* Oct 153: 1149-1155, 1998.

Saunders, A.M.; Strittmatter, W.J.; Schmechel, D.; St. George-Hyslop, P.H.; Pericak-Vance, M.A.; Joo, S.H.; Rosi, B.L.; Gusella, J.F.; Crapper-MacLachlan, D.R.; Alberts, M.J.; Huletta, C.; Crain, B.; Goldgaber, D.; Roses, A.D. Association of apolipoprotein E allele e4 with late-onset familial and sporadic Alzheimer's disease. *Neurology* 43: 1467-1472, 1993

Scacchi, R.; De Bernardini, L.; Mantuano, E.; Vilardo, T.; Donini, L.M.; Ruggeri, M.; Gemma, A.T.; Pascone, R.; Corbo, R.M. DNA polymorphisms of apolipoprotein B and angiotensin I-converting enzyme genes and relationships with lipid levels in Italian patients with vascular dementia or Alzheimer's disease. *Dement Geriatr Cogn Disord* 9: 186-190, 1998.

Scacchi, R.; Gambina, G.; Ruggeri, M.; Martini, M.C.; Ferrari, G.; Silvestri, M.; Schiavon, R.; Corbo, R.M. Plasma levels of apolipoprotein E and genetic markers in elderly patients with Alzheimer's disease. *Neurosci Lett* 259: 33-36, 1999.

Schmidt, R.; Schmidt, H.; Fazekas, F. Vascular risk factors in dementia. *J Neurol* 247: 81-87, 2000.

Schunkert, H. Polymorphism of the angiotensin-converting enzyme gene and cardiovascular disease. *J Mol Med* 75: 867-875, 1997.

Scott, W.K.; Grubber, J.M.; Abou-Donia, S.M.; Church, T.D.; Saunders, A.M.; Roses, A.D.; Pericak-Vance, M.A.; Conneally, P.M.; Small, G.W.; Haines, J.L. Further evidence linking late-onset Alzheimer disease with chromosome 12. *JAMA* 281: 513-514, 1999.

Scott, W.K.; Yamaoka, L.H.; Bass, M.P.; Gaskell, P.C.; Conneally, P.M.; Small, G.W.; Farrer, L.A.; Auerbach, S.A.; Saunders, A.M.; Roses, A.D.; Haines, J.L.; Pericak-Vance, M.A. No genetic association between the LRP receptor and sporadic or late-onset familial Alzheimer disease. *Neurogenetics* 1: 179-183, 1998.

Seshadri, S.; Beiser, A.; Selhub, J.; Jacques, P.F.; Rosenberg, I.H.; D'Agostino, R.B.; Wilson, P.W.; Wolf, P.A. Plasma homocysteine as a risk factor for dementia and Alzheimer's disease. *N Engl J Med* 346: 476-483, 2002,

Serrato, M.; Marian, A.J. A variant of human paraoxonase/arylesterase (HUMPONA) gene is a risk factor for coronary artery disease. J *Clin Invest* 96: 3005-3008, 1995.

Sherrington, R.; Rogaev, E.I.; Liang, Y.; Rogaeva, E.A.; Levesque, G.; Ikeda, M.; Chi, H.; Lin, C.; Li, G.; Holman, K.; Tsuda, T.; Mar, L.; Foncin, J.F.; Bruni, A.C.; Montesi, M.P.; Sorbi, S.; Rainero, I.; Pinessi, L.; Nee, L.; Chumakov, I.; Pollen, D.; Brookes, A.; Sansosu, P.; Polinsky, R.J.; Wasco, W.; Da Silva, H.A.R.; Haines, J.L.; Pericak-Vance, M.A.; Tanzi, R.E.; Roses, A.D.; Fraser, P.E.; Rommens, J.M.; St George-Hyslop, P. Cloning of a novel gene bearing missense mutations in early onset familial Alzheimer's disease. *Nature* 375: 754-760, 1995.

Siest, G.; Pillot, T.; Regis-Bailly, A.; Leininger-Muller, B.; Steinmetz, J.; Galteau, M.M.; Visvikis, S. Apolipoprotein E: an important gene and protein to follow in laboratory medicine. *Clin Chem* 41: 1068-1086, 1995.

Simons, M.; Keller, P.; De Strooper, B.; Beyreuther, K.; Dotti, C.G.; Simons, K. Cholesterol depletion inhibits the generation of beta-amyloid in hippocampal neurons. *Proc Natl Acad Sci USA* 95: 6460-6464, 1998.

Singh, R.; Barden, A.; Morl, T.; Beilin, L. Advanced glycation end-products: a review. *Diabetologia* 44: 129-146, 2001.

Singleton, A.B.; Gibson, A.M.; McKeith, I.G.; Ballard, C.G.; Edwardson, J.A.; Morris, C.M. Nitric oxide synthase gene polymorphisms in Alzheimer's disease and dementia with Lewy bodies. *Neurosci Lett* 303: 33-36, 2001.

Skoog, I. Vascular aspects in Alzheimer's disease. *J Neural Transm Suppl* 59: 37-43, 2000.

Skoog, I.; Lernfelt, B.; Landahl, S.; Palmertz, B.; Andreasson, L.A.; Nilsson, L.; Persson, G.; Oden, A.; Svanborg, A. 15-year longitudinal study of blood pressure and dementia. *Lancet* 347: 1141-1145, 1996.

Skoog, I. A review on blood pressure and ischaemic white matter lesions. *Dement Geriatr Cogn Disor* 9(suppl 1): 13-9, 1998.

Slooter, A.J.; Houwing-Duistermaat, J.J.; van Harskamp, F.; Cruts, M.; Van Broeckhoven, C.; Breteler, M.M.; Hofman, A.; Stijnen, T.; van Duijn, C.M. Apolipoprotein E genotype and progression of Alzheimer disease. The Rotterdam Study. *J Neurol* 246: 304-308, 1999.

Slooter, A.J., Tang, M.X., van Duijn, C.M., Stern, Y., Ott, A., Bell, K., Breteler, M.M., Van Broeckhoven, C., Tatemichi, T.K., Tycko, B., Hofman, A., Mayeux, R. Apolipoprotein E ε4 and the risk of dementia with stroke. A population- based investigation. *JAMA* 277: 818-821, 1997.

Smith, G.E.; Bohac, D.L.; Waring, S.C.; Kokmen, E.; Tangalos, E.G.; Ivnik, R.J.; Petersen, R.C. Apolipoprotein E genotype influences cognitive 'phenotype' in patients with Alzheimer's disease but not in healthy control subjects. *Neurology* 50: 355-362, 1998.

Snowdown, D.A.; Greiner, L.H.; Mortimer, J.A.; Riley, K.P.; Greiner, P.A.; Markesbery, W.R. Brain infarction and the clinical expression of Alzheimer disease. The Nun Study. *JAMA* 277: 813-817, 1997.

Sodeyama, N.; Yamada, M.; Itoh, Y.; Suematsu, N.; Matsushita, M.; Otomo, E.; Mizusawa, H. No association of paraoxonase gene polymorphism with atherosclerosis or Alzheimer's disease. *Neurology* 53: 1146-1148, 1999.

Solfrizzi, V.; Panza, F.; D'Introno, A.; Colacicco, A.M.; Capurso, C.; Basile, A.M.; Capurso, A. Lipoprotein(a); apolipoprotein E genotype; and risk of Alzheimer's disease. *J Neurol Neurosurg Psychiatry* 72: 732-736, 2002.

Solfrizzi, V.; Panza, F.; Capurso, A. The role of diet in cognitive decline. *J Neural Transm* 110: 95-110, 2003.

Song, Y.Q.; Rogaeva, E.; Premkumar, S.; Brindle, N.; Kawarai, T.; Orlacchio, A.; Yu, G.; Levesque, G.; Nishimura, M.; Ikeda, M.; Pei, Y.; O'Toole, C.; Duara, R.; Barker, W.; Sorbi, S.; Freedman, M.; Farrer, L.; St George-Hyslop, P. Absence of association between Alzheimer disease and the -491 regulatory region polymorphism of APOE. *Neurosci Lett* 250: 189-192, 1998.

Sparks, D.L.; Scheff, S.W.; Liu, H.; Landers, T.M.; Coyne, C.M.; Hunsaker, J.C. 3rd. Increased incidence of neurofibrillary tangles (NFT) in non-demented individuals with hypertension. *J Neurol Sci* 131: 162-169, 1995.

Sparks, D. Coronary artery disease; hypertension; ApoE; and cholesterol: a link to Alzheimer disease? *Ann NY Acad Sci* 826: 128-146, 1997.

Sparks, D.L.; Scheff, S.W.; Liu, H.; Landers, T.; Danner, F.; Coyne, C.M.; Hunsaker, J.C. 3rd. Increased density of senile plaques; but not neurofibrillary tangles; in non-demented individuals with the apolipoprotein E4 allele: comparison to confirmed Alzheimer disease patients. *J Neurol Sci* 138: 97-104, 1996.

Stewart, R.; Prince, M.; Mann, A. Vascular risk factors and Alzheimer's disease. *Aust NZ J Psych* 33: 809-813, 1999.

Strittmatter, W.J.; Weisgraber, K.H.; Huang, D.Y.; Dong, L.M.; Salvesen, G.S.; Pericak-Vance, M.; Schmechel, D.; Saunders, A.M.; Goldgaber, D.; Roses, A.D. Apolipoprotein E: high avidity binding to ß-amyloid and increased frequency of type 4 allele in late-onset familial Alzheimer's disease. *Proc Natl Acad Sci USA* 90: 1977-1981, 1993.

Taddei, K.; Clarnette, R.; Gandy, S.E.; Martins, R.N. Increased plasma apolipoprotein E (apoE) levels in Alzheimer's disease. *Neurosci Lett* 223: 29-32, 1997.

Tatsuguchi, M.; Furutani, M.; Hinagata, J.; Tanaka, T.; Furutani, Y.; Imamura, S.; Kawana, M.; Masaki, T.; Kasanuki, H.; Sawamura, T.; Matsuoka, R. Oxidized LDL receptor gene (OLR1) is associated with the risk of myocardial infarction. *Biochem Biophys Res Commun* 303: 247-250, 2003.

Tatemichi, T.K.; Paik, M.; Bagiella, E.; Desmond, D.W.; Stern, Y.; Sano, M.; Hauser, W.A.; Mayeux, R. Risk of dementia after stroke in a hospitalized cohort: results of a longitudinal study. *Neurology* 44: 1885-1891, 1994.

Tedde, A.; Nacmias, B.; Cellini, E.; Bagnoli, S.; Sorbi, S. Lack of association between NOS3 polymorphism and Italian sporadic and familial Alzheimer's disease. *J Neurol* 249: 110-111, 2002.

Tiret, L.; Rigat, B.; Visvikis, S.; Breda, C.; Corvol, P.; Cambien, F.; Soubrier, F. Evidence; from combined segregation and linkage analysis; that a variant of the angiotensin I-converting enzyme (ACE) gene controls plasma ACE levels. *Am J Hum Genet* 51: 197-205, 1992.

Tooyama, I.; Kawamata, T.; Akiyama, H.; Kimura, H.; Moestrup, S.K.; Gliemann, J.; Matsuo, A.; McGeer, P.L. Subcellular localization of the low density lipoprotein receptor-related protein (alpha 2-macroglobulin receptor) in human brain. *Brain Res* 691: 235-238, 1995.

Urakami, K.; Wada-Isoe, K.; Wakutani, Y.; Ikeda, K.; Ji, Y.; Yamagata, K.; Kowa, H.; Okada, A.; Adachi, Y.; Nakashima, K. Lipoprotein(a) phenotypes in patients with vascular dementia. *Dement Geriatr Cogn Disord* 11: 135-138, 2000.

Uwabo, J.; Soma, M.; Nakayama, T.; Kanmatsuse, K. Association of a variable number of tandem repeats in the endothelial constitutive nitric oxide synthase gene with essential hypertension in Japanese. *Am J Hypertens* 11: 125-128, 1998.

Wakutani, Y.; Kowa, H.; Kusumi, M.; Yamagata, K.; Wada-Isoe, K.; Adachi, Y.; Takeshima, T.; Urakami, K.; Nakashima, K. Genetic analysis of vascular factors in Alzheimer's disease. *Ann N Y Acad Sci* 977: 232-238, 2002.

Wang, J.C.; Kwon, J.M.; Shah, P.; Morris, J.C.; Goate, A. Effect of APOE genotype and promoter polymorphism on risk of Alzheimer's disease. *Neurology* 55: 1644-1649, 2000.

Warwick Daw, E.; Payami, H.; Nemens, E.J.; Nochlin, D.; Bird, T.D.; Schellenberg, G.D.; Wijsman, E.M. The number of trait loci in late-onset Alzheimer disease. *Am J Hum Genet* 66: 196-204, 2000.

Wavrant-DeVrieze, F.; Perez-Tur, J.; Lambert, J.C.; Frigard, B.; Pasquier, F.; Delacourte, A.; Amouyel, P.; Hardy, J.; Chartier-Harlin, M.C. Association between the low density lipoprotein receptor-related protein (LRP) and Alzheimer's disease. *Neurosci Lett* 227: 68-70, 1997.

Whitehouse, P.J.; Sciulli, C.G.; Mason, R.M. Dementia drug development: use of information systems to harmonize global drug development. Psychopharmacol Bull 33: 129-133, 1997.

Wilson, R.S.; Schneider, J.A.; Barnes, L.L.; Beckett, L.A.; Aggarwal, N.T.; Cochran, E.J.; Berry-Kravis, E.; Bach, J.; Fox, J.H.; Evans, D.A.; Bennett, D.A. The apolipoprotein E epsilon 4 allele and decline in different cognitive systems during a 6-year period. *Arch Neurol* 59: 1154-1160, 2002.

Wilson, P.W.F.; Myers, R.H.; Larson, M.G.; Ordovas, J.M.; Wolf, P.A.; Schaefer, E.J. Apolipoprotein E alleles; dyslipidemia; and coronary heart disease: the Framingham offspring study. *JAMA* 272: 1666-1671, 1994.

Wolozin, B.; Kellman, W.; Ruosseau, P.; Celesia, G.G.; Siegel, G. Decreased prevalence of Alzheimer disease associated with 3-hydroxy-3-methyglutaryl coenzyme A reductase inhibitors. *Arch Neurol* 57: 1439-1443, 2000.

Wolozin, B.; Kellman, W.; Ruosseau, P.; Celesia, G.G.; Siegel, G. Reduced prevalence of Alzheimer disease associated with 3OH 3 methyl-glutaryl coenzyme: a reductase inhibitor. *Arch Neurol* 57: 1439-1443, 2000.

Woodward, R.; Singleton, A.B.; Gibson, A.M.; Edwardson, J.A.; Morris, C.M. LRP gene and late-onset Alzheimer's disease. *Lancet* 352: 239-240, 1998.

Yaffe, K.; Barret-Connor, E.; Lin, F.; Grady, D. Serum lipoprotein levels; statin use; and cognitive function in older women. *Arch Neurol* 59: 378-384, 2002.

Yang, J.D.; Feng, G.; Zhang, J.; Lin, Z.X.; Shen, T.; Breen, G.; St Clair, D.; He, L. Association between angiotensin-converting enzyme gene and late onset Alzheimer's disease in Han chinese. *Neurosci Lett* 295: 41-44, 2000.

Yoshitake, T.; Kiyohara, Y.; Kato, I.; Ohmura, T.; Iwamoto, H.; Nakayama, K.; Ohmori, S.; Nomiyama, K.; Kawano, H.; Ueda, K. Incidence and risk factors of vascular dementia and Alzheimer disease in a defined elderly Japanese population: the Hisayama Study. *Neurology* 45: 1161-1168, 1995.

Zill, P.; Engel, R.; Hampel, H.; Behrens, S.; Burger, K.; Padberg, F.; Stubner, S.; Moller, H.J.; Ackenheil, M.; Bondy, B. Polymorphisms in the apolipoprotein E (APOE) gene in gerontopsychiatric patients. *Eur Arch Psychiatry Clin Neurosci* 251: 24-28, 2001.

Zenker, G.; Koltringer, P.; Bone, G.; Niederkorn, K.; Pfeiffer, K.; Jurgens, G. Lipoprotein(a) as a strong indicator for cerebrovascular disease. *Stroke* 17: 942-945, 1986.

Zuliani, G.; Ble', A.; Zanca, R.; Munari, M.R.; Zurlo, A.; Vavalle, C.; Atti, A.R.; Fellin, R. Genetic polymorphisms in older subjects with vascular or Alzheimer's dementia. *Acta Neurol Scan*d 103: 304-308, 2001.

Zurutuza, L.; Verpillat, P.; Raux, G.; Hannequin, D.; Puel, M.; Belliard, S.; Michon, A.; Pothin, Y.; Camuzat, A.; Penet, C.; Martin, C.; Brice, A.; Campion, D.; Clerget-Darpoux, F.; Frebourg, T. APOE promoter polymorphisms do not confer independent risk for Alzheimer's disease in a French population. *Eur J Hum Genet* 8: 713-716, 2000.

In: Cognitive Science Compendium, Volume 1
Editor: Miao-Kun Sun
ISBN: 978-1-60456-597-3

Chapter 3

ERP STUDIES OF RECOGNITION MEMORY: DIFFERENTIAL EFFECTS OF FAMILIARITY, RECOLLECTION, AND EPISODIC PRIMING

David Friedman*
Cognitive Electrophysiology Laboratory, New York State Psychiatric Institute, New York, New York 10032

ABSTRACT

Dual process theorists postulate that recognition memory performance is subserved by two processes labeled familiarity (relatively automatic, reflecting the retrieval of the item per se without contextual details) and recollection (effortful, reflecting retrieval of specific contextual details surrounding the initial episode). Some theorists have also argued that these two processes have unique brain correlates. In an attempt to further understand the nature of the distinction between these two hypothesized constructs, event-related brain potentials (ERPs) were recorded in four conditions that were compared and contrasted: episodic priming, remember and know judgments, and misses. Like recognition judgments based on familiarity, episodic priming is thought to reflect an automatic process; remember and know judgments are closely allied with recollection and familiarity, respectively; and misses (old items called new) have been assumed to reflect an "implicit" retrieval process. An episodic memory (EM) or old/new effect between 200 and 300 ms was the earliest to distinguish old from new items, and did not differ in scalp topography across conditions. This EM effect could be the earliest to reflect familiarity-based recognition. Between 300 and 500 ms, the topographies of the ERPs associated with the episodic priming effect, remember and know judgments and misses all differed reliably. The scalp topography associated with the 300-500 ms window, previously assumed to be associated with familiarity, differed from the 500-700 ms window, the putative correlate of recollection. These data imply that familiarity-based processing may occur quite early. Further, the results suggest that within a 300-500 ms

* Address correspondence to: David Friedman, Ph.D., Cognitive Electrophysiology Laboratory, 1051 Riverside Drive, Unit 6, New York, NY 10032, Phone: 212-543-5476, Fax: 212-543-6002, Email: dfl2@columbia.edu

interval three different cognitive processes can contribute to recognition memory performance.

Keywords: Episodic memory, recollection, familiarity, episodic priming, event-related potentials

INTRODUCTION

Behavioral Studies of Recollection and Familiarity

How humans remember facts and events has long fascinated the lay public as well as the scientific community. This is attested to by the long history of research into the workings of memory, perhaps formally beginning with the classic studies of Ebbinghaus in the 1800s (Ebbinghaus, 1964), and continuing with the increasingly large number of papers that have appeared in the recent cognitive neuroscience and behavioral literatures (for review examples see, Johnson, 1995; McGaugh, 2000; Friedman and Johnson, 2000; Schacter, 1999; Gabrieli, 1998; Rugg and Wilding, 2000; Wilding and Sharpe, 2002). Despite about a century of formally documented memory research, there are still unanswered questions. Several of these concern the neurocognitive underpinnings of recognition memory judgments. For example, we have all had the experience of knowing that we have previously experienced an event, but were unable to remember specific details that surrounded that particular event. One fairly common example of this phenomenon is seeing a person who looks so familiar that you are certain that you have met him/her before. However, that "recognition" response, though apparently relatively automatic and seeming to "pop into mind," may not be accompanied by any other "contextual" details, e.g., where you met the person (was it at a party, or another type of gathering?), when you met the person (was it recently or has some time elapsed?) or other specific details, such as the person's name and occupation. These latter details may only be retrieved after effortful search for the context or source of that initial recognition judgment. The exact nature of the cognitive processes involved in these types of judgments as well as their neuroanatomical underpinnings is the subject of a great deal of current research and considerable debate.

Nonetheless, on the basis of these kinds of introspective experiences, cognitive theorists have postulated that at least two behavioral processes contribute to recognition memory judgments. These processes and the models used to account for them have recently been exhaustively well-reviewed by Yonelinas (2002). The first of these, labeled "familiarity," is thought to be relatively fast-acting and automatic, reflecting the retrieval of item or content information devoid of contextual details. The second, labeled "recollection," is thought to be effortful and slower-acting, indicating the retrieval of "source" information, i.e., contextual details such as the perceptual or emotional attributes in which the initial episode was experienced. Evidence concerning the speed with which these different types of information accrue has been provided by Johnson et al., (1994) and McElree et al., (1999). These reaction time data indicate that the products of recollection (i.e., source information) are available some 200 ms after the products of familiarity (i.e., item information) have become available.

Several investigators of these phenomena have used the remember/know (R/K) paradigm originally advanced by Tulving (1985) to provide support for the distinction between familiarity and recollection (for reviews see Gardiner, 2001 and Yonelinas, 2002). In this paradigm, following a study phase, participants are asked to make old/new recognition judgments. For any item attracting a judgment of old, the participant is asked to make an additional R or K judgment. They are to state "K" if they know the item was on the study list, but cannot retrieve specific details about the item's original occurrence, and "R" if they know that the item was on the study list and, in addition, can retrieve contextual details (whether idiosyncratic, inherent in the studied item, or experimentally induced). On the basis of a number of dissociations between experimental manipulations such as levels of processing (Craik and Lockhart, 1972), and the frequency with which old items attract R or K responses, several authors have argued that R and K are surrogates, respectively, for recollection and familiarity (but see Gardiner et al., 1999).

Other Mechanisms Potentially Contributing to Recognition Memory

In addition to familiarity and recollection, other mechanisms could influence recognition memory performance. For example, a perceptual implicit memory mechanism might be responsible for the brain's automatic retrieval of a previously experienced event in the absence of conscious awareness about that episode. In fact, some investigators have suggested that this implicit memory mechanism and familiarity can be equated (e.g., Parkin et al., 2001; Mandler, 1980). On the other hand, others have argued that this is not the case, with familiarity and implicit memory making distinct contributions to recognition memory performance (Donaldson et al., 2001; Wagner and Gabrieli, 1998). This latter hypothesis is supported by data indicating that residual recognition memory performance of amnesiacs with damage to the medial temporal lobes is not likely due to familiarity engendered by an Aimplicit@ memory mechanism (Knowlton and Squire, 1995; Reed et al., 1997). Moreover, amnesic individuals can often perform as well as controls on measures of perceptual implicit memory while failing to recognize these same items as having been previously studied (Cermak et al., 1988; Schacter et al., 1994).

Finally, in some retrieval situations, another mechanism could potentially influence recognition memory performance. Episodic priming can occur for newly-learned associations created in the context of study episodes for subsequent recognition memory tasks (McKoon and Ratcliff, 1979), as in the task described in this report. In this paradigm, two unassociated nouns occurred in a studied sentence and were subsequently tested for recognition memory. During the memory test, the nouns were presented sequentially in two-noun pairs. They were presented either in their original noun 1 - noun 2 sequence (i.e., from the same studied sentence) or from two different studied sentences. Like its perceptual priming counterpart, episodic priming refers to a facilitation (either in reaction time or accuracy) in processing the second noun of the pair from the same sentence compared to the second noun of a pair from a different sentence. Similar to current theorizing about retrievals based on familiarity and implicit mechanisms, episodic priming effects are thought to be relatively automatic (McKoon and Ratcliff, 1979). However, at the current state of knowledge, it is not known whether episodic priming reflects an implicitly- or familiarity-based mechanism, or an independent retrieval mechanism. Attempting to determine which of these alternatives holds is one of the purposes of this report.

Event-Related Potential (ERP) Studies of Recollection, Familiarity, Implicit Memory, and Episodic Priming

Recollection versus Familiarity

As stated earlier, based on the premise that the behavioral processes of recollection and familiarity reflect unique states of awareness, some authors have suggested that they are subserved by distinct or at least partially non-overlapping neural networks (e.g., Yonelinas, 2002). Evidence for this hypothesis has come from the study of ERPs (e.g., Curran, 2000; Friedman and Johnson, 2000; Johnson et al., 1998; Rugg et al., 1998a), as well as other techniques, such as fMRI (Eldridge et al., 2000; Henson et al., 1999), and neuropsychological studies of amnesiacs (Yonelinas et al., 1998). Two aspects of the ERP waveform, one relatively early (~300-500 ms) and one subsequent to the early region (~500-800 ms), have been the most consistently observed to be modulated by experimental manipulations thought to engender reliance on, respectively, familiarity and recollection. The earlier of these two has been labeled the medial prefrontal episodic memory (EM) effect, because its scalp distribution is characterized by a focus of positive activity over medial prefrontal scalp. The later of the two has been labeled the left parietal EM effect, as its topography is focused over the left temporo-parietal scalp when words are used as stimuli (Friedman and Johnson, 2000). In general, the sequential time course of these two EM effects fits rather well with the behavioral data showing that, as mentioned earlier, the products of familiarity-based processing become available about 200 ms before the products of recollection-based processing begin to influence retrieval (Johnson et al., 1994; McElree et al., 1999).

A few methods have been used in an attempt to disentangle the effects of familiarity and recollection on the ERP waveform. Because some theorists argue that familiarity reflects a global matching process between the to-be-retrieved item and its previously studied counterpart (e.g. Curran, 2000), foil or lure items that are similar but not identical to previously studied items are expected to generate a strong familiarity signal. Therefore, the ERPs elicited by the highly similar lure and the previously studied item should both contain a similar-amplitude medial prefrontal EM effect. This prediction has been shown to be the case with, among others, the ERPs associated with studied words and semantically similar lures (Nessler et al., 2001), and studied pictures and those same pictures presented in reversed orientation at test (Curran and Cleary, 2003). In these same studies, the left parietal EM effect was larger to previously studied items relative to the lures, consistent with the notion that the left parietal EM effects reflects the retrieval of specific features of the study episode, i.e., recollection.

Another method for assessing whether the medial prefrontal and parietal EM effects are differentially sensitive to familiarity and recollection is to use variables that have been shown to have dissociative effects in behavioral studies. For example, using the remember/know paradigm, Trott and colleagues (Trott et al., 1999) showed that, for young adult subjects, the medial prefrontal EM effect was similar in magnitude regardless of whether the retrieved item attracted a remember or a know judgment; on the other hand, only the ERPs elicited by items associated with a remember judgment showed a robust parietal EM effect (see also Duzel et al., 1997, Mark and Rugg, 1998, and Smith, 1993).

A third technique that has been employed in this endeavor is the source memory paradigm (for reviews see Friedman and Johnson, 2000, and Rugg and Wilding, 2000). In this design, participants are tested via old/new recognition memory and, for any item judged old, an additional source judgment is solicited. As the hallmark of a recollective response is the retrieval of contextual information from the initial learning experience, the ERPs associated with correctly recognized old items whose sources are correctly retrieved are assumed to be comprised of signatures related to recollection (contextual memory) as well as familiarity (item memory). By contrast, ERPs associated with correctly recognized items whose sources are misattributed, are assumed to be based on familiarity only. By and large, ERP investigators using this design have reported results in line with these assumptions, i.e., the ERPs presumed to be associated with recollection-based processing show larger parietal EM effects than those assumed to be based on familiarity, whether the sources are temporal (Trott et al., 1999), gender of voice (Wilding and Rugg, 1996) or spatial location (Van-Petten et al., 2000). Moreover, in the Trott et al. (1999) study, the medial prefrontal effect was of similar magnitude in the ERPs associated with correct and incorrect source retrievals, suggesting a link to the familiarity component of recognition memory.

The directed forgetting paradigm has also been employed to inform the ERP distinction between recollection and familiarity (Ullsperger et al., 2000). In this design, subjects are instructed during study to remember some words (TBR) and forget others (TBF). During recognition, subjects are instructed to decide whether the item is old or new regardless of the TBR or TBF status of the item In support of the association of the medial prefrontal EM effect with familiarity, Ullsperger et al. (2000) found that correctly recognized TBR and TBF words contained equal magnitude medial prefrontal EM effects. By contrast, only the TBR words were associated with a reliable left parietal EM effect.

Recently, however, the results of a few investigations have cast doubt on the association of the medial prefrontal EM effect with familiarity. Curran and Dien (2003), Duarte et al.,. (2004) and Tsivilis et al., (2001) have all observed EM effects that could reflect a familiarity-based retrieval process, and these effects have been observed to occur earlier than those associated with the medial prefrontal effect. For example, Tsivilis et al., (2001) reported an unpredicted finding that the amplitude of an early EM effect (between 100 and 300 ms), with a fronto-polar scalp topography, responded in a fashion that was more consistent with a familiarity-based effect than the medial prefrontal EM component. During the encoding phase of their study, pictures of common objects were "placed" in a background scene. During test, same (object, background) pairings, rearranged pairings (both object and background were old, but from different pairings), old object/new background, new object/old background, and new object/new background pairings were presented. The subject's task was to judge the previous occurrence of the objects without regard to their associated contexts. Tsivilis et al., (2001) found that the medial prefrontal effect was reliably more positive in association with correctly recognized same and rearranged pairs compared to the other 3 pairings. Because it was absent in response to the old/new pairing, and performance for this pairing did not differ from that to rearranged pairs, the authors argued that this EM effect could not have been a direct correlate of familiarity. Rather, the authors argued that an earlier EM effect (100-300 ms) could have reflected familiarity, as its amplitude was more positive (compared to new/new pairings) for any pairing that had at least one previously studied object or background.

During encoding, Duarte et al., (2004) asked their subjects to study pictures of common objects under two instructional regimes presented in separate encoding blocks, animacy (living/non-living), or manipulability (can the object be manipulated). During the test phase subjects made trinary remember, know, or new judgments via reaction time. For all items given either remember or know judgments, a subsequent source judgment (which task was the old item associated with, animacy or manipulability?) was also solicited. During retrieval, an EM effect measured between 150 and 300 ms, with a fronto-polar scalp distribution, was sensitive to familiarity, i.e., it was of equal magnitude in the R and K waveforms, suggesting a link with familiarity-based retrievals. In addition, the 300-450 ms window was also associated with familiarity-based responding. Interestingly, these two regions of the waveform did not differ topographically. Hence, unlike the Tsivilis et al., (2001) data, the Duarte et al., (2004) finding is equivocal with respect to the two regions reflecting unique mnemonic functions.

In a very recent investigation, Curran and Dien (2003) also observed an early EM effect between 176 and 260 ms characterized by a fronto-polar scalp distribution, similar to those reported by Tsivilis et al., (2001) and Duarte et al., (2004). During encoding, Curran and Dien (2003) presented words either auditorally or visually and then, during the test phase, tested all words visually, yielding auditory-visual and visual-visual study-test pairings. Based on the premise that familiarity is an amodal process rather than perceptually specific, it was predicted that the medial prefrontal EM effect (300-500 ms) would be of similar magnitude regardless of the type of study-test pairing. This prediction was borne out. However, the earlier EM effect (176-260 ms) was perceptually specific as it was observed only following visual study. Taken as a whole, then, the data on the early EM effect are conflicting and further study is clearly warranted. Hence, in the current investigation, the data of Trott et al., (1999) were re-examined using an early window (200-300 ms) to determine if a reliable EM effect would be observed and, if so, whether it would differ across conditions.

Implicit Memory

Although familiarity effects have been clearly evident in several ERP studies of recognition memory, the same cannot be said for implicit memory effects, with which they have been equated by some theorists (see above). Attempts to find an ERP correlate of implicit memory retrieval have, for the most part, been unconvincing. One reason for the lack of compelling evidence is that many investigators have assumed that, because a task is "indirect" (i.e., does not necessarily require conscious access to previous episodes, e.g., repetition priming), retrievals are based upon an implicit mechanism. This is, however, not the case, as explicit retrieval strategies cannot be ruled out in the majority of these studies (Schacter et al., 1989). In an experiment that attempted to control for this difficulty, and one particularly germane to the current investigation, Rugg et al., (1998a) had subjects encode verbal items either shallowly (first and last letters alphabetical?) or deeply (incorporate word into a sentence). They were then tested for retention of these items using a standard old/new recognition procedure. Because perceptual implicit memory effects are known to be insensitive to depth of processing manipulations (Roediger and McDermott, 1993), whereas direct memory measures are, this investigation provided a sounder, dissociative method for determining if ERP EM effects would be consistent with implicit memory retrieval (see also Paller, 1998; Paller and Kutas, 1992 for similar methods). With respect to implicit memory, the major finding was that for shallowly encoded words, missed items at test elicited a reliable ERP EM effect between 300 and 500 ms with a scalp topography that differed from that associated with correctly recognized shallowly studied words, i.e., it could have reflected an "implicit" retrieval. By contrast, the 300-500 ms EM effects associated with deeply and shallowly encoded correctly recognized old items did not differ in scalp distribution. Therefore, the authors concluded that this latter EM effect could be a correlate of familiarity. Hence, between 300 and 500 ms, two functionally and topographically dissociable EM effects were observed by Rugg et al., (1998a). In showing unique scalp topographies for retrievals assumed to be based on an implicit memory mechanism compared to those based on familiarity, the Rugg et al., (1998a) data, like those based on amnesic patients (Knowlton and Squire, 1995; Reed et al., 1997), suggest that the two most likely make unique contributions to recognition memory.

Episodic Priming

As described earlier, in similar fashion to familiarity- and implicitly-based retrievals, episodic priming has been associated with automatic processes, but whether or not episodic priming differs from these other mnemonic processes has not been the subject of ERP investigations. The ERP episodic priming effect described by Trott et al., (1999) was interpreted by them to have impacted the N400 component (Kutas and Hillyard, 1980b), a brain potential that has typically been associated with semantic priming tasks. During the test phase of the Trott et al., study, the N400 to a second noun that followed a first noun from the same studied sentence was reliably smaller than a noun 2 that followed a noun 1 from a different studied sentence, i.e., demonstrating an episodic priming effect on the N400. The Trott et al., (1999) data extended the circumstances under which N400 amplitude can be modulated to a situation in which new associations were encoded during a single study session. Further, although controversial, there is good evidence that the N400 effect may reflect relatively automatic, non-conscious processes (Luck et al., 1996; Deacon et al., 2000). On these bases, Trott et al., concluded that their episodic priming effect could have reflected an automatic, non-conscious process contributing to the faster retrieval of the second noun in a same-sentence pairing.

The Current Study

Here I concentrate only on the young adult data and report analyses that were not possible to pursue due to space limitations and the age-related focus of the original investigation (Trott et al., 1999). There were four goals: 1) to determine if the familiarity and recollection components of dual-process theories of recognition memory (Mandler, 1980) have unique ERP signatures; 2) to assess whether the ERP familiarity effect is similar to or different from an "implicitly-based" response presumably reflected by the ERPs associated with unrecognized or "missed" old items (Rugg et al., 1998a); 3) because episodic priming effects are thought to be relatively automatic (McKoon and Ratcliff, 1979), it was of interest to determine if their ERP correlates were similar to either familiarity- or implicitly-based ERP effects, as these are also thought to be automatic and not necessarily accessible to consciousness; 4) in view of the findings by Curran and Dien (2003), Tsivilis et al., (2001) and Duarte et al., (2004), the current data were re-examined to determine if there was evidence for a fronto-polar EM effect that might be an earlier manifestation of a brain event reflecting familiarity-driven processes.

To assess differences among these conditions, scalp topography analyses were performed. These analyses were based on the premise that differences in scalp distribution among conditions of the experiment would suggest differences in the intracranial configuration underlying those different conditions (see Johnson, 1993 for a review and caveats; and see Urbach and Kutas, 2002 for counter arguments). Reliable changes in scalp topography between, for example, the putative ERP indices of familiarity and recollection, would provide evidence for those processes reflecting unique states of awareness subserved by different neural networks. As stated earlier, recent evidence from the neuropsychological literature suggests that familiarity should not be equated with an "implicit" memory mechanism (Knowlton and Squire, 1995; Reed et al., 1997). To the extent that a miss (an old item labeled new) reflects an implicitly-driven retrieval (Rugg et al., 1998a) while a know judgment represents a familiarity-based retrieval, these two waveforms should differ reliably in scalp distribution. Although, as stated above, episodic priming effects are thought to be relatively automatic, it is not known whether these reflect implicitly- or familiarity-based mechanisms. Hence, comparison of the scalp distributions associated with these conditions would allow preliminary conclusions to be drawn. In addition, some evidence suggests that an EM effect between 100 and 300 ms with a fronto-polar topography is the earliest manifestation of a familiarity-based retrieval. This EM effect has been shown to be distinct topographically from the subsequent medial prefrontal EM effect in at least one investigation (Tsivilis et al., 2001). Therefore, an assessment of topographic similarity or difference between these two EM effects was also performed.

Methods and Materials

Complete methodological details are available in Trott et al., (1999) and only a brief description will be provided here.

Subjects

These were the original 16 young adult women (age range = 21-28 yrs.) whose data are presented in Trott et al., (1999). All subjects signed informed consent and the study was approved by the New York State Psychiatric Institute's Institutional Review board.

Experimental Design and Procedures

The paradigm used was based on that originally published by McKoon and Ratcliff (1979). It was modified by Howard et al., (1986) to investigate the episodic priming of new associations in young and older adults. We employed Howard et al.'s modification, adding our own changes to accommodate ERP recording. The design was also based on the Remember/Know (R/K) distinction originally advanced by Tulving (1985).

Study Phase

During each study phase, participants viewed a series of sentences embedded in two distinctly demarcated lists (labeled *List 1* and *List 2*). The target sentences were presented on a computer screen one word at a time for 300 ms with an inter-stimulus interval (ISI) of 1200 ms. Immediately after the presentation of the List 1 sentences, the List 2 sentences were presented. There were 8 target sentences in each list, yielding a total of 128 study sentences for the entire experiment (i.e., 8 sentences per list by 2 lists by 8 study-test sessions = 128). To assess episodic priming, each sentence contained two unassociated nouns. The determination that the nouns were unassociated had been obtained from a normative study performed with a separate group of 20 young adult subjects.

Test Phase[1]

Each test trial consisted of the sequential presentation of two nouns each displayed for a duration of 300 ms with an ISI of 2000 ms. Subjects made speeded and accurate old/new judgments to each noun via button press. After the presentation of a two-noun sequence, if either noun had been judged old, it was re-presented, and participants made an R/K judgment, followed by a temporal source judgment (i.e., List). These latter two judgments were untimed, but generally took between 2-4 s each. The sequence of events during study and test is depicted in Figure 1.

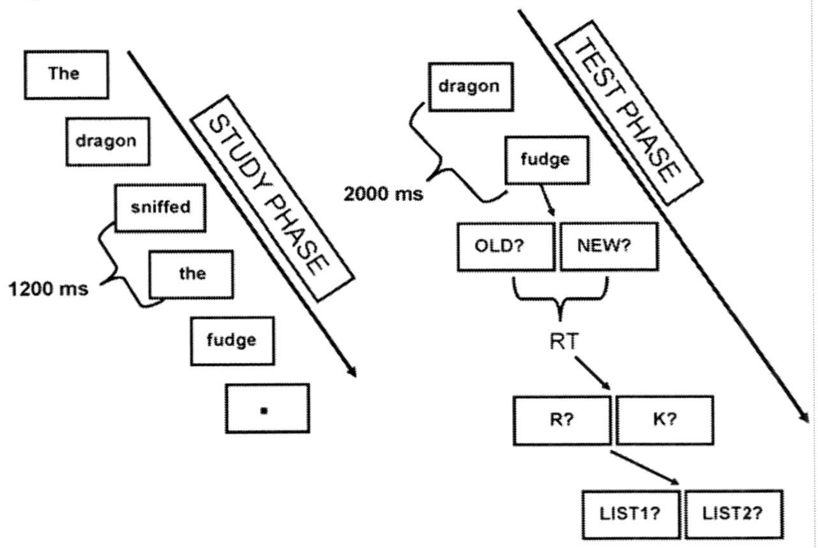

Figure 1: Schematic diagram of the experimental protocol. The diagram depicts a single sentence presented during the study phase and an example of a test pair in which both nouns come from the same sentence (Same condition). During the study phase, each word (including the period indicating the end of the sentence) was presented for 300 ms with an ISI of 1200 ms. During the test phase, the nouns were presented for 300 ms, separated by an ISI of 2000 ms.

[1] List membership served as the temporal contextual attribute for a subsequent source memory judgment during the test phase. This aspect of the data will not be dealt with (see (Trott et al., 1999) for complete details).

During the test phase "forward" episodic priming was assessed by combining pairs of nouns so that they had either been 1) presented together in the same sentence (labeled Same), or 2) presented in two different sentences (labeled Diff). Other combinations of new and old nouns were also employed, but the Same and Diff conditions were critical for assessing the presence of episodic priming. Each test phase contained the 32 target nouns from the study sentences (16 from each list), plus 32 foil nouns, yielding a total of 32 pairs presented during each test block. Therefore, a total of 512 nouns was presented across all 8 test blocks.

EEG Recording

The electroencephalogram (EEG) was recorded from 30 scalp sites using an Electro cap (Electrocap International, Inc.) for scalp placements and disposable Ag/AgCl electrodes for the face and mastoids. All electrodes, including the mastoids, were referred to the nosetip. The EEG data were digitally re-referred to an averaged mastoid reference, resulting in data available from a total of 28 active scalp sites. Horizontal and vertical electrooculogram (EOG) were recorded bipolarly with electrodes placed, respectively, at the outer canthi of both eyes and above and below the right eye. EOG and EEG were recorded continuously with a 5.3 s time constant (30-Hz upper cutoff), and digitized at 100 Hz. Test-phase data were epoched off-line with 100 ms pre- and 1900 ms post-stimulus intervals. Eye-movement artifacts were corrected off-line (Gratton et al., 1983).

Data Analyses

With the exception of the ERPs associated with misses, and the measurement of an early 200-300 ms interval, all other amplitude analyses have been reported in Trott et al., (1999). Hence, only the statistical analyses on the raw amplitude data associated with misses and the early 200-300 ms interval will be reported. The key analyses in the present report focus on the comparisons of the topographies associated with the episodic priming, Remember/Know and miss data. These had not been the subject of the Trott et al., report. To compare and contrast the conditions of the experiment, the data were normalized using the root mean square vector method (McCarthy and Wood, 1985; Ruchkin et al., 1999). This procedure effectively removes amplitude information allowing for distribution comparisons across different samples and/or conditions. A reliable difference in scalp distribution results in a significant Electrode Location by Condition interaction in the ANOVA performed on the normalized data. The data from all 28 electrode locations were entered into the normalized analyses. In order to pinpoint where on the scalp the topographic differences were localized, post-hoc testing consisted of assessing differences among conditions at each electrode site.

Maps of surface potential (SP) and current source density (CSD) activity were computed using a third order spherical spline interpolation algorithm (Perrin et al., 1989). SP and CSD maps provide a different picture of electrical activity, because the two mapping methods differentially weight the contribution of subcortical activity. Voltage maps are derived from the ERP amplitudes obtained at each electrode site, and thus provide a means of visualizing the summed activity from *all* active brain areas, both cortical and subcortical. By contrast, before CSD maps are calculated, the amplitudes are spatially filtered with an algorithm that removes the activity volume conducted from subcortical and distant cortical areas. The resulting maps provide a spatially-sharpened, reference-free display of positive and negative current densities that emphasize cortical differences (Nunez and Pilgreen, 1991; Picton et al., 1995). In addition to yielding different pictures of brain activity, another reason for including both types of maps is that the vast majority of investigators of memory-related phenomena have used only SP maps to display their data, and thus the SP maps depicted in the results section provide an opportunity for comparison with these previous results.

The SPSS Version 11.5 for Windows (repeated measures ANOVA) computer program was used for all analyses. The Greenhouse-Geisser epsilon (ε) correction (Jennings and Wood, 1976) was used where appropriate. Uncorrected degrees of freedom are reported along with the epsilon value; the *P* values reflect the epsilon correction. Significant main effects and interactions were followed-up, where appropriate, with simple effects tests and/or post-hoc analyses using the Tukey Honestly Significant Difference (HSD) test.

RESULTS

Unless otherwise stated, all results reported in the text below reflect a P value of < .05.

Behavioral Data

All of the behavioral data have been reported in Trott et al., (1999), and are described briefly here. Collapsed across nouns 1 and 2, the mean hit rate was 0.72, with a mean false alarm rate of 0.08, leading to a corrected recognition score of 0.63 (Snodgrass and Corwin, 1988). The mean proportion of hits that attracted a remember judgment was 0.50; for know judgments, this value was 0.22.

The episodic priming (EP) effect (noun 2 only) was obtained by subtracting the reaction times associated with a noun 2 that was preceded by a noun 1 from the same studied sentence (Same condition; M RT = 926 ms) from a noun 2 that was preceded by a noun 1 from a different studied sentence (Diff condition; M = 980 ms). This reliable facilitation in reaction time (RT) was, therefore, 54 ms. Accuracy, as measured by sensitivity (Snodgrass and Corwin, 1988), was also significantly higher in the Same (M d_L = 4.1) than in the Diff (M d_L = 3.6) condition.

ERP Waveforms

Figure 2 depicts the unsubtracted waveforms from the four conditions of the experiment, while the corresponding difference waveforms are depicted in Figure 3. For Figure 2, the 4 conditions are, respectively, episodic priming (Figure 2, column 1, Figure 3, column 1), remember (Figure 2, column 2; Figure 3, column 2), know (Figure 2, column 2; Figure 3, column 2), and misses (Figure 2, column 3; Figure 3, column 3). The data are depicted at 4 scalp sites (Fz, Cz, Pz, O1) where these effects appeared largest. The waveforms are pictured for a period of 1000 ms and encompass the three regions (early EM effect, 200-300 ms;

medial prefrontal EM effect, 300-500 ms; left parietal EM effect, 500-800 ms) focused on in the current paper.

To set the stage for the new topographical analyses that are the subject of this report, the reliable amplitude findings from the Trott et al., (1999) investigation will first be briefly summarized. Following this, the new amplitude analyses of the data associated with misses will be detailed. Then, following Duarte et al., (2004) and Tsivilis et al., (2001), new analyses of an early EM effect (200-300 ms) not described in Trott et al., (1999) will be presented. Finally, the topographical analyses that compared directly the various conditions of the experiment will be described in detail.

Summary of (Trott et al., 1999) Amplitude Findings

All of the descriptions below are based on the statistical analyses originally reported in Trott et al., (1999)[2,3], and reflect comparisons that returned significant amplitude differences among the conditions depicted in each of the three columns of waveforms in Figure 2. The first column of ERPs shows that the waveforms associated with a noun 2 that was correctly recognized are more positive than the ERPs associated with correctly rejected foil nouns (i.e., show a robust EM effect between approximately 200 and 500 ms). However, the ERP associated with a noun 2 that was embedded in a different sentence than its noun 1 counterpart shows greater negativity (i.e., enhanced N400) than a noun 2 that came from the same sentence as its noun 1 associate (i.e., this region of the waveform demonstrates robust episodic priming). Note also that only the two early regions of the waveforms show this effect – there is no difference between these conditions for the subsequent positivity. Nonetheless, there is a robust EM effect for this later region of the waveform, as both of the ERPs associated with old items show greater positivity than the corresponding correctly rejected foil nouns. Note also that the waveforms associated with the two old items diverge from those associated with new items beginning at about 200 ms, some 100 or so ms prior to the maximum of the medial prefrontal EM effect.

The second column of waveforms in Figure 2 depicts the data associated with remember and know judgments (collapsed across nouns 1 and 2). During the earliest region (200-300 ms), as well as the region of the medial prefrontal EM effect (300-500 ms), the R and K waveforms do not differ, but both are more positive than the ERPs elicited by correctly rejected new nouns (i.e., show robust EM effects). However, during the 500-800 ms region, only the ERPs associated with R judgments show a robust EM effect. This can be clearly seen in the difference means depicted in Figure 3, column 2.

The third column of waveforms in Figure 2 depicts the ERPs (collapsed across nouns 1 and 2) associated with correctly recognized old items (hits – collapsed across R and K), misses and correctly rejected foil nouns. Note that, in the 200 - 300 and 300 – 500 ms

[2] Because of slight differences in the latency windows used by Trott et al. (1999) and those used here, all of the analyses reported in Trott et al. (i.e., R vs. K vs. new; Same vs. Diff) were re-computed using 300-400, 400-500, 500-600 and 600-700 ms averaged voltage intervals. These analyses support, in all respects, the ordering of conditions as reported in Trott et al. (1999).

[3] With the exception of the data for misses, which were not reported in the Trott et al. (1999) paper, the mean number of trials are as reported in the original manuscript (Trott et al., 1999). For the miss data reported here, the mean number (±SD) of trials entering the averages was 50.9 (33.2).

intervals, the ERPs associated with missed items do not appear to differ markedly from those associated with hits, but both are more positive than the ERPs associated with correctly rejected nouns, hence appearing to demonstrate robust EM effects. For the subsequent region of the waveform, however, the ERPs associated with misses and new items do not differ and both are less positive than the ERPs associated with hits. These effects can be observed in the difference waveforms of Figure 3, column 3.

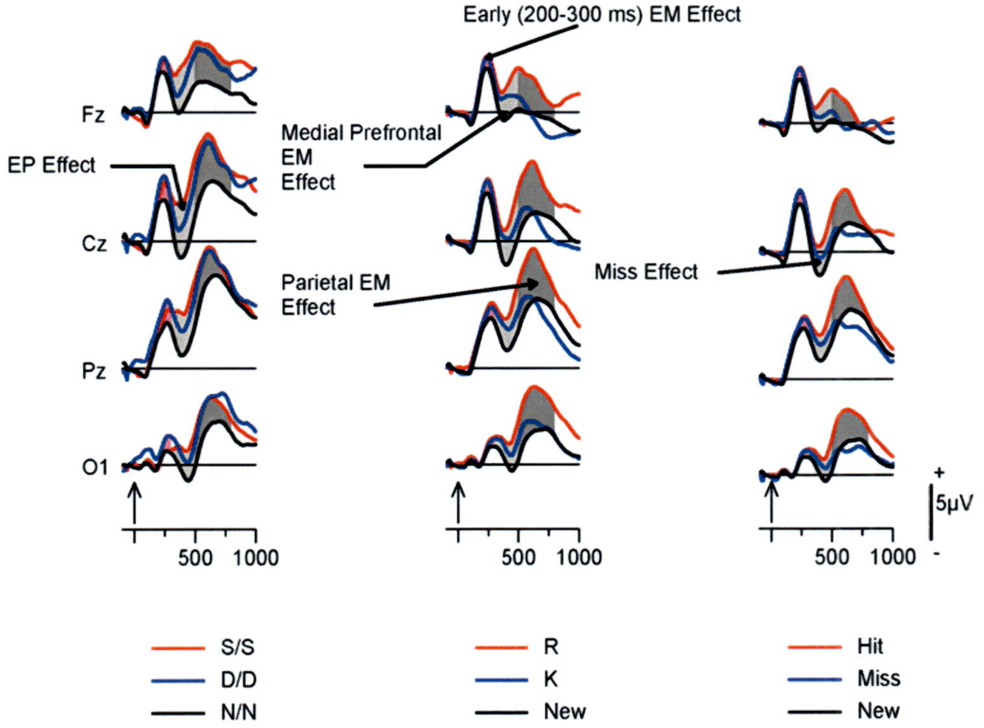

Figure 2: Grand mean ERPs averaged across subjects for the episodic priming (column 1; noun 2 only), remember/know (column 2; collapsed across nouns 1 and 2), and miss (column 3; collapsed across nouns 1 and 2) conditions of the experiment. Arrows mark stimulus onset with time lines every 500 ms. The ERP EM effects of interest are denoted by arrows. They are, an early (200-300 ms) effect, larger in all classes of old items relative to new items; episodic priming (300-500 ms, column 1), in which the ERPs elicited by noun 2 ERPs in the Diff condition are more negative than those in the Same condition; the medial prefrontal EM effect (300-500 ms, column 2), in which the ERPs associated with R and K judgments do not differ but both are less negative than those associated with new items; the parietal EM effect (measured between 500 and 700 ms, column 2), in which only the ERPs associated with R judgments show a robust EM effect; and the miss effect (300-500 ms, column 3), in which the ERPs associated with missed (i.e., old) trials are more positive than their new counterparts. Pink shading marks the EM effect between 200 and 300 ms; light gray shading marks the EM effect between 300 and 500 ms, and dark gray shading between 500 and 800 ms. Abbreviations for the episodic priming conditions, which are labeled for their noun 1/noun 2 pairings, are S/S = Same; D/D = Diff; and N/N = New).

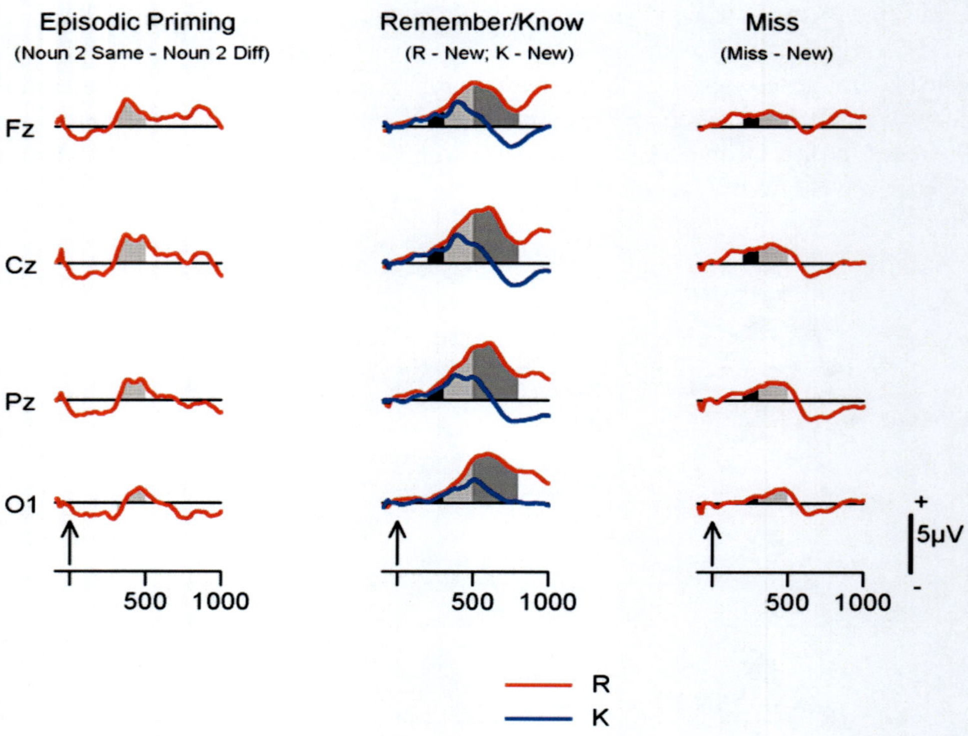

Figure 3: Difference waveforms corresponding to the ERPs shown in Figure 2. Column 1 shows the difference ERPs obtained by subtracting the ERPs associated with a correctly recognized noun 2 that came from a different sentence than the preceding noun from the ERPs elicited by a noun 2 that came from the same sentence as the preceding noun. Column 2 shows the difference waveforms computed by subtracting the ERPs associated with correct rejections from those associated with remember (red line) or know (blue line) judgments. Column 3 shows the difference waveforms obtained by subtracting the ERPs associated with correct rejections from those associated with missed trials. Arrows mark stimulus onset with 500-ms time markers. Black shading reflects the 200-300 ms time period, light gray, the 300-500 ms time period, and dark gray the 500-800 ms time period.

Raw Amplitude Analyses of the Misses

The data (based on 16 subjects) depicted in the third column of Figure 2 were analyzed in a Condition (hit, miss, correct rejection) by Bin (200-800 ms in 100 ms increments) by Electrode Location (Fz, Cz, Pz) ANOVA. These scalp sites were chosen as these relatively early EM effects (200-300 and 300-500 ms) did not show any asymmetries and were largest at the midline. As expected, the Condition main effect was reliable ($F(2,30) = 7.99$, $P < .006$, $\epsilon = 0.69$), indicating greater amplitude for hit trials. The main effect of Bin was also significant ($F(5,75) = 4.75$, $P < .01$, €o€ $= 0.41$). Both main effects were modified by the Condition by Bin ($F(10,150) = 6.65$, $P < .004$, $\epsilon = 0.19$) and Condition by Bin by Electrode Location ($F(20,300) = 6.21$, $P < .0001$, $\epsilon = 0.23$) interactions. Post-hoc testing of the triple interaction revealed that the ERPs associated with hits and misses were more positive between 200 and 300 ms at Fz and Cz than those associated with correct rejections. These comparisons remained reliable for the 300-400 ms bin. Between 400 and 500 ms, with the exception of Fz, miss ERPs were again more positive than those associated with correct rejections but, in this latency range, the ERPs associated with hits were now more positive than those associated with correct rejections and misses. For the remaining windows (600-800 ms, i.e., within the range of the parietal EM effect), the miss and correct rejection ERPs no longer differed, and hit trials were more positive than those associated with misses and correct rejections.

To summarize, the ERPs associated with misses show a reliable EM effect that is active between ~ 200 and 500 ms, suggesting a very early onset similar to those observed by Duarte et al., (2004) and Tsivilis et al., (2001). However, there is no reliable EM effect for the later region (left parietal EM effect) of the waveforms associated with misses.

Raw Amplitude Analyses of the ERPs Associated with R and K Judgments and Episodic Priming between 200 and 300 ms

For the waveforms associated with remember and know judgments, the data were analyzed in a Condition (R, K, New) by Electrode (Fz, Cz, Pz) ANOVA. This analysis resulted in a reliable main effect of Condition only ($F(2,30) = 6.98$, $P < 0.004$, $\epsilon = 0.96$). Paired comparisons indicated that the R and K waveforms were reliably more positive than the ERPs associated with new items, but did not differ from each other.

The averaged voltages measured between 200 and 300 ms associated with the episodic priming conditions (noun 2 only) were subjected to a Condition (Same, Diff, New) by Electrode (Fz, Cz, Pz) ANOVA. This ANOVA returned a reliable main effect of Condition only ($F(2,30) = 3.34$, $P < 0.05$, $\epsilon = 0.96$). Paired comparisons indicated that the ERPs associated with the Diff condition were significantly more positive than those associated with new items; the averaged voltages associated with the Same condition were marginally more positive than the ERPs associated with new items ($P < 0.065$). However, the Diff and Same ERPs did not differ.

In sum, the analyses of the 200-300 ms region of the waveform suggests the presence of a robust EM effect similar to that observed by Curran and Dien (2003), Duarte et al., (2004) and Tsivilis et al., (2001).

Scalp Topography Analyses

Figure 4 depicts the surface potential (SP) and current source density (CSD) topographic maps based on the difference means presented in Figure 3. The maps were computed for the 200-300, 300-400 and 400-500 ms intervals. For the episodic priming conditions, there are no maps depicted for the 200-300 ms region because there was no difference for the two priming conditions during that latency window (see Figures 2 and 3). Note that, in the majority of the maps depicted in Figure 4, the SP data show widespread activity with a single focus. By contrast, the CSD maps are much more differentiated, suggesting the presence of multiple (cortical) generators that contribute to the data recorded at the scalp.

The first set of topographic ANOVAs compared the early 200-300 ms window across conditions. This analysis contrasted the normalized averaged voltage data computed from the old minus new difference waveforms associated with the second noun when primed, and the R, K and miss conditions. The old minus new averaged voltage differences computed on the noun 2 waveforms were included as a check on the reliability of the results, as these data formed a subset of the remember and know waveforms. The Condition (noun 2 old-new; R-new; K-new; miss-new) by Electrode (all 28 locations) ANOVA failed to return a significant Condition by Electrode interaction ($F = 1.68$, $P > 0.10$), suggesting that the scalp distributions of the early 200-300 ms EM effect did not differ across these conditions.

Based on the results of Tsivilis et al., (2001) and Duarte et al., (2004), it was of great interest to determine if the early 200-300 ms EM effect differed topographically from the subsequent 300-500 ms EM effect. Therefore, the second set of topographic analyses compared the 200-300 ms window with the 300-400 and 400-500 ms windows for each condition separately using the data from all 28 scalp locations. For the noun 2 old-new ($F(54,810) = 40.32$, $P < 0.0001$, $\epsilon = 0.02$), and the miss-new ($F(54,810) = 32.15$, $P < 0.0001$), $\epsilon = 0.03$) ANOVAs, the Window (200-300, 300-400, 400-500) by Electrode interactions were reliable. For the remember-new differences, the interaction was marginally significant ($P < 0.07$), while for the know-new analysis, the interaction failed to reach the appropriate level of significance ($P. > 0.10$). In all cases, post-hoc testing indicated that the distribution associated with the early 200-300 ms EM effect was more frontally oriented than those associated with the 300-400 and 400-500 ms windows. The results of these analyses suggest that the early (200-300 ms) EM effect might be distinct from the subsequent 300-500 ms medial prefrontal EM effect, the putative correlate of familiarity-based recognition.

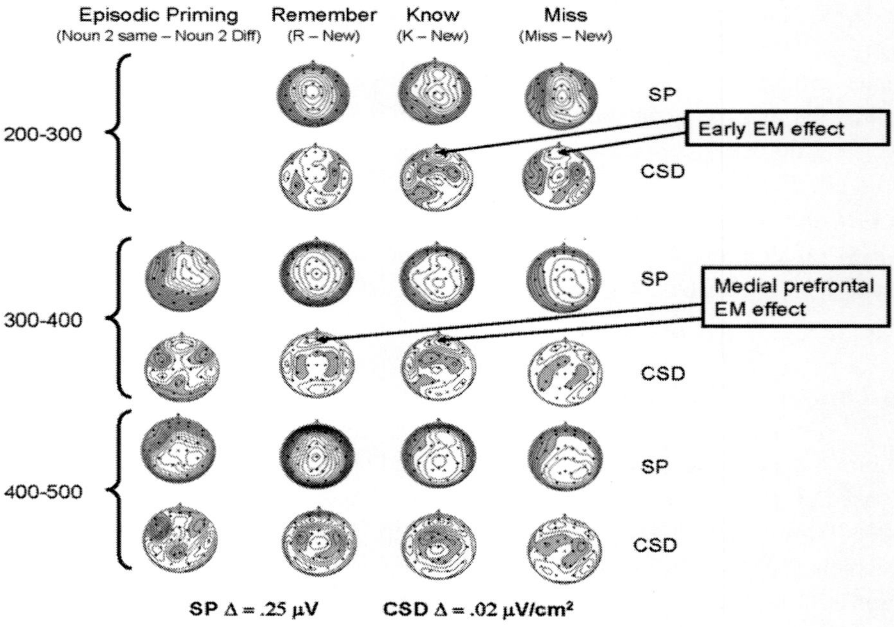

Figure 4. Surface potential (SP) and current source density (CSD) maps computed on the Episodic Priming, R, K and Miss difference waveforms between 200 and 500 ms. The maps are based on the waveforms illustrated in Figure 3. The maps in this and the remaining figures were computed using the average reference (Scherg, 1990). Separation between isopotential lines is 0.25 μV for the SP data and .02 $\mu V/cm^2$ for the CSD data. Shaded areas represent negative values, unshaded areas positive values. Filled dots represent the 28 electrode sites.

Table 1: Results of the ANOVAs on the normalized difference mean data between 300 and 700 ms. Tabled values are for the Condition by Electrode Interactions.

Analysis:	EP[**] vs. Miss	EP vs. R	EP vs. K	R[$] vs. Miss	K[$$] vs. Miss	R vs. K
300-400 ms						
F-ratio	**26.97**	**4.87**	1.95	1.90	<1	<1
df	27,351	27,351	27,351	27,405	27,405	27,405
Epsilon (ε)	0.09	0.12	0.12	0.13	0.11	0.11
***P*-value**	<.0001	<.005	>.10	>0.10	>0.10	>0.05
400-500 ms						
F-ratio	**16.13**	**5.93**	**2.95**	**4.61**	1.08	<1
df	27,351	27,351	27,351	27,405	27,405	27,405
Epsilon (ε)	0.06	0.06	0.13	0.12	0.16	0.15
***P*-value**	<.0001	<.01	<.04	<.005	>0.10	>0.10
500-600 ms[*]						
F-ratio						1.00
df						27,405
Epsilon (ε)						0.13
***P*-value**						>0.10
600-700 ms[*]						
F-ratio						2.05
df						27,405
Epsilon (ε)						0.11
***P*-value**						>0.10

Notes. [*] The difference waveforms for the EP and Miss ERPs did not contain a left parietal EM effect between 500 and 700 ms. Therefore, there are no tabled values for these windows for the first 5 comparisons (cells with gray fill).

[**] Episodic priming

[$] Remember; [$$] Know.

Bolded F-ratios are significant.

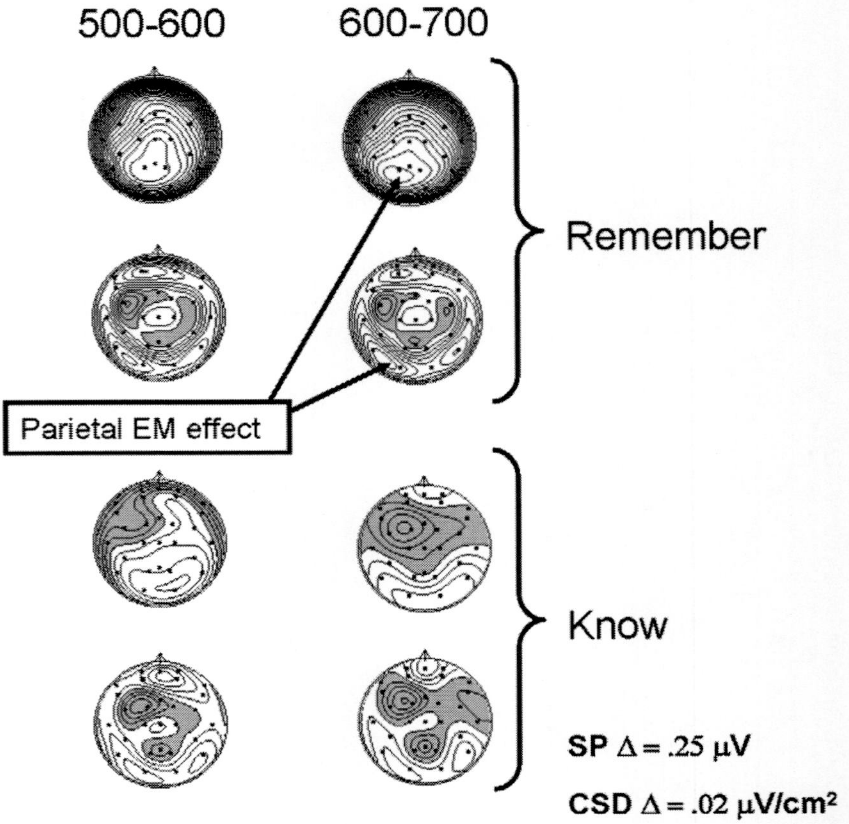

Figure 5: Surface potential (SP) and current source density (CSD) maps computed on the R and K difference waveforms between 500 and 700 ms (see Figure 3, second column of waveforms). Separation between isopotential lines is 0.25 μV for the SP data and .02 μV/cm^2 for the CSD data. Shaded areas represent negative values, unshaded areas positive values. Filled dots represent the 28 electrode sites.

The third set of topographic ANOVAs performed on the averaged voltages between 300 and 700 ms are presented in Table 1. Data from 14 rather than 16 subjects contributed to all of the comparisons involving episodic priming due to the small numbers of trials in the episodic priming ERPs of 2 participants (see Trott et al., 1999 for complete details). The table shows that between 300 and 500 ms, reliable differences in scalp distribution were obtained for all comparisons with the exceptions of K versus Miss and R versus K. For the 300-400 ms interval, post-hoc tests indicated that the distribution of the waveforms associated with episodic priming was more frontally oriented than those associated with missed trials and R judgments. Post-hoc testing of the 400-500 ms interval revealed that, relative to the topography associated with misses, which demonstrated a right posterior scalp distribution, the topography associated with episodic priming was right lateralized over temporal scalp (see Figure 4). For the EP versus Know comparison, post-hoc tests indicated that the distribution associated with Know judgments was more frontally oriented than that associated with episodic priming. A similar pattern held for the EP versus R comparison. For the R versus Miss comparison between 400 and 500 ms, post-hoc testing indicated that the R topography was more frontally oriented with a somewhat left-sided preponderance compared to the distribution associated with misses.

To briefly summarize, these analyses suggest topographic (and presumably functional) distinctions between the episodic priming, putatively "implicitly-based" response to misses, and the presumed familiarity-based responses to old items associated with remember and know judgments.

Figure 5 depicts the SP and CSD distributions associated with R and K judgments between 500 and 700 ms, the time period during which the left parietal EM effect was active. As for Figure 4, the CSD maps are highly differentiated compared to the SP maps, showing the presence of frontal as well as bilateral, parietally-oriented positive scalp current densities, the latter with a strong left-sided preponderance. Between 500 and 700 ms, the R versus K comparisons did not yield significant Condition (R vs. K) by Electrode Location (all 28 scalp sites) interactions on the normalized data.

Additional analyses comparing the scalp topographies of the waveforms associated with R and K were also undertaken (Figures 4 and 5) to determine if the medial prefrontal and the left parietal EM effects were different topographically. These Latency Window by Electrode Location ANOVAs (Table 2) on the normalized difference mean data compared the 300-400 and 400-500 ms windows (reflecting the medial prefrontal EM effect) with the 500-600 and 600-700 ms regions (reflecting the left parietal EM effect). For the R waveforms, post-hoc testing showed that for the 300-400 vs. 500-600, 300-400 vs. 600-700 and 400-500 vs. 500-600 ms comparisons the distributions associated with the later windows were more posteriorly focused than those associated with the early windows. For the K waveforms, none of the comparisons was reliable, although the direction of the differences was the same as for the R waveforms. This could have resulted from a lack of power due to the smaller number of trials comprising the ERP averages associated with K judgments. In summary, these data suggest that the early window typically associated with familiarity (300-500 ms) differs in scalp distribution from the later window that has been associated with recollection (500-700 ms).

Table 2: Results of the ANOVAs on the difference mean data normalized within the R and K conditions. Tabled values are for the Latency Window by Electrode Location interactions.

REMEMBER	*300-400 vs. 500-600*	*300-400 vs. 600-700*	*400-500 vs. 500-600*	*400-500 vs. 600-700*
F-ratio	**3.88**	**4.18**	**3.89**	1.34
df	27,405	27,405	27,405	27,405
epsilon	0.094	0.07	0.08	0.07
P-value	<0.02	<0.03	<.03	>0.10
KNOW				
F-ratio	1.57	1.70	<1	1.64
df	27,405	27,405	27,405	27,405
epsilon	0.12	0.10	0.12	0.11
P-value	>0.10	>0.10	>0.10	>0.10

Notes. **Bolded** F-ratios are significant.

Overall Summary of Findings

The first goal of this series of analyses was to determine if the familiarity and recollective components of dual-process theories of recognition memory (Mandler, 1980) have unique ERP signatures. To the extent that the 300-500 ms window reflects familiarity and the 500-700 ms region reflects recollection, the current data suggest that the two recognition memory aspects do have unique ERP signatures (but see Discussion below). That is, the data imply that the two processes are the reflection of at least partially non-overlapping neural generators, or an amplitude difference in one or more brain regions common to both sets of processes. The second goal was to determine whether the ERP "familiarity effect" between 300 and 500 ms is similar to or different from an "implicitly-based" response presumably reflected by the ERPs associated with unrecognized or "missed" old items. Here again, the answer appears to be yes. When compared to correctly rejected new items, misses were associated with a robust EM effect between 300 and 500 ms, suggesting that the brain appreciated the previous occurrence of these events despite the lack of a correct behavioral response. Moreover, this robust "miss" EM effect, which was right-lateralized with a posterior focus, differed topographically from the EM effect associated with correctly recognized old items attracting either R or K judgments. This latter, medial prefrontal EM effect was characterized by a frontally-oriented scalp focus (Figure 4). In similar fashion to the conclusions of Rugg et al., (1998a), these data imply that the 300-500 ms "familiarity-based" and 300-500 ms "implicitly-based" EM effects are different.

The third goal was to determine if the ERP correlates of episodic priming were similar to either familiarity- or implicitly-based ERP effects. The current data indicate that the episodic priming effect is distinct from both the 300-500 ms familiarity- and implicitly-based EM effects, suggesting that it makes a unique contribution to recognition memory. The final goal was to assess the evidence for a fronto-polar EM effect that might be the earliest manifestation of a brain event reflecting familiarity-driven processes (Curran and Dien, 2003; Duarte et al., 2004; Tsivilis et al., 2001). In answer to this query, there was some evidence for a very early EM effect between 200 and 300 ms that could reflect the earliest brain event signaling a familiarity-based retrieval.

DISCUSSION

Recollection and Familiarity

Several previous studies have produced data that suggest a strong association between the left parietal EM effect and recollection ((Friedman and Johnson 2000; Wilding and Nobre, in press). Some of these investigations have shown the parietal EM effect to be larger in conjunction with remember compared to know judgments (Duzel et al., 1997; Mark and Rugg, 1998; Smith, 1993; Trott et al., 1999). Though subjective, a remember judgment presumably reflects the retrieval of contextual information, and hence more information about the initial episode than that reflected by a know judgment. Similarly, as described in the Introduction, Ullsperger et al., (2000) reported that, during the test phase of a directed forgetting paradigm, the ERPs associated with to-be-remembered words contained a larger magnitude parietal EM effect compared to those that were associated with to-be-forgotten words. Consistent with these reports, Wilding (Wilding, 2000) has demonstrated convincingly that the parietal EM effect is larger the greater the amount of contextual information retrieved, suggesting that this EM effect reflects the recollection, on a continuum, of source-related information (see also Tsivilis et al., 2001). Moreover, a number of source memory studies have shown that the parietal EM effect is larger to correctly recognized old items in association with correct compared to incorrect source judgments, regardless of the type of information that must be retrieved -- gender of voice (Wilding and Rugg, 1996; Rugg et al., 1998b), modality (Wilding et al., 1995), list membership (Trott et al., 1999), spatial location (Van-Petten et al., 2000) and the second word given the first word of a previously associated word pair (Donaldson and Rugg, 1998).

On the other hand, the association of the medial prefrontal EM effect (300-500 ms) with the familiarity component of recognition memory has not been as consistently observed, although there is a growing body of evidence that is concordant with this hypothesis. The assumption made in the majority of these studies is that, based upon their intra-experimental history, the two (or more) classes of old items should be equally familiar. If this is, in fact, the case and the medial prefrontal EM effect indeed reflects familiarity, then it should be of equal magnitude in the ERPs evoked by these different classes of old items (e.g., hits associated with remember and know judgments; hits attracting correct and incorrect source judgments etc.). The results of some investigations support this interpretation (for example, the current study; Curran, 2000; Curran and Cleary, 2003; Mecklinger, 2000; Nessler et al., 2001; Ullsperger et al., 2000; for a review see Rugg and Yonelinas, 2003). However, although several investigators have recorded ERP data that might speak to this issue, they have not always considered and/or analyzed their data within the latency range of the medial prefrontal EM effect (e.g., Spencer et al., 2000; Wilding and Rugg, 1996; Van-Petten et al., 2000). Nonetheless, in some of these aforementioned studies, the published data do suggest, on visual inspection, an association of the 300-500 ms EM effect with familiarity-based processes.

By contrast with the studies reviewed above, Tsivilis et al., (2001) published data suggesting that the 300-500 ms medial prefrontal effect may not reflect familiarity *per se*. As described in the Introduction, Tsivilis and co-workers argued that an earlier EM effect, with a fronto-polar scalp topography, may be a more likely candidate of familiarity-based processes. As also detailed in the Introduction, an early EM effect similar to that of Tsivilis et al., (2001) was reported by Duarte et al., (2004) and Curran and Dien (2003). Unlike the current data, the EM effect measured between 300 and 450 ms by Duarte and colleagues (Duarte et al., 2004), which demonstrated a fronto-polar distribution, was larger in the ERPs associated with R compared to K responses. However, although appearing to differ functionally in terms of their relations to R and K judgments, the 150-300 ms EM effect and the one measured between 300 and 450 ms did not differ topographically, counter to the data reported by Tsivilis et al., (2001). Moreover, the early EM effect observed by Curran and Dien (2003), which also demonstrated a fronto-polar scalp topography between 176 and 260 ms, was only reliable following visual-visual study test pairings and not following auditory-visual pairings. Thus, unlike the medial prefrontal EM effect measured between 300 and 500 ms by Curran and Dien (2003), which showed an amodal pattern of response consistent with familiarity-based responding, the 176-260 ms effect demonstrated a response pattern more consistent with repetition priming than with familiarity.

The current 200-300 ms EM effect is similar to the Tsivilis et al., (2001), Curran and Dien (2003) and Duarte et al., (2004) data, as the ERPs elicited by old items associated with R and K judgments, hits and misses and episodically primed or unprimed items did not differ in magnitude (all differed reliably from new items). These findings suggest that this EM effect could be the earliest ERP manifestation of a familiarity-based retrieval. Topographic analyses indicated that, in two of the four comparisons performed (noun 2 old-new; misses-correct rejections), the scalp distributions of this early effect were more frontally oriented than those of the 300-400 and 400-500 ms windows. However, for the other two comparisons (Remember; Know), the Latency Window by Electrode interactions were marginally reliable or not significant. Hence, caution must be exercised when considering the notion that this early EM effect is topographically unique compared to the 300-500 ms region of the waveform, which has been thought to be the putative correlate of familiarity. Further, at this stage of knowledge, the inconsistencies among the published reports concerning this early, fronto-polar EM effect, warrants additional study of this phenomenon and the experimental circumstances under which it occurs.

ERP scalp topographic data can also inform the distinction between recollection and familiarity. To date, however, the results have proven disappointing. Investigators who have compared directly the ERPs associated with hits whose sources were and were not correctly attributed (Wilding and Rugg, 1996; Trott et al., 1999), or the ERPs associated with correctly recognized items given R or K judgments (Smith, 1993; Trott et al., 1999), have reported that the scalp distributions were not reliably different. On the other hand, Duzel et al., (1997) reported that the ERPs to hits associated with R and K judgments were different. However, Duzel et al., (1997) failed to normalize their data (McCarthy and Wood, 1985; Ruchkin et al., 1999), rendering their strong conclusion concerning differences in topography equivocal.

However, the methods used by the investigators cited above may not have been optimal. Typically, the remember/know or source correct/incorrect waveforms were compared using either the 300-500 ms or the 500-800 ms window. However, if the 300-500 ms region is the putative correlate of familiarity, one would not expect a difference in scalp topography between these conditions. Similarly, if the left parietal EM effect is the putative correlate of recollection, which, based on ERP evidence (Wilding, 2000) can be a graded as compared to an all-or-none, threshold process (Yonelinas, 2002), comparison of the scalp distributions for these different conditions should yield no evidence of differences in scalp distribution. In order to demonstrate that the scalp-recorded ERP signatures of familiarity and recollection might receive contributions from different sets or configurations of intracranial generators requires comparing the distributions of the medial prefrontal effect with that of the parietal effect. When this has been done, the evidence suggests that these two regions of the waveform may indeed receive contributions from different brain regions (the current data; Curran, 2000).

To summarize at this juncture, the current data and literature reviewed suggest a strong association between the left parietal EM effect and recollection-based processing. Similarly, a body of evidence has begun to accrue suggesting that the medial prefrontal EM effect active between 300 and 500 ms may be associated with familiarity-driven processing. There are, however, contra-indications to this argument (e.g., Tsivilis et al., 2001; Yovel and Paller, in press see also below).

Implicit Retrieval Effects

In the current study, another EM effect, present only in the ERPs associated with missed items, occurred within the same time window as the medial prefrontal EM effect. However, this EM effect had a right-lateralized, posterior parietal scalp distribution, distinct from those associated with the medial prefrontal and left parietal EM effects. The scalp distribution of this EM effect is reminiscent of that observed by (Rugg et al., 1998a), who, as reviewed earlier, demonstrated a reliable EM effect in association with missed items. The Rugg et al., (1998a) topography associated with misses is depicted in Figure 6 along with the scalp topography associated with misses from the current study. Visual inspection indicates a good deal of similarity between the two distributions. Hence, similar to the conclusion reached by Rugg et al., (1998a), the current data suggest that this EM effect is a potential candidate for a correlate of implicit memory retrieval. As suggested by Wilding and Sharpe (2002), the data from previous studies have also indicated this possibility, but the presence of these effects has not always been noted or statistically analyzed.

The current methodology differs considerably from the dissociative design employed by Rugg et al., (1998a). Nonetheless, the reliable distinction between the ERPs associated with misses and correct rejections at posterior scalp sites appears to satisfy the definition of implicitly-based retrieval. To my knowledge, this and the Rugg et al., (1998a) papers are the only two to have noted this distinction. Thus, further replication will be required before these data can be accepted with confidence.

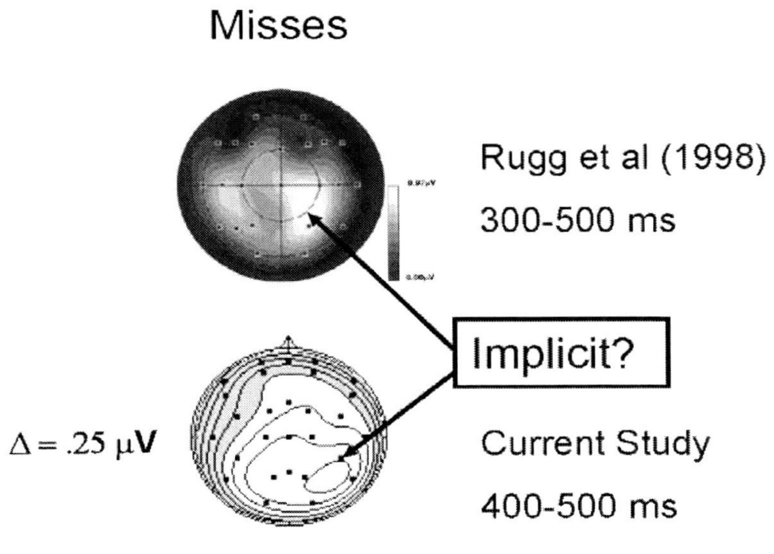

Figure 6: *Top Row*. Topographic surface potential grayscale maps from Rugg et al., (1998a). The data are based on the difference waveforms computed by subtracting the ERPs associated with missed shallowly-studied old items from the ERPs elicited by correctly rejected new items. The map has been adapted from Rugg et al., (1998a) and reproduced by permission of Nature Publishing Group and the author. The microvolt scale ranges from 0.0 to +0.97 μV. *Bottom Row*. The topographic map associated with the difference waveforms computed by subtracting correctly rejected new items from misses from the current study. Isopotential lines are separated by 0.25 μV. Shaded areas represent negative activity, unshaded areas positive activity. Filled dots represent the 28 electrode sites.

Because of the differences in scalp topography between the miss EM effect and the 300-500 ms medial prefrontal EM effect associated with correct recognition, these data suggest that familiarity- and implicitly-based retrievals are not identical. That is, the kind of processing responsible for preserved perceptually-based implicit memory performance in amnesic individuals is not likely the same as familiarity that operates in explicit or direct recognition memory tests. Similarly, as described earlier, Curran and Dien (2003) asked participants to study words presented visually and auditorally and then tested all words visually. Based on the premise that familiarity is an amodal process, the prediction was that the medial prefrontal EM effect would be as large for visual-visual as it would for auditory-visual, study-test pairings. This prediction was borne out, arguing against the medial prefrontal EM effect reflecting a perceptual implicit memory mechanism. Hence, taken as a whole, the current data and those of Curran and Dien (2003) join a number of behavioral studies (Wagner et al., 1997; Wagner and Gabrieli, 1998) as well as neuropsychological investigations (Fleischman et al., 1995; Gabrieli et al., 1995); Manns et al., 2003) in supporting the hypothesis that familiarity-based and implicitly-based perceptual memory are not identical.

Conceptually-Driven Implicit Retrieval Effects

Alternatively, it could be the case that, while the medial prefrontal EM effect is not likely to reflect a perceptual implicit memory mechanism, it might be associated with conceptually-based implicit memory processes, as has been suggested by Paller et al., (2003) and Yovel and Paller (in press). The basis for this argument is as follows. Repeating a previously studied item in recognition memory test phases induces a reduction of the N400 component (between ~ 300 and 500 ms), which is an aspect of the medial prefrontal EM effect. The N400 is strongly associated with semantic processing, i.e., it is conceptually based (Kutas and Hillyard, 1980a). Further, while some amnesic patients show deficits in familiarity-based processing during recognition memory testing, some of these same patients show intact conceptual priming. For example, Olichney et al., (2000) presented amnesiacs and controls with a category name which was followed either by a congruous or incongruous target exemplar. Congruent and incongruent targets were presented either once, twice or three times. Amnesiacs showed a robust incongruity effect (i.e., larger N400 to incongruous relative to congruous exemplars), which was just as large as that recorded in the controls. By contrast, the performance of the amnesiacs during recognition, free- and cued-recall memory testing was markedly deficient relative to the controls. Hence, because the vast majority of investigators of ERP memory-related phenomena have used words as stimuli, the mechanisms involved in the conceptual priming shown by the amnesiacs and controls in the Olichney et al., (2000) experiment may overlap those reflected in the reduction of the N400 that comprises one aspect of the medial prefrontal EM effect. Hence, rather than reflecting familiarity per se, the medial prefrontal EM effect may be associated with conceptual priming mechanisms.

On the other hand, some very recent data suggest that the medial prefrontal EM effect may be more a reflection of familiarity than conceptually-driven processing. For example, Curran et al., (2002) and Curran (2003) described analyses of ERP data recorded during recognition memory testing of objects labeled "blobs" that could not have easily been associated with semantic labels. The medial prefrontal effect, albeit of small magnitude, was statistically reliable, suggesting that familiarity rather than conceptual processing was responsible for this EM effect. Similarly, Curran and Friedman (2004) demonstrated that the medial prefrontal EM effect survived retention intervals as long as 24 hours. By contrast with previous investigations that had initially questioned the relevance of this EM effect to mnemonic function (Rugg and Nagy, 1989), the Curran and Friedman (2004) data suggest that it could reflect a familiarity-based response that contributes to long-term memory.

Arguably, the episodic priming that resulted in the current paradigm must have been due to a "conceptual" mechanism similar to that involved in associative priming, as the two nouns in each studied sentence were *unassociated* prior to the experiment (see also below). As described earlier, comparison of the topographies of the medial prefrontal EM effect and the episodic priming effect revealed reliable differences. Based on this comparison, it is not likely that the medial prefrontal EM effect is based solely on conceptually-driven processes. Although this may be a plausible interpretation, the current design did not include an encoding manipulation, such as levels of processing, which would have comprised "conceptual" and "perceptual" levels (for example, semantic encoding versus orthographic encoding, respectively) that may have shown dissociative effects on the putative ERP correlates of recollection and familiarity (Rugg et al., 1998a). This would have allowed a more forceful interpretation. Clearly, whether the medial prefrontal EM effect receives contributions from conceptual- as well as familiarity-based processes is an open question, one deserving of further research effort.

Episodic Priming

Although a fair amount of knowledge concerning the brain correlates of repetition priming has accrued (Henson and Rugg, 2003), virtually nothing is known about the physiological correlates of episodic priming. Further, after an initial series of behavioral studies of episodic priming in the late 1970s and early 1980s (e.g., McKoon and Ratcliff, 1979), there have been very few additional investigations of this phenomenon in the last 10 to 20 years (Schrijnemakers and Raaijmakers, 1997). Hence, there is very little basis for comparison of the present physiological findings with previous results.

In ERP studies of semantic priming, presentation of a prime followed by a target (e.g., nurse-doctor) produces a priming effect on the N400 component (Kutas and Hillyard, 1980b) in the ERP waveform elicited by the target -- N400 is larger or more negative when the target is semantically unrelated compared to when it is highly related to the prime (Kutas and Hillyard, 1984). Applying this logic in the current investigation, when a target noun 2 was preceded by a noun 1 from the same studied sentence (i.e., it was episodically primed), the N400 component was smaller than when a target noun was preceded by a noun 1 from a different studied sentence (Figure 2). The topography of this effect (Figure 4) was right-lateralized and focused over posterior temporal scalp. By contrast, the results of a recent study of the N400 semantic priming effect with a larger number of electrode sites than used previously, showed that the N400 priming effect was symmetrically distributed with a centro-parietal midline scalp focus (Rossell et al., 2003), consistent with previous investigations of the topography of the N400 semantic priming effect based on smaller sets of scalp locations. Although by no means definitive, these data suggest that the episodic and semantic priming effects are not likely to be similar. However, firmer conclusions await assessment of the two effects with the same subjects in the same experimental paradigm.

Some authors have argued that episodic priming effects reflect retrieval from semantic memory (Tulving, 1983), whereas others have argued that the presence of this effect reflects retrieval from an episodic store (McKoon and Ratcliff, 1986). Still others have advanced the hypothesis that episodic priming effects reflect the influence of both the semantic and episodic systems (Schrijnemakers and Raaijmakers, 1997). The current experiment was not designed to test these alternatives. However, the data do suggest that within a relatively early time window (300-500 ms), the scalp distribution of the ERPs associated with episodically primed pairs differed from those associated with correctly recognized and unrecognized old items.

Conclusions

Between 300 and 500 ms, at least three different cognitive processes can contribute to recognition memory performance. These are, an episodically-based priming mechanism, as well as "familiarity-like," and implicitly-based retrieval mechanisms. The ERP effects reflecting implicit and episodic priming are not likely to be similar because they were shown to be topographically distinct. Moreover, evidence from patients with well-circumscribed lesions of the medial temporal lobe shows that some of these patients are impaired on measures of associative priming, but not on measures of perceptual item priming (Yang et al., 2003). These three early EM effects also differed reliably from the subsequent left parietal EM effect, the putative correlate of recollection-based retrieval. The current picture of the medial prefrontal EM effect as reflecting familiarity may have to be revised if data similar to those presented by Tsivilis et al., (2001) begin to accrue. Importantly, work with non-human primates indicates that neurons responding to repeated complex visual stimuli do so within 100 ms of their presentation (Xiang and Brown, 1998). This finding has been interpreted as reflecting a familiarity-based brain response (Brown and Bashir, 2002). If this is so, those data would suggest that the medial prefrontal effect occurs too late in time to reflect directly a familiarity-based decision. Consistent with this conclusion, an early prefrontally-oriented EM effect did differentiate old from new items between 200 and 300 ms post-stimulus, and could be the earliest manifestation of familiarity-based recognition. However, considerably more work needs to be done to bolster the evidence for this phenomenon. Furthermore, additional work will be required to confirm the distinction between episodic priming and implicitly-based retrieval mechanisms. Nonetheless, the current data indicate that the temporal precision of the ERP coupled with the judicious use of scalp topographic information can provide a more complete temporo-spatial picture of mnemonic function by the human brain.

Acknowledgments

The author thanks Mr. Charles L. Brown, III for computer programming and Mr. Jeff Cheng for technical assistance. The author thanks Drs. Yael Cycowicz, Monica Fabiani, Helen Gaeta, Walter Ritter, Joan G. Snodgrass, and Charlotte Trott for their invaluable contributions to the research reported here and for discussions concerning these data. I thank Dr. Axel Mecklinger for his helpful comments on an earlier draft of this manuscript, and give special thanks to Dr. Doreen Nessler who made many helpful comments and suggestions on previous drafts of this manuscript. Preparation of this paper was supported in part by grants AG05213 and AG09988 from the NIA, grant HD14959 from NICHD, and by the New York State Department of Mental Hygiene.

REFERENCES

Brown MW and Bashir ZI. Evidence concerning how neurons of the perirhinal cortex may effect familiarity discrimination. *Philosophical Transactions of the Royal Society of London B: Biological Sciences* 357: 1083-1095, 2002.

Cermak LS, Blackford SP, O'Connor M, and Pratt-Bleich R. The implicit memory ability of a patient with amnesia due to encephalitis. *Brain and Cognition* 7: 145-156, 1988.

Craik FIM and Lockhart S. Levels of processing: A framework for memory research. *Journal of Verbal Learning and Verbal Behavior* 11: 671-684, 1972.

Curran T. Brain potentials of recollection and familiarity. *Memory and Cognition* 28: 923-938., 2000.

Curran T. ERP correlates of familiarity-based recognition. *Psychophysiology* 40: S14, 2003.

Curran T and Cleary AM. Using ERPs to dissociate recollection from familiarity in picture recognition. *Cognitive Brain Research* 15: 191-205, 2003.

Curran T and Dien J. Differentiating amodal familiarity from modality-specific memory processes: An ERP study. *Psychophysiology* 40: 979-988, 2003.

Curran T and Friedman WJ. ERP old/new effects at different retention intervals in recency discrimination tasks. *Cognitive Brain Research*, 18: 107–120, 2004.

Curran TE, Tanaka JW, and Weiskopf DM. An electrophysiological comparison of visual categorization and recognition memory. *Cognitive, Affective, & Behavioral Neuroscience* 2: 1-18, 2002.

Deacon D, Hewitt S, Yang C, and Nagata M. Event-related potential indices of semantic priming using masked and unmasked words: evidence that the N400 does not reflect a post-lexical process. *Cognitive Brain Research* 9: 137-146, 2000.

Donaldson DI, Petersen SE, and Buckner RL. Dissociating memory retrieval processes using fMRI: evidence that priming does not support recognition memory. *Neuron* 31: 1047-1059, 2001.

Donaldson DI and Rugg MD. Recognition memory for new associations: electrophysiological evidence for the role of recollection. *Neuropsychologia* 36: 377-395, 1998.

Duarte A, Ranganath C, Winward L, Hayward D, and Knight RT. Dissociable neural correlates for familiarity and recollection during the encoding and retrieval of pictures: an ERP study. *Cognitive Brain Research*, 18: 255–272, 2004.

Duzel E, Yonelinas AP, Mangun GR, Heinze H-J, and Tulving E. Event-related brain potential correlates of two states of conscious awareness in memory. *Proceedings of the National Academy of Sciences* 94: 5973-5978, 1997.

Ebbinghaus H. *Memory: a contribution to experimental psychology*. New York: Dover Publications, 1964.

Eldridge LL, Knowlton BJ, Furmanski CS, Bookheimer SY, and Engel SA. Remembering episodes: a selective role for the hippocampus during retrieval. *Nature Neuroscience* 3: 1149-1152, 2000.

Fleischman DA, Gabrieli JDE, Reminger S, Rinaldi J, Morrell F, and Wilson R. Conceptual priming in perceptual identification for patients with Alzheimer's disease and a patient with right occipital lobectomy. *Neuropsychology* 9: 187-197, 1995.

Friedman D and Johnson R. Event-related potential (ERP) studies of memory encoding and retrieval: A selective review. *Microsc Res Tech* 51: 6-28., 2000.

Gabrieli JD. Cognitive neuroscience of human memory. *Annual Review of Psychology* 49: 87-115, 1998.

Gabrieli JDE, Fleischman DA, Keane MM, Reminger SL, and Morrell F. Double dissociation between memory systems underlying explicit and implicit memory in the human brain. *Psychological Science* 6: 76-82, 1995.

Gardiner JM. Episodic memory and autonoetic consciousness: a first-person approach. *Philosophical Transactions of the Royal Society of London Series B: Biological Sciences* 356: 1351-1361, 2001.

Gardiner JM, Ramponi C, and Richardson-Klavehn A. Response deadline and subjective awareness in recognition memory. *Consciousness and Cognition* 8: 484-496, 1999.

Gratton G, Coles MGH, and Donchin E. A new method for off-line removal of ocular artifact. Electroencephalography and clinical *Neurophysiology* 55: 468-484, 1983.

Henson R, Rugg MD, Shallice T, Josephs O, and Dolan RJ. Recollection and Familiarity in Recognition Memory: An Event-Related Functional Magnetic Resonance Imaging Study. *Journal of Neuroscience* 19: 3962-3972, 1999.

Henson RN and Rugg MD. Neural response suppression, haemodynamic repetition effects, and behavioural priming. *Neuropsychologia* 41: 263-270, 2003.

Howard DV, Heisey JG, and Shaw RJ. Aging and the Priming of Newly Learned Associations. *Developmental Psychology* 22: 78-85, 1986.

Jennings JR and Wood CC. The ε-adjustment procedure for repeated measures analyses of variance. *Psychophysiology* 13: 277-278, 1976.

Johnson MK, Kounios J, and Reeder JA. Time-course studies of reality monitoring and recognition. *Journal of experimental Psychology: Learning, Memory, and Cognition* 20: 1409-1419, 1994.

Johnson R. Event-related potential insights into the neurobiology of memory systems. In: *Handbook of Neuropsychology*, edited by Boller F and Grafman J. Amsterdam: Elsevier, 1995, p. 135-163.

Johnson R. On the neural correlates of the P300 component of the event-related potential. *Psychophysiology* 30: 90-97, 1993.

Johnson R, Jr., Kreiter K, Russo B, and Zhu J. A spatio-temporal analysis of recognition-related event-related brain potentials. *International Journal of Psychophysiology* 29: 83-104, 1998.

Knowlton BJ and Squire LR. Remembering and knowing: Two different expressions of declarative memory. *Journal of Experimental Psychology: Learning, Memory, and Cognition* 21: 699-710, 1995.

Kutas M and Hillyard SA. Brain potentials during reading reflect word expectancy and semanti association. *Nature* 307: 161-163, 1984.

Kutas M and Hillyard SA. Reading between the lines: Event-related brain potentials during natural sentence processing. *Brain and Language* 11: 354-373, 1980a.

Kutas M and Hillyard SA. Reading senseless sentences: Brain potentials reflect semantic incongruity. *Science* 207: 203-205, 1980b.

Luck S, Vogel E, and Shapiro K. Word meanings can be accessed but not reported during the attentional blink. *Nature* 383: 616-618, 1996.

Mandler G. Recognizing: The judgement of previous occurrence. *Psychological Review* 87: 252-271, 1980.

Manns JR, Hopkins RO, Reed JM, Kitchener EG, and Squire LR. Recognition memory and the human hippocampus. *Neuron* 37: 171-180, 2003.

Mark RE and Rugg MD. Age effects on brain activity associated with episodic memory retrieval: An electrophysiological study. *Brain* 121: 861-873, 1998.

McCarthy G and Wood CC. Scalp distributions of event-related potentials: An ambiguity associated with analysis of variance models. *Electroencephalo-graphy and clinical Neurophysiology* 62: 203-208, 1985.

McElree B, Dolan PO, and Jacoby LL. Isolating the contributions of familiarity and source information to item recognition: a time course analysis. *Journal of Experimental Psychology: Learning, Memory and Cognition* 25: 563-582, 1999.

McGaugh JL. Memory-a Century of Consolidation. *Science* 287: 248-251, 2000.

McKoon G and Ratcliff R. Automatic activation of episodic information in a semantic memory task. *Journal of Experimental Psychology: Learning, Memory, and Cognition* 12: 108-115, 1986.

McKoon G and Ratcliff R. Priming in Episodic and Semantic Memory. *Journal of Verbal Learning and Verbal Behavior* 18: 463-480, 1979.

Mecklinger A. Interfacing mind and brain: a neurocognitive model of recognition memory. *Psychophysiology* 37: 565-582, 2000.

Nessler D, Mecklinger A, and Penney TB. Event related brain potentials and illusory memories: the effects of differential encoding. *Cognitive Brain Research* 10: 283-301., 2001.

Nunez PL and Pilgreen KL. The spline-Laplacian in clinical neurophysiology: A method to improve EEG spatial resolution. *Journal of Clinical Neurophysiology* 8: 397-413, 1991.

Olichney JM, Van Petten C, Paller KA, Salmon DP, Iragui VJ, and Kutas M. Word repetition in amnesia: Electrophysiological measures of impaired and spared memory. *Brain* 123: 1948-1963, 2000.

Paller KA. Brain potentials associated with perceptual priming vs explicit remembering during the repetition of visual word-form. *Neuropsychologia* 36, 1998.

Paller KA and Kutas M. Brain potentials during memory retrieval provide neurophysiological support for the distinction between conscious recollection and priming. *Journal of Cognitive Neuroscience* 4: 375-391, 1992.

Paller KA, Yovel G, Boehm SG, and Florczak SM. Investigating the experience of familiarity by simulating the butcher-on-the-bus phenomenon. *Psychophysiology* 40: S14, 2003.

Parkin AJ, Ward J, Squires EJ, Furbear H, Clark A, and Townshend J. Data-driven recognition memory: a new technique and some data on age differences. *Psychonomic Bulletin and Review* 8: 812-819, 2001.

Perrin F, Pernier J, Bertrand O, and Echallier JF. Spherical splines for scalp potential and current density mapping. *Electroencephalography and clinical Neurophysiology* 72: 184-187, 1989.

Picton TW, Lins OG, and Scherg M. The recording and analysis of event-related potentials. In: *Handbook of Neuropsychology*, edited by Jr. RJ and Baron JC. Amsterdam: Elsevier, 1995, p. 3-73.

Reed JM, Hamann SB, Stefanacci L, and Squire LR. When amnesic patients perform well on recognition memory tests. *Behavioral Neuroscience* 111: 1163-1170, 1997.

Roediger HLI and McDermott KB. Implicit memory in normal human subjects. In: *Handbook of Neuropsychology*, edited by Boller F and Grafman J. Amsterdam: Elsevier Science Publishers, B.V., 1993, p. 63-131.

Rossell SL, Price CJ, and Nobre AC. The anatomy and time course of semantic priming investigated by fMRI and ERPs. *Neuropsychologia* 41: 550-564, 2003.

Ruchkin DS, Johnson R, Jr., and Friedman D. Scaling is necessary when making comparisons between shapes of event- related potential topographies: a reply to Haig et al., *Psychophysiology* 36: 832-834., 1999.

Rugg MD, Mark RE, Walla P, Schloerscheidt AM, Birch CS, and Allan K. Dissociation of the neural correlates of implicit and explicit memory. *Nature* 392: 595-598, 1998a.

Rugg MD and Nagy ME. Event-related potentials and recognition memory for words. Electroencephalography and clinical *Neurophysiology* 72: 395-406, 1989.

Rugg MD, Schloerscheidt AM, and Mark RE. An electrophysiological comparison of two indices of recollection. *Journal of Memory and Language* 39: 47-69, 1998b.

Rugg MD and Wilding EL. Retrieval processing and episodic memory. *Trends in Cognitive Sciences* 4: 108-115, 2000.

Rugg MD and Yonelinas AP. Human recognition memory: a cognitive neuroscience perspective. *Trends in Cognitive Sciences* 7: 313-319, 2003.

Schacter DL. The seven sins of memory. Insights from psychology and cognitive neuroscience [In Process Citation]. *American Psychologist* 54: 182-203, 1999.

Schacter DL, Church B, and Treadwell J. Implicit memory in amnesic patients: Evidence for spared auditory priming. *Psychological Science* 5: 20-25, 1994.

Schacter DL, J.Bowers, and Booker J. Intention, awareness, and implicit memory: THe retrieval intentionality criterion. In: *Implicit Memory: Theoretical Issues*, edited by Lewandowsky S, Dunn JC and Kirsner K. Hillsdale, N.J.: Lawrence Earlbaum Associates, 1989, p. 47-65.

Scherg M. Fundamentals of dipole source analysis. In: *Auditory evoked magnetic fields and potentials*, edited by H.M. Basel: Karger, 1990, p. 40-69.

Schrijnemakers JM and Raaijmakers JG. Adding new word associations to semantic memory: evidence for two interactive learning components. *Acta Psychologica* 96: 103-132, 1997.

Smith ME. Neurophysiological manifestations of recollective experience during recognition memory judgments. *Journal of Cognitive Neuroscience* 5: 1-13, 1993.

Snodgrass JG and Corwin J. Pragmatics of measuring recognition memory: applications to dementia and amnesia. *Journal of Experimental Psychology: General* 117: 34-50., 1988.

Spencer KM, Abad EV, and Donchin E. On the search for the neurophysiological manifestation of recollective experience. *Psychophysiology* 37: 494–506., 2000.

Trott CT, Friedman D, Ritter W, Fabiani M, and Snodgrass JG. Episodic priming and memory for temporal source: event-related potentials reveal age-related differences in prefrontal functioning. *Psychol Aging* 14: 390-413., 1999.

Tsivilis D, Otten LJ, and Rugg MD. Context effects on the neural correlates of recognition memory. an electrophysiological study. *Neuron* 31: 497-505., 2001.

Tulving E. *Elements of Episodic Memory*. New York: Oxford University Press, 1983.

Tulving E. Memory and consciousness. *Canadian Psychologist* 26: 1-12, 1985.

Ullsperger M, Mecklinger A, and Muller U. An electrophysiological test of directed forgetting: the role of retrieval inhibition. *Journal of Cognitive Neuroscience* 12: 924-940., 2000.

Urbach TP and Kutas M. The intractability of scaling scalp distributions to infer neuroelectric sources. *Psychophysiology* 39: 791-808, 2002.

Van-Petten C, Senkfor AJ, and Newberg WM. Memory for drawings in locations: Spatial source memory and event-related potentials. *Psychophysiology* 37: 551-564, 2000.

Wagner A, Gabrieli J, and Verfaellie M. Dissociations between familiarity processes in explicit recognition and implicit perceptual memory. *J Exp Psychol Learn Mem Cogn* 23: 305-323, 1997.

Wagner AD and Gabrieli JD. On the relationship between recognition familiarity and perceptual fluency: evidence for distinct mnemonic processes. *Acta Psychologica* 98: 211-230, 1998.

Wilding EL. In what way does the parietal ERP old/new effect index recollection? *International Journal of Psychophysiology* 35: 81-87, 2000.

Wilding EL, Doyle MC, and Rugg MD. Recognition memory with and without retrieval of context: An event-related potential study. *Neuropsychologia* 33: 1-25, 1995.

Wilding EL and Nobre AC. Task-dependency of the electrophysiological correlates of episodic retrieval. *Neuropsychologia*, in press.

Wilding EL and Rugg MD. An event-related potential study of recognition memory with and without retrieval of source. *Brain* 119: 889-905, 1996.

Wilding EL and Sharpe H. Episodic memory encoding and retrieval: Recent insights from event-related postentials. In: *The Cognitive Electrophysiology of Mind and Brain,* edited by Zani A and Proverbio AM. New York: Academic Press, 2002, p. 169-196.

Xiang JZ and Brown MW. Differential neuronal encoding of novelty, familiarity and recency in regions of the anterior temporal lobe. *Neuropharmacology* 37: 657-676, 1998.

Yang J, Weng X, Guan L, Kuang P, Zhang M, Sun W, Yu S, and Patterson K. Involvement of the medial temporal lobe in priming for new associations. *Neuropsychologia* 41: 818-829, 2003.

Yonelinas AP. The Nature of Recollection and Familiarity: A Review of 30 Years of Research. *Journal of Memory and Language* 46: 441-517, 2002.

Yonelinas AP, Kroll NE, Dobbins I, Lazzara M, and Knight RT. Recollection and familiarity deficits in amnesia: convergence of remember-know, process dissociation, and receiver operating characteristic data. *Neuropsychology* 12: 323-339, 1998.

Yovel G and Paller KA. The neural basis of the butcher-on-the-bus phenomenon: When a face seems familiar but is not remembered. *NeuroImage*, in press.

In: Cognitive Science Compendium, Volume 1
Editor: Miao-Kun Sun
ISBN: 978-1-60456-597-3

Chapter 4

INFANTILE AMNESIA: A CROSS-CULTURAL INVESTIGATION

Qi Wang*
Cornell University, U.S.A.
Martin A. Conway
University of Durham, England
Yu-bo Hou
Peking University, China

ABSTRACT

Chinese, British, and Caucasian American participants were asked to recall as many childhood memories (of events occurring below the age of 5 years), including their earliest memory, as they could in a five-minute period. Chinese participants recalled fewer memories and had earliest memory 6 months later than either of the Western groups. Women provided more memories than men. The cultural differences in the period of childhood amnesia are explained in terms of cultural self-construals and related cultural narrative practices. It is proposed that cultures with an autonomous self-construal which prioritizes elaborative memory talk have earlier childhood memories and shorter periods of infantile amnesia than cultures with a relational self-construal.

Keywords: culture, infantile amnesia, memory, self

Adults are unable to access many autobiographical memories from the earliest years of life. This phenomenon was first referred to by Freud (1905/1949) as *Infantile amnesia,* which "turns everyone's childhood into something like a prehistoric epoch" (p. 54). Studies with Western samples have empirically demonstrated this phenomenon, showing that most adults

* Please address correspondence to Qi Wang, Department of Human Development, Cornell University, MVR Hall, Ithaca, NY 14853. Tel: (607) 255-9376. E-mail: qw23@cornell.edu.

cannot remember events that occurred before 2 to 3 years of age, forget at an accelerated rate experiences occurring before age 5 (Wetzler & Sweeney, 1986), and yet recall memories from age 5 onward with sharply increasing accessibility and elaboration (for review see Conway, 1990, and Pillemer & White, 1989, and see Rubin, 2000, for recent findings). In spite of the empirical evidence, there is no generally agreed explanation. Instead, currently there are many conjectures and theories which variously emphasize social/cultural/linguistic factors (e.g. Fivush & Hamond, 1990; Nelson, 1996; Pillemer, 1998; Reese, 2002; Wang et al., 1998), self/cognitive factors (e.g. Conway & Pleydell-Pearce, 2000; Howe & Courage, 1997; Usher & Neisser, 1993; Waldfogel, 1948), and neurobiological changes (e.g. Conway & Pleydell-Pearce, 2000; Newcombe, et al., 2000).

It seems unlikely, however, that any single account will provide a full explanation of infantile amnesia because so many interrelated changes in development take place during the first five years of life. Thus, the factors featured in all the above proposals may collectively contribute towards lowering the accessibility of memories from this period. Although the nature and timing of these factors in determining the period of childhood amnesia is yet clear, there are, nonetheless, certain landmarks. Howe and Courage (1997) propose that the development of a 'cognitive' self, structured around the distinction between "I" and "Me", during the period of 2 to 3 years of age is critical in providing an organizational structure within which memories can be represented. This does not mean that memories are not encoded before this period (see Bauer, 2002, for review of infant memory) but rather that they can now be organized in a way that makes them more accessible to retrieval. However, the age 2 to 3 also sees the emergence of language as a fully developing system and this too may have a profound effect on the accessibility of memories by rendering them verbally accessible (Nelson, 1996). Moreover, the emergence of language ushers in a whole new range of linguistic interactions with others and this too may impact on the way in which memories are represented and later accessed. In particular, the frequency and nature of rehearsal of memories with a parent may have important consequences for the later accessibility of memories (Bauer, 2002; Fivush & Hamond, 1990). There may well, of course, also be complex interactions between the developing self, emerging language abilities, and verbally dominated social interactions (e.g., Reese, 2002). All of these may also be influenced by cultural practices which act to facilitate or attenuate the rate of memory development (Wang, 2003).

A cross-cultural approach provides a unique means to investigate the collective effect of these variables on infantile amnesia. Indeed research has already established that when asked to recall their earliest childhood memory, Asians, including native Koreans and Chinese and overseas Asians, report events dating from more than 6 months later than do Europeans and Caucasian Americans, who remember earliest events occurring, on average, at age 3.5 (MacDonald et al., 2000; Mullen, 1994; Wang, 2001). The cultural variation in the age of earliest memory is thought to stem from different cultural conceptions of selfhood (Markus & Kitayama, 1991; Mullen, 1994; Pillemer, 1998; Wang, 2001, 2003). An autonomous self-construal that focuses on the individual as a centrally important agent authoring life experiences, like that in many Western cultures, may drive the early emergence of an organized, articulated, durable memory system for events that happened to "me". In contrast, a relational self-construal that focuses on community rather than agency may de-emphasize individuality and promote social integration and dependence. Under such influences the development of a structured autobiographical memory system would not be prioritized and this might lead to a longer period of childhood amnesia.

The nature of the self, its focus (agentic versus communal), and concomitant development may then influence the age of first memories typically accessible in a culture. The aim of the present research is to determine whether these factors have a more widespread influence on a range of earliest memories, rather than just on the first memory. Thus, we used an exhaustive-search method in which participants recalled as many childhood memories as they could within a limited timeframe. This might be thought of as a type of 'autobiographical memory fluency' test in which the total number of memories is a measure of fluency of access to, in the present case, childhood memories. This method allows us to examine the earliest age of recall (thus replicating previous cross-cultural studies) and assess adults' ability to access memories from the period of infantile amnesia more generally. Also, compared with the method of recalling only one memory by each participant in previous studies on the earliest memory, listing all memories may help to trigger very early memories and so raise accessibility.

We chose to focus on three culture groups who differ in the degree that they value individuality and an autonomous self: U.S., England, and China. American culture puts a prime value on personal autonomy and self sufficiency, qualities that are also emphasized but less strongly so in British culture and the least in Chinese culture that promotes interdependence and a relational self (Hofstede, 1980; Hus, 1970). Based on the theorization of cultural self-construal and autobiographical remembering, we predicted that the Chinese group would show less accessibility to childhood events and report later-dated earliest memory than either of the Western groups, particular the U.S. group. In addition, we examined gender effects on the recall of childhood events. We expected that, in support of previous findings (Davis, 1999; Mullen, 1994; Thorne, 1995; Wang, 2001), women would recall a greater number of childhood memories than men.

METHOD

Participants

The sample consisted of 302 participants, 101 from the U.S. (all Caucasian, 76 females, 25 males), 104 from England (87 females, 17 males), and 97 from China (47 females, 50 males).[4] All were college students. They took part in the study for partial course credit.

Procedure

The study was conducted in participants' native languages. Participants met with an experimenter in small groups and each received an instruction sheet describing the task. They were asked to recall childhood memories of events that occurred when they were under 5 years of age (including 5). They were told that they must recall as many memories as they

[4] We originally included a sample of Russian participants. We dropped it because the sample size was too small to warrant analysis.

could in a 5-minute period, and write on a provided sheet a word or short phrase whenever a memory was formed in order to remind themselves later of which memory they had brought to mind. The instruction emphasized that participants should respond quickly with the first memory that came to mind and not reject memories. Participants were told that later they would go back and rate each memory on several dimensions. When they were ready to begin, the experimenter started the timer.

After the time was up, participants received rating sheets (one for each memory) and a written rating instruction. On each rating sheet, participants first dated the memory, putting down how old they were to the nearest month when the recalled event occurred. They then used 5-point scales to rate the following: 1) how often they had thought and/or talked about the memory before; 2) how personally important the recalled experience was to them; 3) how detailed and clear the memory was; 4) how emotionally intense the experience was; 5) how negative or positive the experience was (from 1"very negative" to 5 "very positive"); and 6) the source of the memory (from 1 "only know about it from others" to 5 "remember it completely on your own").[5] Finally, participants provided demographic information.

RESULTS

The entire sample generated a total of 2805 memories, of which 2695 were dated. The 110 memories (US 27, England 34, China 49) with no age estimates were excluded from analysis pertinent to retrieval curve.

Retrieval Curve

A core result of interest is the retrieval curve from the first five years of life. Figure 1 illustrates the mean number of memories plotted in one year time bins for each culture group. All groups showed similar linear trends indicating age-related increase in memory accessibility. Trend analyses revealed that the linear fit r-square was U.S. $r^2 = .86$, England $r^2 = .88$, and China $r^2 = .87$. The retrieval curve of the entire data set also showed a linear fit with $r^2 = .87$. We performed a mixed model analysis on the number of memories reported, with culture, gender, and memory age being fixed factors and subject being a random factor. Significant main effects of culture and age emerged, F (2, 287) = 47.20, $p < .0001$, F (4, 1156) = 147.59, $p < .0001$, respectively, qualified by a Culture x Age interaction, F (8, 1156) = 21.52, $p < .0001$. As illustrated in Figure 1, although all culture groups recalled more memory events that occurred in the later years of the infantile amnesia period, the rate of increase in reported memories varied across cultures. Further focused analyses contrasting each pair of cultures revealed significant differences, U.S.-China F (1, 189) = 92.55, $p < .0001$, England-China, F (1, 189) = 33.84, $p < .0001$, and U.S.-England F (1, 196) = 11.52, $p = .0008$. On average, U.S. participants (M = 12.24, SD = 5.47) recalled the greatest number of childhood memories, followed by participants from England (M = 9.83, SD = 4.45), and lastly China (M = 5.68, SD = 1.68). In addition, a significant gender effect emerged, F (1, 287) =

[5] Participants also did other ratings and completed an attachment questionnaire at the end. Relevant results address separate research questions and are discussed elsewhere.

8.87, p = .003, whereby women (M = 10.20, SD = 5.12) in all cultures recalled more childhood memories than did men (M = 7.25, SD = 3.97) across the 5-year period.

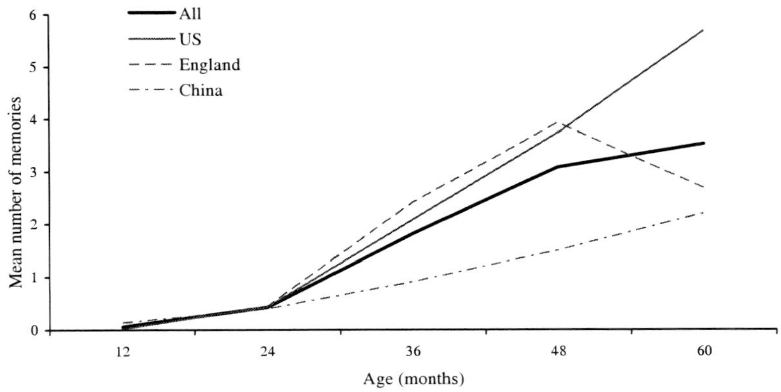

Figure 1. Memory retrieval curves as a function of culture and age of the memory.

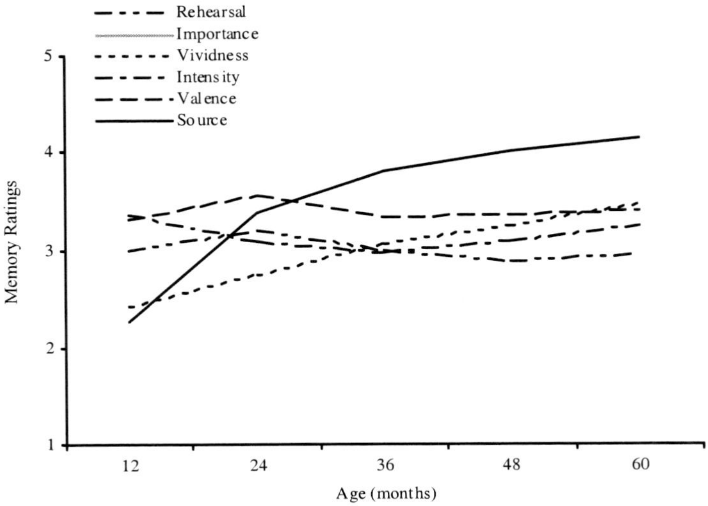

Figure 2. Memory ratings as a function of age of the memory.

Age at Earliest Memory

In most earliest-memory studies, participants are instructed to report what they remember for themselves and not to include 'memories' they *only* knew from external sources. Because the present study aimed to promote rapid free recall of memories no constraints were placed on the types of memories that could be brought to mind. Inevitably then our sample of first memories included "only-know-it-from-others" (OKIFO) memories (Ulric Neisser, personal communication, 2003). To control for this we took a rating of source of memory (other versus self) and so were able to partial out OKIFO memories (US 12, England 15, China 33). The average age at first memory, with OKIFO memories excluded, was U.S. $M = 32.37$ ($SD = 9.32$), England $M = 31.00$ ($SD = 7.34$), and China $M = 37.60$ ($SD = 12.01$). A 3 x 2 (Culture x Gender) ANOVA showed a main effect of culture, $F\,(2, 281) = 8.15$, $p = .0004$, in which the Chinese group drew their first memories from a reliably later age than did either of the Western groups who, in turn, did not differ, Tukey-Kramer HSD, $p < .05$ (note that the same pattern was present when OKIFO memories were included).

Memory Ratings

Although various characteristics of the memories were assessed by the ratings a problem with these data related to the fact that different participants recalled different numbers of memories. Thus, making comparisons between means is problematic because different participants contributed different numbers of ratings to the mean ratings. Instead of detailed analyses, i.e. by culture and gender, we report only the overall means to provide some perspective on the characteristics of memories recalled.

Overall means were: Rehearsal 2.95 (SD = 0.91), Importance 2.94 (SD = 0.97), Vividness 3.19 (SD = 1.02), Emotional Intensity 3.09 (SD = 0.93), Valence 3.37 (SD = 0.85), and Source of the memory 3.91 (SD = 0.99). The childhood memories were then ones that had received moderate amount of rehearsal. They were of moderate personal importance and slightly above average vividness. These events were of moderate emotional intensity and were mildly positive in nature. Finally, the source of the memories was predominantly the rememberer rather than an outside agency, i.e. the memories were, largely just that: memories. Figure 2 illustrates the memory ratings as a function of the age of the memory. Simple regressions using memory age as a predictor for the ratings (with subject being a random factor) showed that with age, memories became more vivid, $F\,(1, 719) = 50.16$, $p < .0001$, more emotionally intense, $F\,(1, 719) = 4.56$, $p = .03$, and more remembered on one's own, $F\,(1, 719) = 62.56$, $p < .0001$.

Discussion

The present study is to our knowledge the first cross-cultural investigation of memory accessibility for the period of infantile amnesia. All culture groups found it extremely difficult, if not impossible, to retrieve memory information from the first 2 to 3 years of life. They also all showed increasing ability to recall events occurring in the later part of the 5-year period. However, U.S. participants recalled the greatest number of childhood memories and showed the greatest age-linked increase in memory accessibility, followed by British participants, and then Chinese. In addition, consistent with previous findings (MacDonald et al., 2000; Mullen, 1994; Wang, 2001), Chinese participants' earliest childhood memories were approximately 6 months later than their Western counterparts. Thus, the period of infantile amnesia differs in extent across cultures, with Chinese showing the longest period and Caucasian Americans the shortest and the British group falling in between.

If we assume that cognitive development has universal aspects that are common to all three cultures, the differences we have observed must largely be due to cultural influences. Thus, for instance, a cognitive self (Howe & Courage, 1997) emerges, as indicated by mirror self-recognition, around 18 to 24 months in children across different cultures (Kagan, 1989; Priel & de Schonen, 1986) and develops in the same manner with further cognitive progress (Harter, 1998). The same is almost certainly the case for language comprehension and production (Brown, 1973). And, perhaps, it might also be reasonably assumed that the development of the neurological substrate that mediates autobiographical remembering (cf. Conway et al., 2002), especially the frontal lobes, has a common largely genetically determined time course.

In addition to the universal development of cognitive/linguistic/ neurobiological processes that mediate the formation, retention, and retrieval of childhood memories are cultural influences that determine the extent and way in which these processes are deployed. Developmental studies provide converging evidence that the formation of the self entails constructive processes embedded in a myriad of daily exchanges between children and socialization agents from early in life (Haight, 1999; Miller et al.,1997; Mullen & Yi, 1995; Shweder et al., 1998; Wang et al., 2000). Children then develop a self-construal that integrates the framework of their respective cultures (Wang, 2004), which further influences cognitive processes by channeling or directing these processes in specific ways. An autonomous self-construal may prioritize the early formation of a unique, detailed and articulate personal history, resulting in earlier first memories and more childhood memories generally. In contrast, a relational self-construal may prioritize instead the retention of knowledge, not necessarily in the form of autobiographical memories, that is related to experiences with predominately interpersonal themes (cf., Cohen & Gunz, 2002; Mullen, 1994; Schrauf, 2000; Wang, 2001).

A factor that seems intertwined with the influence of cultural self-construal is the narrative practices employed by parents in different cultures. Elaborative memory conversations are found to have facilitative effects on the development of autobiographical memory in children (e.g., Fivush & Hamond, 1990; Nelson, 1996). Mirroring the respective cultural emphasis on autonomy versus relatedness, U.S. parents often initiate lengthy, elaborate memory conversations that focus on the child's roles, feelings, and predilections, whereas Chinese parents tend to discuss the past with less elaboration and emphasize social rules and discipline (Wang et al., 2000). There is also evidence that U.S. parents focus more on the child's personal rights and feelings during narrative interactions than British parents (Dunn & Brown, 1991). It is possible then that the cultural practice of elaborate, child-centered memory talk played a role in facilitating access to more and earlier childhood memories in our U.S. group relative to the Chinese and British groups. This account could also explain gender effect. There is evidence that parents often have more elaborate memory conversations with their daughters than with their sons (e.g., Fivush, 1998). Such narrative practices, if widespread, might account for at least part of the advantage that women showed over men in the present study in the greater number of childhood memories they were able to rapidly access. Our findings of gender differences are consistent with past work and support the notion that gender-differentiated socialization affects the accessibility of autobiographical memories (Davis, 1999; Thorne, 1995; Wang, 2001).

Notably, there may be many other sociocultural factors that contribute to individual and cultural variations in the accessibility of childhood memories (see Wang, 2003, for a review). For example, Western cultures, especially in the U.S., value adventures, changes and independent exploration (Hsu, 1970), which may create abundant memorable events for individuals to dwell upon. In contrast, Chinese culture emphasizes constancy, predictability and behavioral inhibition, which may render a life with few stories. Obviously, more empirical cross-cultural research is called for.

In conclusion, the pattern of findings in the present study demonstrated reliable and strong cultural differences in the amount of childhood memories that could be rapidly accessed and in the age of the earliest memory. These differences may have arisen through the operation of two interrelated factors. First, the self-construal prevalent in a culture acts to direct cognitive processes in the acquisition, retention, and retrieval of memory. Having an autonomous versus a relational self may affect whether and to what extent cognitive processes and resources are channeled into the early development of an articulate, detailed personal history. Second, culture-specific narrative practices may shape memory development in ways that support the predominant self-construal. Elaborative, child-centered memory talk acts to promote autobiographical remembering in independently-oriented cultures where the development of autobiographical memory is prioritized. Memory in these cultures serves to differentiate the self from others and so provides an important means of constructing a unique individual identity, even at early ages.

REFERENCES

Bauer, P.J. (2002). Early Memory Development. In U. Goswami (Ed.), *Blackwell Handbook of Childhood Cognitive Development* (pp. 127-146). Oxford, United Kingdom: Blackwell Publishers.

Brown, R. (1973). *A first language: The early stages*. MA, Cambridge: Harvard University Press.

Cohen, D. & Gunz, A. (2002). As seen by the other... Perspectives on the self in the memories and emotional perceptions of Easterners and Westerners. *Psychological Science, 13*, 55-59

Conway, M.A. (1990). *Autobiographical Memory: An Introduction*. Buckingham: Open University Press.

Conway, M. & Pleydell-Pearce, C. W. (2000). The construction of autobiographical memories in the self-memory system. *Psychological Review, 107, 2*, 261-288.

Conway, M.A., Pleydell-Pearce, C.W., Whitecross, S. Sharpe, H. (2002). Brain imaging autobiographical memory. *The Psychology of Learning and Motivation, 41*, 229-264.

Davis, P. J. (1999). Gender differences in autobiographical memory for childhood emotional experiences. *Journal of Personality and Social Psychology, 76,* 498–510.

Dunn, J. & Brown, J. (1991). Becoming American or English? Talking about the social world in England and the United States. In M. H. Bornstein (Ed.), *Cross-cultural approaches to parenting* (pp. 155-172). Hilllsdale, NJ: Laurence Erlbaum Associates.

Eacott, M. J. & Crawley, R. A. (1999). Childhood amnesia: On answering questions about very early life events. *Memory, 7, 3*, 279-292.

Fivush, R. (1998). Gendered narratives: Elaboration, structure, and emotion in parent-child reminiscing across the preschool years. In C. P. Thompson & D. J. Herrmann (Eds.), *Autobiographical memory: Theoretical and applied perspectives* (pp. 79-103). Mahwah, NJ: Lawrence Erlbaum Associates, Inc.

Fivush, R. & Hamond, N. R. (1990). Autobiographical memory across the preschool years: Toward reconceptualizing childhood amnesia. In R. Fivush & J. Hudson (Eds.), *Knowing and remembering in young children* (pp. 223-248). New York: Cambridge University Press.

Freud, S. (1949). *Three essays on the theory of sexuality* (J. Strachey, Trans.). London: Imago Publishing Company, Ltd.

Greenfield, P. (1997). Culture as process: Empirical methodology for cultural psychology. In J. W. Berry, Y. H. Poortinga, & J. Pandey (Eds.), *Handbook of cross-cultural psychology: Vol. 1, Theory and method* (pp. 301-346). Boston: Allyn & Bacon.

Harley, K. & Reese, E. (1999). Origins of autobiographical memory. *Developmental Psychology, 35, 5,* 1338-1348.

Harter, S. (1998). The development of self-representations. In W. Damon (Ed.), *Handbook of child psychology, vol. (5th ed.)* (pp. 553-617). New York: Wiley.

Hofstede, G. (1980). *Culture's consequences: International differences in work-related values*. Beverly Hills, CA: Sage Publications.

Howe, M. L. & Courage, M. L. (1997). The emergence and early development of autobiographical memory. *Psychological Review, 104, 3,* 499-523.

Hsu, F. L. K. (1970). *Americans and Chinese: Purpose and fulfillment in great civilizations.* New York: The Natural History Press.

Kagan, J. (1989). *Unstable ideas: Temperament, cognition, and self.* Cambridge, MA: Harvard University Press.

MacDonald, S., Uesiliana, K., Hayne, H. (2000). Cross-cultural and gender differences in childhood amnesia. *Memory, 8, 6,* 365-376.

Markus, H.R. & Kitayama, S. (1991). Culture and the self: Implications for cognition, emotion, and motivation. *Psychological Review, 98, 2,* 224-253.

Mullen, M. K. (1994). Earliest recollections of childhood: A demographic analysis. *Cognition, 52, 1,* 55-79.

Mullen, M. K., & Yi, S. (1995). The cultural context of talk about the past: Implications for the development of autobiographical memory. *Cognitive Development, 10*, 407-419.

Neisser, U. (1962). Cultural and cognitive discontinuity. In T. E. Gladwin & W. C. Sturtevant (Eds.), *Anthropology and human behavior* (pp. 54-71). Washington, DC: Anthropological Society of Washington.

Nelson, K. (1996). *Language in cognitive development: The emergence of the mediated mind.* New York: Cambridge University Press.

Newcombe, N. S., Drummey, A. B., Fox, N. A., Lie E., & Ottinger-Alberts, W. (2000). Remembering early childhood: How much, how, and why (or why not). *Current Directions in Psychological Science, 9, 2*, 55- 58.

Pillemer, D. B. (1998). *Momentous events, vivid memories.* Cambridge, MA: Harvard University Press.

Pillemer, D. B. & White, S. H. (1989). Childhood events recalled by children and adults. In H.W. Reese (Ed.), *Advances in child development and behavior (Vol. 21)* (pp. 297-340). New York: Academic Press.

Reese, E. (2002). A model of the origins of autobiographical memory. In J. Fagen & H. Hayne (Eds.), *Progress in infancy research* (Vol. 2). Mahwah, NJ: Lawrence Erlbaum Associates Inc.

Rubin, D. C. (2000). The distribution of early childhood memories. *Memory, 8, 4*, 265-269.

Schrauf, R. W. (2000). Bilingual autobiographical memory: Experimental studies and clinical cases. *Culture & Psychology, 6, 4*, 387-417.

Thorne, A. (1995). Developmental truths in memories of childhood and adolescence. *Journal of Personality, 63*, 139-163.

Usher, J. A. & Neisser, U. (1993). Childhood amnesia and the beginnings of memory for four early life events. *Journal of Experimental Psychology: General, 122, 2*, 155-165

Waldfogel, S. (1948). The frequency and affective character of childhood memories. *Psychological Monographs*, 62, 1-34.

Wang, Q. (2001). Cultural effects on adults' earliest childhood recollection and self-description: Implications for the relation between memory and the self. *Journal of Personality and Social Psychology, 81*, 2, 220-233.

Wang, Q. (2003). Infantile amnesia reconsidered: A cross-cultural analysis. *Memory, 11, 1, 65-80.*

Wang, Q. (2004). The emergence of cultural self-construct: Autobiographical memory and self-description in American and Chinese children. *Developmental Psychology, 40, 1*, 3-15.

Wang, Q., Leichtman, M. D., & Davies, K. (2000). Sharing memories and telling stories: American and Chinese mothers and their 3-year-olds. *Memory, 8, 3,* 159-177.

Wang, Q., Leichtman, M. D., & White, S. H. (1998). Childhood memory and self-description in young Chinese adults: The impact of growing up an only child. *Cognition, 69, 1,* 75-105.

Wetzler, S. E., & Sweeney, J. A. (1986). Childhood amnesia: An Empirical demonstration. In D. C. Rubin (Ed.), *Autobiographical memory*, (pp. 202-221). Cambridge: Cambridge University Press.

AUTHOR NOTE

Qi Wang was supported by Cornell University Hatch Award. Martin A. Conway was supported by a grant from the Economic and Social Research Council, Changes in Autobiographical Memory During Adolescence, R000239395. The authors thank both bodies for their support. This work is part of a larger project jointly conducted by the two authors into cultural differences in autobiographical memory.

We thank Ulric Neisser, Nora Newcombe, and Peter Gordon for helpful comments on earlier versions of this article. We also thank Abie Li, Melissa McDermott, Erin Spillane, and Christine Tang for their assistance. Special thanks go to the participants who made this study possible.

In: Cognitive Science Compendium, Volume 1
Editor: Miao-Kun Sun
ISBN: 978-1-60456-597-3

Chapter 5

THE NDRG GENE FAMILY AND HUMAN NERVOUS SYSTEM DISORDERS

***Cathy Mitchelmore*[1*]*, Stine Büchmann-Møller*[2] *and Niels A. Jensen*[3]**

[1] Department of Life Sciences and Chemistry, Roskilde University, P.O. Box 260, DK-4000 Roskilde, Denmark

[2]Department of Biology, Institute of Cell Biology, Swiss Federal Institute of Technology, ETH-Hönggerberg, 8093-Zürich, Switzerland

[3]Medical Biotechnological Center, University of Southern Denmark, Winslowparken 25, DK-5000 Odense C, Denmark

ABSTRACT

The N-Myc downstream-regulated gene (NDRG) gene family consists of four highly related genes. The proteins encoded by the NDRG gene family all contain a putative α/β-hydrolase fold, which is typically present in enzymes. However, their functions are little understood. The NDRG gene family has recently come into focus because two members – NDRG1 and NDRG2 – have been linked to human nervous system disorders. NDRG1 was shown to be mutated in a peripheral neuropathy, called hereditary motor and sensory neuropathy-Lom (HMSNL). In this disorder, affected individuals bearing two copies of the mutated NDRG1 allele develop a severe disability in early adulthood, with symptoms including muscle weakness, gait disorder and deafness. Neuropathologically, HMSNL is characterized by defective myelination and axonal loss in the peripheral nervous system. Another member of this gene family, NDRG2, was identified as a gene whose protein product is increased in the brains of Alzheimer's disease (AD) patients and is associated with pathological lesions in AD. In contrast to NDRG2, the highly related NDRG4 gene is downregulated in AD. We describe here the NDRG gene family members and discuss their possible roles in human nervous system disorders.

* Corresponding author. Email: mitch@ruc.dk

Keywords: Alzheimer's disease, hereditary motor and sensory neuropathy-Lom, α/β-hydrolase fold, neurodegeneration, pathology

THE NDRG GENE FAMILY

The NDRG gene family encodes four highly related proteins: NDRG1, NDRG2, NDRG3 and NDRG4. There is 57% to 65% identity between each of these proteins. Furthermore, the identity between each human NDRG protein and the corresponding mouse protein is 92-97%, showing that these proteins have been well conserved during evolution (Zhou et al., 2001). The central region of these proteins, which is particularly well conserved, contains a putative α/β-hydrolase fold, shaded in Figure 1 (Shaw et al., 2002). The α/β-hydrolase fold is present in a diverse range of enzymes which show little sequence homology, including carboxypeptidases, lipases and acetylcholine esterases. Typically, this domain consists of an α/β-sheet of eight β-strands connected by α-helices. Three conserved amino acids make up the catalytic triad: an active serine located in a nucleophile elbow, an acidic amino acid (usually aspartate) and an invariant histidine (Ollis et al., 1992). However, only one of these positions in the catalytic triad (the aspartate) is conserved in the NDRG proteins, suggesting that enzymatic activity is either lacking in the NDRG proteins or is atypical (Shaw et al., 2002).

NDRG1 is the best characterized member of the NDRG family. The mouse NDRG1 homolog was originally identified as a gene repressed by N-myc during early development (Shimono et al., 1999). NDRG1 is also repressed in transformed cells and behaves as a tumor suppressor gene in colon, prostate and breast cancers (Bandyopadhyay et al., 2004; Kurdistani et al., 1998; Segawa et al., 2002; vanBelzen et al., 1997). During terminal differentiation in mouse embryos and in cell lines induced to differentiate, the NDRG1 gene is upregulated (Kurdistani et al., 1998; Piquemal et al., 1999; Shimono et al., 1999; vanBelzen et al., 1997; Xu et al., 1999). Furthermore, NDRG1 is upregulated in response to a range of cellular stresses, for example following treatment of cells with homocysteine, nickel or hypoxia, and can be induced by androgens and activators of protein kinase C (Kokame et al., 1996; Kurdistani et al., 1998; Lin and Chang, 1997; Park et al., 2000; Piquemal et al., 1999; Ulrix et al., 1999; vanBelzen et al., 1997; Xu et al., 1999; Zhou et al., 1998).

There are several indications that, although the NDRG members are highly related, they serve different functions. Their amino acid sequences are less well conserved outside the central region (Figure 1). In the C-terminal region of NDRG1, but not in the other NDRG proteins, there is a triple repeat of a 10-amino-acid sequence of unknown function. Alternative protein isoforms of NDRG2, NDRG3 and NDRG4 have been described, in which a short stretch of amino acids is inserted as a result of alternative splicing. These insertions are located in the less-conserved terminal regions of the proteins and could contribute to protein function. There are also differences in the expression pattern and regulation of the different family members (Table 1). Whereas NDRG1 is ubiquitously expressed, NDRG2, NDRG3 and NDRG4 are more restricted in their expression (Qu et al., 2002; Zhao et al., 2001; Zhou et al., 2001). A further difference is that, although NDRG1 is reciprocally expressed in relation to N-Myc during mouse embryo development, the other NDRG members are not subjected to downregulation by N-myc (Okuda and Kondoh, 1999; Shimono et al., 1999).

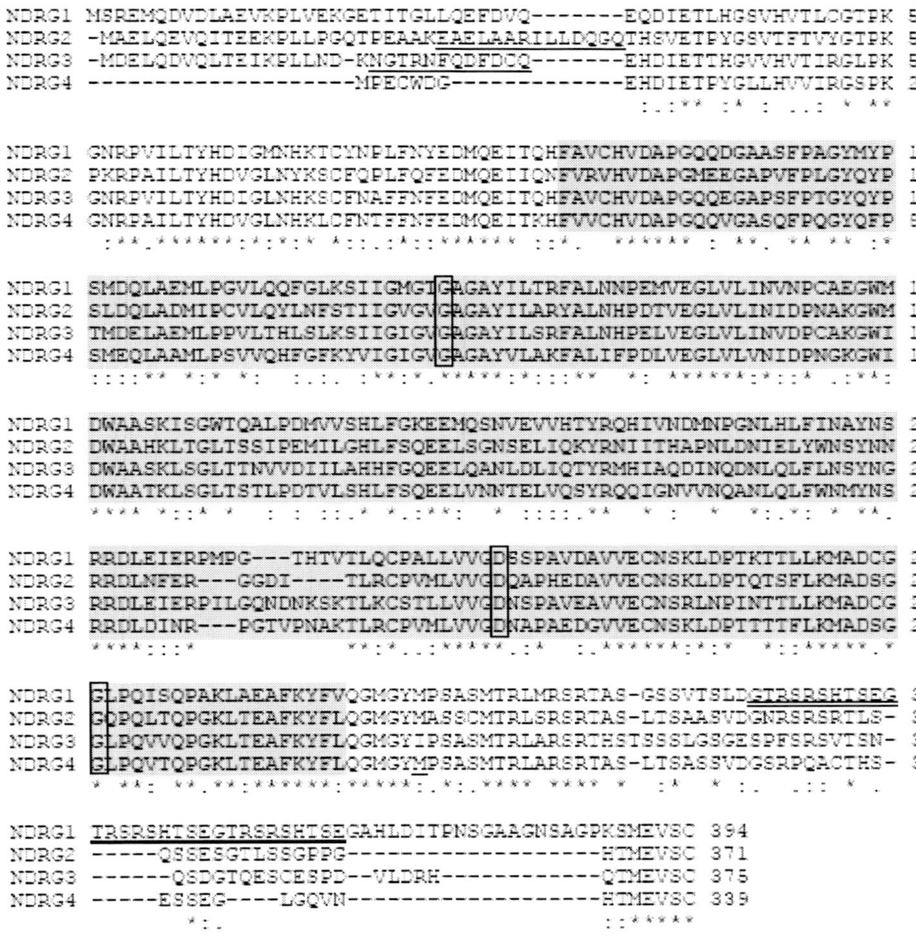

Figure 1. Alignment of the amino acid sequences of human NDRGs in single-letter code. The NDRG proteins are highly conserved, particularly in the central region that is structurally similar to the α/β-hydrolase fold (shaded). Of the three residues that form the catalytic triad in typical α/β-hydrolase fold proteins (boxed), only the aspartic residue is conserved in the NDRG protein family. Additional protein isoforms of NDRG2, NDRG3 and NDRG4 exist which differ in their amino acid sequence outside of the α/β-hydrolase fold. The longer isoforms of NDRG2 and NDRG3 are shown; the shorter forms lack the underlined amino acids. The brain isoform of NDRG4 is shown; in a less common variant, the methionine at position 289 (underlined) is replaced by 14 amino acids. A third isoform, present in both heart and brain, contains a different N-terminal sequence. Only the NDRG1 protein contains a 3-times repeat of the sequence GTRSRSHTSE in the C-terminal region of the protein (double-underlined)

Table 1. The NDRG gene family

NDRG (alternative names)	Chromosome	Protein	Expression	Regulation	Disease association	References
NDRG1 (RTP, Drg1, rit42,Cap43, Ndr1, TDD5, Proxy-1)	8q24.3	NDRG1: 394 aa	Ubiquitous expression in tissues, but with stronger expression in kidney and moderate in heart, placenta, lung and brain. Subcellular: mainly cytoplasmic in myelinating Schwann cells and in rat brain oligodendrocytes. Not detected in CNS neurons.	Upregulated during differentiation of tumor cell lines. Downregulated by N-Myc and in transformed cells. Induced by stress-inducing compounds (homocysteine, tunicamycin, nickel, forskolin and hormones) in cells. Regulated similarly to myelin proteins during nerve development, nerve injury and nerve remyelination in rats.	Mutations in NDRG1 have been found in patients with hereditary motor and sensory neuropathy-Lom (HMSNL), a form of Charcot-Marie-Tooth disease. Downregulated in various cancers.	(Berger et al., 2004; Hunter et al., 2003; Kalaydjieva et al., 2000; Kokame et al., 1996; Kurdistani et al., 1998; Piquemal et al., 1999; Shimono et al., 1999; vanBelzen et al., 1997; Xu et al., 1999; Zhou et al., 1998)
NDRG2 (SYLD, KIAA1248, Ndr2)	14q11.2	NDRG2: 357 aa NDRG2-ins: 371 aa	Strong expression in skeletal muscle, brain and heart. Moderate expression in liver, kidney, trachea, adrenal and salivary gland. Subcellular: cytoplasmic, perinuclear in normal pyramidal neurons but accumulates in neurites when overexpressed.	Induced by aldosterone in kidney and distal colon. Downregulated in glioblastoma cell lines. Upregulated during differentiation of monocyte-derived dendritic cells.	Upregulated in AD, where NDRG2 protein is present in senile plaques and dystrophic neurons. Downregulated in high-grade glioblastoma compared to low-grade glioma and normal brain tissue.	(Boulkroun et al., 2002; Choi et al., 2003; Deng et al., 2003; Mitchelmore et al., 2004; Qu et al., 2002; Zhou et al., 2001)

Table 1. The NDRG gene family (Continued)

NDRG (alternative names)	Chromosome	Protein	Expression	Regulation	Disease association	References
NDRG3 (Ndr3)	20q11.21-23	NDRG3a: 375 aa NDRG3b: 363 aa	Strong expression in prostate, testis and ovary. Moderate expression in brain, kidney and heart.			(Qu et al., 2002; Zhao et al., 2001; Zhou et al., 2001)
NDRG4 (Bdm1, Smap8)	16q21-q22.3	NDRG4-B: 339 aa NDRG4-Bvar: 352 aa NDRG4-H: 371 aa.	NDRG4-H is expressed in both brain and heart, whereas NDRG4-B is restricted to the brain. Little or no expression in other tissues. Subcellular: cytoplasmic expression in neurons of the CNS. Distributed in mitochondria and ER. Not detected in peripheral neurons or in human cancer cell lines	Upregulated in cell lines during neuronal differentiation. Induced by homocysteine in vascular smooth muscle cells.	Downregulated in AD brain tissue compared to normal adult brain.	(Nakada et al., 2002; Nishimoto et al., 2003; Ohki et al., 2002; Yamauchi et al., 1999; Zhou et al., 2001)

NDRG2 and NDRG4 Are Widely Expressed in the Brain

The NDRG2 gene is widely expressed throughout the mammalian brain, where it has been reported to be expressed in both neuronal and glial cells (Deng et al., 2003; Mitchelmore et al., 2004; Nichols, 2003). In the human entorhinal cortex and hippocampus, NDRG2 protein appears to be present mainly in neuronal cells (Mitchelmore et al., 2004). In the human hippocampus, NDRG2 protein expression is highest in the pyramidal neurons (CA1 and CA3 subregions) but is absent from the granule neurons of the dentate gyrus (Mitchelmore et al., 2004). There are two isoforms of NDRG2, which differ by the insertion of 14 amino acids in the long form (Figure 1). Expression of both isoforms is upregulated in the brains of AD patients (Mitchelmore et al., 2004 and see below). During embryonic mouse development, NDRG2 is strongly expressed in the ventricular zone during neurogenesis, in both the brain and spinal cord (Okuda and Kondoh, 1999). Downregulation of NDRG2 expression has been observed in high-grade glioblastoma (Deng et al., 2003).

Outside of the central nervous system, NDRG2 upregulation has been observed during differentiation of monocyte-derived dendritic cells (Choi et al., 2003), and following aldosterone stimulation of epithelial target cells in the kidney and distal colon (Boulkroun et al., 2002).

NDRG4 was first identified by differential display technology as a gene expressed at higher levels in the adult rat brain than in embryonal brains (Yamauchi et al., 1999). The gene is predominantly expressed during early postnatal development and expression of NDRG4 in the brain is widespread (Nakada et al., 2002). NDRG4 expression is mainly restricted to the brain and heart (Zhou et al., 2001). Two isoforms of NDRG4, which differ in their N-terminal sequence, have different expression patterns (Zhou et al., 2001). Recently, NDRG4 has been shown to be induced by homocysteine and to be subjected to mitogen-stimulated phosphorylation in vascular smooth muscle cells (Nishimoto et al., 2003).

NDRG4 is upregulated during nerve growth factor-mediated neuronal differentiation of PC12 cells. In differentiating PC12 cells, the lowering of NDRG4 protein levels through the use of antisense RNA was found to inhibit neurite outgrowth (Ohki et al., 2002). In the brain, NDRG4 expression was localized to neurons by *in situ* hybridisation. Thus, the NDRG4 gene might be involved in neuronal differentiation. The observation that NDRG4 mRNA is downregulated in AD brain tissue may reflect the loss of NDRG4-expressing neurons (Zhou et al., 2001).

NDRG1 Is Mutated in a Hereditary Disorder of the Peripheral Nervous System

A mutation in the NDRG1 gene has recently been linked to hereditary motor and sensory neuropathy-Lom (HMSNL) in gypsies of Romanian origin (Kalaydjieva et al., 2000). HMSNL is an autosomal recessive demyelinating neuropathy characterized by Schwann cell dysfunction and severe axonal loss. Schwann cells are the cells in the peripheral nervous system which are responsible for synthesis of the myelin sheath that surrounds nerve axons. Both motor and sensory nerves are affected in HMSNL. The initial symptoms of this disorder occur in the first decade of life as gait disturbances, followed in the second and third decade by weakness in the upper limbs and deafness (Kalaydjieva et al., 1998). The NDRG1 mutation identified in HMSNL is a nucleotide substitution that results in a premature termination signal at amino acid position 148 (Kalaydjieva et al., 2000). More recently, mutation screening of NDRG1 in an ethnically diverse group of patients with peripheral neuropathies (after exclusion of common causes) revealed the mutation above as well as a new mutation in the NDRG1 gene. The new mutation results in a protein with an altered sequence due to exon-skipping. A patient homozygous for this mutation suffered from severe demyelinating neuropathy with symptoms similar to HMSNL (Hunter et al., 2003).

Since NDRG1 is normally expressed in Schwann cells but not in nerve axons, the symptomology in HMSNL is most likely due to Schwann cell dysfunction leading to demyelination and subsequent axonal degeneration (Kalaydjieva et al., 2000). In rodents, NDRG1 protein is found in the cytoplasm of Schwann cells rather than in the myelin sheath (Berger et al., 2004; Okuda et al., 2004). Presumably, the truncated NDRG1 protein is unable to carry out the normal NDRG1 function in Schwann cells, which may be related to maintenance of the myelin sheaths in peripheral nerves (Okuda et al., 2004). The importance of NDRG1 in Schwann cell function is reinforced by a recent report describing the phenotype resulting from deleting the gene in mice. The NDRG1 knock-out mice suffered from severe demyelination of their peripheral nerves at about 5 weeks of age, leading to progressive muscle weakness, especially in the hind limbs (Okuda et al., 2004). Since the knock-out phenotype is similar to HMSNL, the NDRG1 knock-out mice provide a useful model for the disease.

In the central nervous system of rats, NDRG1 protein appears to be restricted in its expression to oligodendrocytes, where it is mainly cytoplasmic. Futhermore, levels of NDRG1 mRNA parallel that of myelin proteins during development, in both the peripheral and central nervous system, suggesting that NDRG1 may also be involved in myelination in the brain (Berger et al., 2004). In support of this hypothesis, the symptoms of HMSNL include subclinical evidence of slowed conduction of central nervous system neurons (Kalaydjieva et al., 1998).

NDRG2 Is Upregulated in AD

AD is a debilitating disease associated with the progressive loss of memory and cognitive abilities. In early-onset AD, which comprises 5 – 10% of cases, symptoms begin at an early age (around 50 years of age). In many of these patients, genetically inherited mutations have been identified in one of three genes: amyloid precursor protein (APP), presenilin 1 (PS1) or presenilin 2 (PS2) (Price and Sisodia, 1998; Sisodia et al., 1999). However, the majority of AD cases arise sporadically, with a late age of onset (>65 years of age). No clear genetic basis has been discovered for the late-onset form of AD, although there is an increased risk associated with the apolipoprotein ApoE4 allele (Corder et al., 1993; Farrer et al., 1997; Saunders et al., 1993). AD is characterized by the deposition of β-amyloid aggregates in senile plaques as well as by the accumulation of neurofibrillary tangles (NFTs) (Terry et al., 1999). Ultimately, these processes lead to neuronal cell death. The first regions in which cell loss occurs include the entorhinal cortex and the CA1 region of the hippocampus (Gomez-Isla et al., 1996; West et al., 1994). Since these regions are required for learning and memory, this explains why the early symptoms of AD include memory problems (Eichenbaum, 2001; Morris, 1999).

Recent work from our laboratory demonstrated that NDRG2 expression is elevated in late-onset AD and that the NDRG2 protein is closely associated with the hallmarks of AD pathology, plaques and NFTs (Mitchelmore et al., 2004). NDRG2 was identified by a subtractive candidate gene strategy that was designed to pull out genes that were expressed at higher levels in the hippocampus of AD patients compared to unaffected persons. To identify candidate genes that participate in the initiating steps of the neurodegenerative process, post-mortem tissue was obtained from patients in the early stages of the disease, where cognitive defects were still relatively mild. Using RT-PCR and northern blotting, our laboratory demonstrated that the RNA coding for NDRG2 was upregulated at least 2-fold in AD brain tissue. Furthermore, an antibody raised specifically against NDRG2 was used to demonstrate that the protein levels of NDRG2 were also elevated in AD compared to normal tissue (Mitchelmore et al., 2004).

In order to determine whether new AD candidate genes play an active role in the neurodegenerative process, as opposed to being part of a secondary process, it is informative to examine where the corresponding protein is expressed. Genes which are expressed in neurons vulnerable to AD-mediated damage as well as in the pathological lesions themselves, are likely to be strong AD gene candidates. Using immunohistochemical staining of AD tissue, our laboratory showed that NDRG2 is found in neurons in the entorhinal and hippocampal regions, both of which are particularly vulnerable to neuronal loss in AD (Mitchelmore et al., 2004).

Possible Role of the NDRG Gene Family in Neurodegeneration

The four highly related NDRG proteins all harbor a central α/β-hydrolase fold, which is found in a diverse range of enzymes. However, it remains to be demonstrated whether the NDRG proteins have hydrolytic activities (Shaw et al., 2002). Most reports indicate that the NDRG proteins are cytoplasmic, which would be consistent with a role in lysosomal function (Qu et al., 2002; vanBelzen et al., 1997) or maintenance of the cytoskeleton (Kim et al., 2004). As described above, NDRG1 appears to play an important role in normal Schwann cell function, since the absence of this gene in knock-out mice or its mutation in human peripheral neuropathy leads to demyelination of peripheral nerves (Kalaydjieva et al., 2000; Okuda et al., 2004). Recent data suggests that NDRG1 is necessary, but not sufficient, for p53-dependent apoptosis (Stein et al., 2004). Since NDRG1 appears to have some tumor-suppressor properties (Bandyopadhyay et al., 2004; Kurdistani et al., 1998; Segawa et al., 2002; vanBelzen et al., 1997), one of its functions could be in cell cycle checkpoint, perhaps through an interaction with the mitotic spindle (Kim et al., 2004). NDRG1, NDRG2 and NDRG4 are all subject to phosphorylation, which could thereby be a mechanism for regulating the activity or location of these proteins (Agarwala et al., 2000; Burchfield et al., 2004; Murray et al., 2004; Nishimoto et al., 2003).

In normal-appearing hippocampal neurons, NDRG2 immunostaining is localized around the nucleus and is absent from the neuronal processes. However, in dying neurons, NDRG2 protein appears to accumulate in abnormal cellular processes. Interestingly, NDRG2 accumulation in dendritic processes also occurs in transgenic mice which over-express NDRG2 in pyramidal neurons in the hippocampus (Mitchelmore et al., 2004). The upregulation and cellular redistribution of NDRG2 could be involved in neurodegeneration through several mechanisms: 1) through a direct damaging effect on the microtubule network; 2) by triggering the action of other damage-inducing proteins; 3) as a cell survival factor that is recruited, for example, to degrade misfolded proteins; or 4) as part of an indirect response. Since NDRG2 is upregulated in AD whereas NDRG4 is suppressed, it seems likely that NDRG2 plays an active role. This is supported by data showing that NDRG2 protein is present in pyramidal neurons, senile plaques and dying neurons in AD patients.

In summary, the role of the NDRG gene family in nervous system disorders such as AD is poorly understood at the present time. However, the identification of NDRG1 mutations as causative for a severe inherited disorder of the peripheral nervous system, together with the demonstration that NDRG2 is upregulated in AD, and recent indications that these proteins are downregulated in human cancers, indicates that there will be an increasing research focus on the NDRG gene family in the future.

ACKNOWLEDGMENTS

This work was supported by grants from the Lundbeck Foundation, the Novo Nordisk Foundation, the Alfred Benzon Foundation, and the Danish Medical Research Council.

REFERENCES

Agarwala, K.L.; Kokame, K.; Kato, H.; Miyata, T. Phosphorylation of RTP, an ER stress-responsive cytoplasmic protein. *Biochem. Biophys. Res. Commun.* 272: 641-647, 2000.

Bandyopadhyay, S.; Pai, S.K.; Hirota, S.; Hosobe, S.; Takano, Y.; Saito, K.; Piquemal, D.; Commes, T.; Watabe, M.; Gross, S.C.; Wang, Y.; Ran, S.; Watabe, K. Role of the putative tumor metastasis suppressor gene Drg-1 in breast cancer progression. *Oncogene* 23: 5675-5681, 2004.

Berger, P.; Sirkowski, E.E.; Scherer, S.S.; Suter, U. Expression analysis of the N-Myc downstream-regulated gene 1 indicates that myelinating Schwann cells are the primary disease target in hereditary motor and sensory neuropathy-Lom. *Neurobiol. Dis.* 17: 290-299, 2004.

Boulkroun, S.; Fay, M.; Zennaro, M.C.; Escoubet, B.; Jaisser, F.; Blot-Chabaud, M.; Farman, N.; Courtois-Coutry, N. Characterization of rat NDRG2 (N-Myc downstream regulated gene 2), a novel early mineralocorticoid-specific induced gene. *J. Biol. Chem.* 277: 31506-31515, 2002.

Burchfield, J.G.; Lennard, A.J.; Narasimhan, S.; Hughes, W.E.; Wasinger, V.C.; Corthals, G.L.; Okuda, T.; Kondoh, H.; Biden, T.J.; Schmitz-Peiffer, C. Akt mediates insulin-stimulated phosphorylation of Ndrg2- Evidence for cross-talk with protein kinase C theta. *J. Biol. Chem.* 279: 18623-18632, 2004.

Choi, S.C.; Kim, K.D.; Kim, J.T.; Kim, J.W.; Yoon, D.Y.; Choe, Y.K.; Chang, Y.S.; Paik, S.G.; Lim, J.S. Expression and regulation of NDRG2 (N-myc downstream regulated gene 2) during the differentiation of dendritic cells. *FEBS Lett.* 553: 413-418, 2003.

Corder, E.H.; Saunders, A.M.; Strittmatter, W.J.; Schmechel, D.E.; Gaskell, P.C.; Small, G.W.; Roses, A.D.; Haines, J.L.; Pericakvance, M.A. Gene Dose of Apolipoprotein-e Type-4 Allele and the Risk of Alzheimers-Disease in Late-Onset Families. *Science* 261: 921-923, 1993.

Deng, Y.C.; Yao, L.B.; Chau, L.; Ng, S.S.M.; Peng, Y.; Liu, X.P.; Au, W.S.; Wang, J.C.; Li, F.Y.; Ji, S.P.; Han, H.; Nie, X.Y.; Li, Q.; Kung, H.F.; Leung, S.Y.; Lin, M.C.M. N-myc downstream-regulated gene 2 (NDRG2) inhibits glioblastoma cell proliferation. *Int. J. Cancer* 106: 342-347, 2003.

Eichenbaum, H. The hippocampus and declarative memory: cognitive mechanisms and neural codes. *Behav. Brain Res.* 127: 199-207, 2001.

Farrer, L.A.; Cupples, L.A.; Haines, J.L.; Hyman, B.; Kukull, W.A.; Mayeux, R.; Myers, R.H.; Pericakvance, M.A.; Risch, N.; vanDuijn, C.M. Effects of age, sex, and ethnicity on the association between apolipoprotein E genotype and Alzheimer disease - A meta-analysis. *JAMA* 278: 1349-1356, 1997.

Gomez-Isla, T.; Price, J.L.; Mckeel, D.W.; Morris, J.C.; Growdon, J.H.; Hyman, B.T. Profound loss of layer II entorhinal cortex neurons occurs in very mild Alzheimer's disease. *J. Neurosci.* 16: 4491-4500, 1996.

Hunter, M.; Bernard, R.; Freitas, E.; Boyer, A.; Morar, B.; Martins, I.J.; Tournev, I.; Jordanova, A.; Guergelcheva, V.; Ishpekova, B.; Kremensky, I.; Nicholson, G.; Schlotter, B.; Lochmuller, H.; Voit, T.; Colomer, J.; Thomas, P.K.; Levy, N.; Kalaydjieva, L. Mutation screening of the N-myc downstream-regulated gene 1 (NDRG1) in patients with Charcot-Marie-Tooth disease. *Hum. Mutat.* 22: 129-135, 2003.

Kalaydjieva, L.; Gresham, D.; Gooding, R.; Heather, L.; Baas, F.; de Jonge, R.; Blechschmidt, K.; Angelicheva, D.; Chandler, D.; Worsley, P.; Rosenthal, A.; King, R.H.M.; Thomas, P.K. N-myc downstream-regulated gene 1 is mutated in hereditary motor and sensory neuropathy-Lom. *Am. J. Hum. Genet.* 67: 47-58, 2000.

Kalaydjieva, L.; Nikolova, A.; Turnev, I.; Petrova, J.; Hristova, A.; Ishpekova, B.; Petkova, I.; Shmarov, A.; Stancheva, S.; Middleton, L.; Merlini, L.; Trogu, A.; Muddle, J.R.; King, R.H.M.; Thomas, P.K. Hereditary motor and sensory neuropathy - Lom, a novel demyelinating neuropathy associated with deafness in gypsies - Clinical, electrophysiological and nerve biopsy findings. *Brain* 121: 399-408, 1998.

Kim, K.T.; Ongusaha, P.P.; Hong, Y.K.; Kurdistani, S.K.; Nakamura, M.; Lu, K.P.; Lee, S.W. Function of Drg1/Rit42 in p53-dependent mitotic spindle checkpoint. *J. Biol. Chem.* 279: 38597-38602, 2004.

Kokame, K.; Kato, H.; Miyata, T. Homocysteine-respondent genes in vascular endothelial cells identified by differential display analysis - GRP78/BiP and novel genes. *J. Biol. Chem.* 271: 29659-29665, 1996.

Kurdistani, S.K.; Arizti, P.; Reimer, C.L.; Sugrue, M.M.; Aaronson, S.A.; Lee, S.W. Inhibition of tumor cell growth by RTP/rit42 and its responsiveness to p53 and DNA damage. *Cancer Res.* 58: 4439-4444, 1998.

Lin, T.M.; Chang, C.S. Cloning and characterization of TDD5, an androgen target gene that is differentially repressed by testosterone and dihydrotestosterone. *Proc. Natl. Acad. Sci. USA* 94: 4988-4993, 1997.

Mitchelmore, C.; Buchmann-Moller, S.; Rask, L.; West, M.J.; Troncoso, J.C.; Jensen, N.A. NDRG2: a novel Alzheimer's disease associated protein. *Neurobiol. Dis.* 16: 48-58, 2004.

Morris, J.C. Clinical presentation and course of Alzheimer's disease. Terry, R.D.; Katzman, R.; Bick, K.L.; Sisodia, S.S. (Eds.), Alzheimer Disease. pp. 11-24, Lippincott Williams and Wilkins, Philadelphia, 1999.

Murray, J.T.; Campbell, D.G.; Morrice, N.; Auld, G.C.; Shpiro, N.; Marquez, R.; Peggie, M.; Bain, J.; Bloomberg, G.B.; Grahammer, F.; Lang, F.; Wulff, P.; Kuhl, D.; Cohen, P. Exploitation of KESTREL to identify N-myc downstream-regulated gene family members as physiological substrates for SGK1 and GSK3. *Biochem. J.* 384: 477-488, 2004.

Nakada, N.; Hongo, S.; Ohki, T.; Maeda, A.; Takeda, M. Molecular characterization of NDRG4/Bdm1 protein isoforms that are differentially regulated during rat brain development. *Brain Res. Dev. Brain Res.* 135: 45-53, 2002.

Nichols, N.R. Ndrg2, a novel gene regulated by adrenal steroids and antidepressants, is highly expressed in astrocytes. *Ann. N. Y. Acad. Sci.* 1007: 349-356, 2003.

Nishimoto, S.; Tawara, J.; Toyoda, H.; Kitamura, K.; Komurasaki, T. A novel homocysteine-responsive gene, smap8, modulates mitogenesis in rat vascular smooth muscle cells. *Eur. J. Biochem.* 270: 2521-2531, 2003.

Ohki, T.; Hongo, S.; Nakada, N.; Maeda, A.; Takeda, M. Inhibition of neurite outgrowth by reduced level of NDRG4 protein in antisense transfected PC12 cells. *Brain Res. Dev. Brain Res.* 135: 55-63, 2002.

Okuda, T.; Higashi, Y.; Kokame, K.; Tanaka, C.; Kondoh, H.; Miyata, T. Ndrg1-deficient mice exhibit a progressive demyelinating disorder of peripheral nerves. *Mol. Cell Biol.* 24: 3949-3956, 2004.

Okuda, T.; Kondoh, H. Identification of new genes Ndr2 and Ndr3 which are related to Ndr1/RTP/Drg1 but show distinct tissue specificity and response to N-myc. *Biochem. Biophys. Res. Commun.* 266: 208-215, 1999.

Ollis, D.L.; Cheah, E.; Cygler, M.; Dijkstra, B.; Frolow, F.; Franken, S.M.; Harel, M.; Remington, S.J.; Silman, I.; Schrag, J.; Sussman, J.L.; Verschueren, K.H.G.; Goldman, A. The Alpha/Beta-Hydrolase Fold. *Protein Eng.* 5: 197-211, 1992.

Park, H.; Adams, M.A.; Lachat, P.; Bosman, F.; Pang, S.C.; Graham, C.H. Hypoxia induces the expression of a 43-kDa protein (PROXY-1) in normal and malignant cells. *Biochem. Biophys. Res. Commun.* 276: 321-328, 2000.

Piquemal, D.; Joulia, D.; Balaguer, P.; Basset, A.; Marti, J.; Commes, T. Differential expression of the RTP/Drg1/Ndr1 gene product in proliferating and growth arrested cells. *Biochim. Biophys. Acta* 1450: 364-373, 1999.

Price, D.L.; Sisodia, S.S. Mutant genes in familial Alzheimer's disease and transgenic models. *Ann. Rev. Neurosci.* 21: 479-505, 1998.

Qu, X.H.; Zhai, Y.; Wei, H.D.; Zhang, C.G.; Xing, G.C.; Yu, Y.T.; He, F.C. Characterization and expression of three novel differentiation-related genes belong to the human NDRG gene family. *Mol. Cell Biochem.* 229: 35-44, 2002.

Saunders, A.M.; Schmader, K.; Breitner, J.C.S.; Benson, M.D.; Brown, W.T.; Goldfarb, L.; Goldgaber, D.; Manwaring, M.G.; Szymanski, M.H.; Mccown, N.; Dole, K.C.; Schmechel, D.E.; Strittmatter, W.J.; Pericakvance, M.A.; Roses, A.D. Apolipoprotein-E-Epsilon-4 Allele Distributions in Late-Onset Alzheimers-Disease and in Other Amyloid-Forming Diseases. *Lancet* 342: 710-711, 1993.

Segawa, T.; Nau, M.E.; Xu, L.L.; Chilukuri, R.N.; Makarem, M.; Zhang, W.; Petrovics, G.; Sesterhenn, I.A.; Mcleod, D.G.; Moul, J.W.; Vahey, M.; Srivastava, S. Androgen-induced expression of endoplasmic reticulum (ER) stress response genes in prostate cancer cells. *Oncogene* 21: 8749-8758, 2002.

Shaw, E.; Mccue, L.A.; Lawrence, C.E.; Dordick, J.S. Identification of a novel class in the alpha/beta hydrolase fold superfamily: The N-myc differentiation-related proteins. *Proteins* 47: 163-168, 2002.

Shimono, A.; Okuda, T.; Kondoh, H. N-myc-dependent repression of Ndr1, a gene identified by direct subtraction of whole mouse embryo cDNAs between wild type and N-myc mutant. *Mech. Dev.* 83: 39-52, 1999.

Sisodia, S.S.; Thinakaran, G.; Borchelt, D.R.; Wong, P.C.; Lee, M.K.; Price, D.L. Function and dysfunction of the presenilins. Terry, R.D.; Katzman, R.; Bick, K.L.; Sisodia, S.S. (Eds.), Alzheimer Disease. pp. 327-337, Lippincott Williams and Wilkins, Philadelphia, 1999.

Stein, S.; Thomas, E.K.; Herzog, B.; Westfall, M.D.; Rocheleau, J.V.; Jackson, R.S.; Wang, M.; Liang, P. NDRG1 is necessary for p53-dependent apoptosis. *J. Biol. Chem.* 279: 48930-48940, 2004.

Terry, R.D.; Masliah, E.; Hansen, L.A. The neuropathology of Alzheimer disease and the structural basis of its cognitive alterations. Terry, R.D.; Katzman, R.; Bick, K.L.; Sisodia, S.S. (Eds.), Alzheimer Disease. pp. 187-206, Lippincott Williams and Wilkins, Philadelphia, 1999.

Ulrix, W.; Swinnen, J.V.; Heyns, W.; Verhoeven, G. The differentiation-related gene 1, Drg1, is markedly upregulated by androgens in LNCaP prostatic adenocarcinoma cells. *FEBS Lett.* 455: 23-26, 1999.

vanBelzen, N.; Dinjens, S.N.M.; Diesveld, M.P.G.; Groen, N.A.; vanderMade, A.C.J.; Nozawa, Y.; Vlietstra, R.; Trapman, J.; Bosman, F.T. A novel gene which is up-regulated during colon epithelial cell differentiation and down-regulated in colorectal neoplasms. *Lab. Invest.* 77: 85-92, 1997.

West, M.J.; Coleman, P.D.; Flood, D.G.; Troncoso, J.C. Differences in the Pattern of Hippocampal Neuronal Loss in Normal Aging and Alzheimers-Disease. *Lancet* 344: 769-772, 1994.

Xu, B.; Lin, L.; Rote, N.S. Identification of a stress-induced protein during human trophoblast differentiation by differential display analysis. *Biol. Reprod.* 61: 681-686, 1999.

Yamauchi, Y.; Hongo, S.; Ohashi, T.; Shioda, S.; Zhou, C.J.; Nakai, Y.; Nishinaka, N.; Takahashi, R.; Takeda, F.; Takeda, M. Molecular cloning and characterization of a novel developmentally regulated gene, Bdm1, showing predominant expression in postnatal rat brain. *Mol. Brain Res.* 68: 149-158, 1999.

Zhao, W.; Tang, R.; Huang, Y.; Wang, W.; Zhou, Z.X.; Gu, S.H.; Dai, J.L.; Ying, K.; Xie, Y.; Mao, Y.M. Cloning and expression pattern of the human NDRG3 gene. *Biochim. Biophys. Acta* 1519: 134-138, 2001.

Zhou, D.J.; Salnikow, K.; Costa, M. Cap43, a novel gene specifically induced by Ni2+ compounds. *Cancer Res.* 58: 2182-2189, 1998.

Zhou, R.H.; Kokame, K.; Tsukamoto, Y.; Yutani, C.; Kato, H.; Miyata, T. Characterization of the human NDRG gene family: A newly identified member, NDRG4, is specifically expressed in brain and heart. *Genomics* 73: 86-97, 2001.

In: Cognitive Science Compendium, Volume 1
Editor: Miao-Kun Sun
ISBN: 978-1-60456-597-3

Chapter 6

NEUROBEHAVIORAL CHARACTERISTICS OF PRESENILIN TRANSGENIC MICE

R. Lalonde[1,*] R. Jazi[2] and C. Strazielle[2]

[1] Université de Rouen, Faculté de Médecine et de Pharmacie, Bâtiment de Recherche, 22 bld Gambetta, INSERM U614 (IFRMP 23), Salle 1D18, 76183 Rouen Cedex, France

[2] Université Henri Poincaré, Nancy I, Laboratoire de Pathologie Moléculaire et Cellulaire en Nutrition, EMI-INSERM U724

Service de Microscopie Electronique, Faculté de Médecine, 54500 Vandoeuvre-les-Nancy, France

ABSTRACT

When inserted into mouse brain by transgene or knockin methods, the presenilin mutations responsible for familial Alzheimer's disease cause anomalies in calcium homeostasis, potentiate excitotoxicity, and alter the electrophysiologic properties of neurons despite the absence of Aβ plaques. Transgenic or knockin mice with presenilin mutations have been shown to differ from controls in several types of behavior, including basic neurologic function, exploratory activity, motor coordination, spatial learning, and object recognition. In particular, the acquisition of spatial learning in the Morris water maze was slowed down in some *PS1* and *PS2* mutants. Novel object recognition was impaired in the *PS1*/L235P transgenic model. In addition, *PS1* mutants have been shown to have disinhibitory tendencies in the elevated plus-maze. However, the mutants do not appear affected for exploratory activity in the open-field and in the Y-maze and have only mild and subtle deficits in motor coordination tests. These functional effects may be mediated by increased $A\beta_{42}$ concentrations or by the direct effect of mutated *PS1* or *PS2*.

* Correspondence: TEL +33 (0)2 35 148281; FAX +33 (0)2 3514 8237.
E-mail: robert.lalonde@univ-rouen.fr

Keywords: Alzheimer's disease; PS1; PS2; Spatial learning; Motor coordination; Motor activity.

1. Characteristics of Alzheimer's Disease

1.1. Neuropsychology

Probable Alzheimer's disease is diagnosed when a progressive difficulty in remembering new material (anterograde amnesia) appears in combination with deficits of at least one of the following functions: language (aphasia), object use or drawing (apraxia), object recognition (agnosia), or step-by-step planning and reasoning (executive functions) [McKhann et al., 1984]. Progressive anterograde amnesia is sometimes the sole sign for several years [Weintraub and Mesulam 1993]. The memory loss comprises an inability in retaining factual information (declarative memory) combined with a difficulty in word retrieval (fluency and naming) [Graham et al., 2004; Jacobs et al., 1995; Helmes et al., 2002]. Except for the conditioned reflex of the eye-blink response [Woodruff-Pak et al., 1996], sensorimotor learning tasks such as pursuit-rotor [Eslinger and Damasio 1986; Jacobs et al., 1999] are relatively spared, as well as verbal priming [Karlsson et al., 2002].

Psychiatric symptoms are very likely to appear at some point of disease progression, such as apathy, depression, agitation, disinhibition, and hallucinations [Chen et al., 1991; Chung and Cummings 2000; Daffner et al., 1992; Frisoni et al., 1999; Hart et al., 2003; Petry et al., 1988; Senanarong et al., 2004]. Neurologic symptoms usually appear in late stages, including disturbed postural control and gait [Huff et al., 1987; Pettersson et al., 2002], myoclonus [Chen et al., 1991; Förstl et al., 1992], epilepsy, frontal lobe-associated release signs, and parkinsonian rigidity and tremor [Chen et al., 1991; Förstl et al., 1992; Huff et al., 1987].

1.2. Neuropathology

Definite Alzheimer's disease is diagnosed with post-mortem evidence of extracellular neuritic plaques and intracellular neurofibrillary tangles [Perl et al 2000]. The neuropathology also involves granulovacuolar degeneration, marked by intracytoplasmic vesicles 1-5 μm in diameter [Su et al., 2002]. The neuritic plaques contain A-beta-protein (Aβ) and cellular debris, while the Aβ-positive [Perry et al., 1992] neurofibrillary tangles contain hyperphosphorylated tau protein [Perl et al 2000]. Cored plaques are distinguishable from diffuse plaques by their β-pleated structure and the later onset of their appearance [Thal et al., 2000]. While the diffuse plaques contained the long $A\beta_{42}$ form, the mature plaques possessed both $A\beta_{42}$ and $A\beta_{40}$ major forms [Iwatsubo et al., 1994]. The Aβ plaques are surrounded by proliferating microglia and astrocytes as well as activated inflammation-related proteins [Eikelenboom et al., 2002].

Presumably as a result of Aβ and tau accumulation, atrophy appears in widespread brain regions. In comparison to age-matched controls, the grey matter density of Alzheimer brain decreased in temporal, parietal, and prefrontal neocortex as well as hippocampus, amygdala, and insula [Frisoni et al., 2002]. The volume loss was not limited to grey matter, but extended as well to white matter in neocortex [Bartzokis et al., 2003] and hippocampus [Callen et al., 2001; Jack et al., 1998; Laakso et al., 1998]. Volume shrinkage was also observed in associated limbic structures, such as the parahippocampal cortex, the septum, the anterior thalamus, the mamillary bodies, the posterior cingulate, and the orbitofrontal cortex [Callen et al., 2001], as well as entorhinal and perirhinal cortices [Juottonen et al., 1998]. In a multi-stage study, the volume loss appeared early in medial temporal and parietal lobes, progressed to the infero-lateral temporal neocortex, and then spread to the prefrontal cortex [Scahill et al., 2002]. The widespread nature of brain atrophy is documented by findings of grey matter loss in the caudate nucleus [Rombouts et al., 2000] and the vermal region of the cerebellum [Sjöbeck et al., 2001], although these regions contained diffuse but few neuritic Aβ plaques [Moran et al., 1993]. Subcortical structures directly innervating the neocortex are also subject to atrophy, particularly the basal nucleus of Meynert, the dorsal raphe. the locus coeruleus, and to a lesser extent the substantia nigra pars compacta, sources of acetylcholine, 5-hydroxytryptamine (5HT), noradrenaline, and dopamine, respectively [Lyness et al., 2003].

Gouras et al., [2000] hypothesized that in addition to extracellar plaque deposits, the neurodegeneration is caused by intracellular $A\beta_{42}$ accumulation, as seen in multivesicular bodies at pre- and postsynaptic sites of nerve terminals [Takahashi et al., 2002] as well as the cell membrane [Yamaguchi et al., 2000]. Supporting this hypothesis is the finding that non-fibrillar Aβ induced apoptotic cell death in rat cortical neuronal cultures [Pillot et al., 1999]. Although Alzheimer brain is characterized by DNA fragmentation, other morphological markers of apoptosis, such as chromatin condensation, nuclear blebbing, and apoptotic bodies, are rare [Anderson et al., 2000; Broe et al., 2001; Colurso et al., 2003; Lassmann et al., 1995; Troncoso et al., 1996]. The term "aborted apoptosis" has been used to describe this condition, as in certain cells, upstream but not downstream caspases were activated [Raina et al., 2001], although one of the latter (caspase-3) was activated in cells with granulovacuolar degeneration [Su et al., 2002].

1.3. Genetics

Mutations of *APP*, *PS1,* and *PS2* genes have been identified as the cause of the familial form of Alzheimer's disease [Campion et al., 1996, 1999; De Jonghe et al., 1999; Tanzi and Bertram 2001]. Although familial Alzheimer's disease appears in younger individuals than the more common sporadic form, as early as the twenties for some *PS1* mutations [Campion et al., 1996; De Jonghe et al., 1999], and regional brain metabolism appears more severely affected in the former [Mosconi et al., 2003], the neuropathological [Mendez et al., 1992] and neuropsychological [Swearer et al., 1992] features are equivalent. *APP* encodes beta-amyloid precursor protein, the source of $A\beta_{42}$ and $A\beta_{40}$ after successive proteolytic cleavage at N- and C-terminals, catalyzed respectively by β- and γ-secretase. There are three major APP isoforms: APP_{695}, APP_{751}, and APP_{770}. The APP_{695} isoform is mostly found in neurons and the longer forms in glia [Rohan de Silva et al., 1997]. PS1 and PS2 proteins facilitate the amyloidogenic pathway by increasing γ-secretase activity. The identification of these mutations supports the amyloid hypothesis of Alzheimer's disease, stating that amyloid accumulation is the underlying cause of neurofibrillary tangles, cell degeneration, and cognitive deficits [Hardy and Selkoe 2002]. At least, some Alzheimer-related pathologies may also be caused by the direct action of mutated APP [Neve et al 2000] and its C-terminal fragments [Turner et al., 2003].

2. Amyloid Changes in Presenilin Transgenic Mice

2.1. Behaviorally Characterized PS1 Mutants

Presumably because of species differences, no transgenic mouse with a *PS1* or a *PS2* mutation develops Aβ plaques. Despite this limitation, such models offer the possibility of examining the role of mutated presenilins on neurobehavioral functions before plaque development. Those transgenic mice with *PS1* and *PS2* mutations that have so far been behaviorally characterized are presented in Table 1. In the first description of this type, mice with *PS1*/M146L and *PS1*/M146V transgenes driven by the *PDGFβ2* promoter had higher $A\beta_{42}$ but not $A\beta_{40}$ concentrations [Duff et al., 1996]. A second *PS1*/M146L model as well as a *PS1*/L286V one were generated by cDNA insertion into a hamster *PrP* expression vector, enhancing $A\beta_{42}$ concentrations in mice transgenic for human wild-type APP_{695} [Citron et al 1997].

Table 1. Description of behaviorally characterized transgenic or knockin mice expressing mutated *PS1* or *PS2*

Mutation	Type	Promoter	Reference
PS1/A246E	transgenic	human *THY1*	[Qian et al., 1998]
PS1/M146L	transgenic	*PDGFβ2*	[Duff et al., 1996]
PS1/M146V	transgenic	*PDGFβ2*	[Duff et al., 1996]
PS1/M146L	transgenic	hamster *PrP*	[Citron et al., 1997]
PS1/L286V	transgenic	hamster *PrP*	[Citron et al., 1997]
PS1/L235P	transgenic	not mentioned	[Huang et al., 2003]
PS1/M146L	knockin	endogenous control	[Guo et al., 1999,a,b]
PS1/I213T	knockin	endogenous control	[Nakano et al., 1999]
PS2/N141I	transgenic	*NSE*	[Hwang et al., 2002]

The same pattern of increased $A\beta_{42}$ but not $A\beta_{40}$ concentrations was revealed in *PS1*/L235P transgenic mice [Huang et al., 2003]. In addition, human *PS1*/A246E and *PS1*/wild-type transgenes driven by the human *THY1* promoter have been generated on a murine *PS1* knockout background [Qian et al., 1998]. The $A\beta_{42}$ concentrations of the *PS1*/A246E transgenic mice were higher than those of wild-type. An attractive aspect of this construct is the elimination of the potentially confounding influence of endogenous PS1. The same advantage can be gained with knockin (KI) models. The insertion of human *PS1* with the Japanese I213T mutation on the chromosomal site normally occupied by the endogenous gene augmented $A\beta_{42}$ concentrations relative to normal controls [Nakano et al., 1999]. A second KI model is available, this time with the M146L mutation [Guo et al., 1999a,b].

2.2. Behaviorally Characterized PS2 Mutants

Only a single *PS2* mouse model has been behaviorally characterized, consisting of the N141I Volga-German (VG) mutated transgene driven by the *NSE* promoter [Hwang et al., 2002]. As in mutated *PS1* strains, these *PS2*/VG transgenic mice had elevated $A\beta_{42}$ levels but without Aβ plaques. A second model features the same mutation driven by the endogenous *Prp* promoter interbred with *APP* mice with the Swedish mutation driven by the endogenous *thy1* promoter [Richards et al., 2003]. As seen in several *APP*+*PS1* bigenic mice, APP_{751}SWE+*PS2*/VG (TgPS2APP) mouse brains contained Aβ plaques surrounded by activated microglial cells and astrocytes.

3. Neurological Signs in Presenilin Transgenic Mice

Following up on general reviews of Alzheimer-type transgenic mice [German and Eisch 2004; Morgan et al., 2003; Sant'Angelo et al., 2003], we present the consequences of *PS1* and *PS2* transgenesis on multiple aspects of behavior, including specific neurological symptoms, exploratory activity, and motor coordination, as well as spatial and non-spatial learning. Clues as to the neuroanatomical pathways affected by Aβ accumulation can be obtained by observing neurological signs in multi-test batteries, such as the **S**mithKline/**H**arwell/**I**mperial College/**R**oyal **H**ospital/**P**henotype **A**ssessment (SHIRPA) [Rogers et al., 1997].

3.1. Myoclonus and Epilepsy

Myoclonus and epileptic seizures are two neurological symptoms liable to appear in Alzheimer's disease [Chen et al., 1991; Förstl et al., 1992], particularly in patients with *APP* and *PS1* mutations [Campion et al., 1996]. In general, myoclonic twitching appears at a later stage of disease progression than such neurological symptoms as extrapyramidal signs and psychosis [Chen et al., 1991]. One possible cause of myoclonus is damage to the cerebellum, as additional cerebellar signs [Huff et al., 1987] and a reduction in Purkinje cell number [Sjöbeck et al., 2001] have been reported in Alzheimer's disease. Unlike the sporadic form, Aβ plaques accumulated in the cerebellum of patients with hereditary Alzheimer's disease and myoclonus [Ishino et al., 1984; Lemere et al., 1996].

In mice expressing the human *PS1*/A246E transgene, a species-specific form of myoclonus was discovered, namely repetitive jumping against the walls of their home cage [Lalonde et al., 2003]. This abnormal pattern of activity was seen in 3 out of 14 (21%) transgenic mice but in none of the human *PS1*/wild-type controls. However, myoclonic jumping was not observed in *PS1*/I213T KI mice [Lalonde and Strazielle 2005]. Instead, myoclonic jumping was observed in APP_{751}SWE (TgAPP23) transgenic mice with the Swedish mutation causing Aβ plaques [Lalonde et al., 2002, 2005]. It may be hypothesized that $A\beta_{42}$ accumulation in cerebellar and/or 5HT-related pathways is responsible, as this sign was observed in mutant mice with cerebellar damage (such as *Grid2*^{Lc}, unpublished observations) and normal animals injected with 5HT precursors [Pappert et al., 1998].

3.2. Pathological Reflexes

The pathological reflexes seen in Alzheimer's disease include grasping and sucking in response to tactile stimuli, caused by damage to the frontal lobe [Förstl et al., 1992; Huff et al., 1987]. Species-specific pathological reflexes are observable in neurologically impaired mice during tail-lifting tests. When slowly lowered toward a horizontal surface, most mice display a visual or a tactile placing response by extending all four paws in anticipation of ground contact. In contrast, some neurologically impaired mice display flexion or even clasping of the two hindpaws together in mid-air. However, this sign did not distinguish *PS1*/A246E mice from *PS1* wild-type [Lalonde et al., 2003] or *PS1*/I213T KI mice from wild-type [Lalonde and Strazielle 2005].

4. Exploration and Anxiety in Presenilin Transgenic Mice

4.1. Open-Field and Y-Maze

The analysis of exploratory behavior is relevant with regard to the changes in mood and affect displayed by Alzheimer patients [Chung and Cummings 2000]. One of their most common neuropsychiatric symptom is apathy, often accompanied by dysphoria, social withdrawal, and depression [Chung and Cummings 2000; Daffner et al., 1992; Frisoni et al., 1999; Hart et al., 2003; Petry et al., 1988; Senanarong et al., 2004]. Conversely, agitation, restlessness, disinhibition, and inappropriate euphoria may be predominant [Chung and Cummings 2000; Frisoni et al., 1999; Hart et al., 2003; Senanarong et al., 2004]. These opposite signs are often found in the same patient at different times, as indicated by the significant correlations existing between agitation on one hand and apathy, anxiety, disinhibition, and euphoria, on the other [Senanarong et al., 2004].

The most commonly used methods for assessing motor activity in mice are open-field [Walsh and Cummins 1976] and Y-maze [King and Arendash 2002] tests. In non-automated versions of these tests, a trained observer can measure segments crossed, maze arms entered, and rears. Photocell activity monitors dissociate ambulatory from stereoptyped movements and their greater spatial resolution permit estimation of smaller-amplitude movements than non-automated versions. The effects of presenilin transgenesis on horizontal motor activity are summarized in Table 2. To our knowledge, the results for single *PS1* mutants are entirely negative.

For example, the activity of *PS1*/M146L/*PDGFβ2* transgenic mice was unchanged relative to non-transgenic controls in the Y-maze [Holcomb et al., 1998, 1999] and in the open-field [Sadowski et al., 2004]. Moreover, no effect in the open-field was observable in *PS1*/M146L/*Prp* [Janus et al., 2000], *PS1*/L286V [Vaucher et al., 2002], and *PS1*/A246E [Dineley et al., 2002] transgenic mice as well as *PS1*/I213T KI mice [Lalonde and Strazielle 2005] relative to controls. The same result was obtained when the *PS1*/A246E transgenic model was compared to human wild-type *PS1* [Lalonde et al., 2003].

Table 2. Activity levels of *PS1* transgenic (Tg) or knockin (KI) mutants in the open-field or Y-maze relative to normal controls or to human *PS1* wild-type for the *PS1*/A246E Tg

Mutation	Activity levels	Reference
PS1/M146L/*PDGFβ2* Tg	normal (open-field and Y-maze)	[Holcomb et al., 1998, 1999; Sadowski et al., 2004]
PS1/M146L/*Prp* Tg	normal (open-field)	[Janus et al., 2000]
PS1/L286V Tg	normal (open-field)	[Vaucher et al., 2002]
PS1/A246E Tg	normal (open-field)	[Dineley et al., 2002; Lalonde et al., 2003]
PS1/I213T KI	normal (open-field)	[Lalonde and Strazielle 2005]

4.2. Spontaneous Alternation

Spontaneous alternation occurs whenever mice choose different arms on successive trials in a two-choice maze (either T- or Y-shaped), indicating a willingness to explore novel stimuli [Dember and Fowler 1958]. The spontaneous alternation test is sensitive to lesions of widespread cortico-subcortical circuits, notably the limbic system [Lalonde 2002]. A failure to alternate may mimick the diminished curiosity seen in Alzheimer patients [Chung and Cummings 2000; Daffner et al., 1992; Frisoni et al., 1999].

APP_{695}SWE and APP_{695}SWE+*PS1*/M146L mice, both harboring Aβ plaques, were impaired in the Y-maze spontaneous alternation test [Arendash et al., 2004; Holcomb et al., 1998, 1999]. APP_{751}SWE+*PS2*/VG bigenic mice were also impaired for Y-maze alternation [Richards et al., 2003]. Paradoxically, the deficit was found before (4 months of age) but not after (8, 12, and 16 months) Aβ plaque formation. This result underlines the importance of the age factor in this test, as old mice lack curiosity and are thereby less susceptible to transgene-mediated effects [Lalonde 2002].

In contrast to these Aβ plaque-containing models, *PS1*/M146L/*PDGFβ2* mice were not impaired for Y-maze spontaneous alternation [Holcomb et al., 1998, 1999]. Likewise, no deficit was found in *PS1*/A246E [Lalonde et al., 2003] and *PS1*/I213T KI [Lalonde and Strazielle 2005] mice in the T-maze. Thus, single *PS1* models do not appear to be vulnerable to open-field and spontaneous alternation tests that estimate brain activation and the motivation to explore novel stimuli.

4.3. Elevated Plus-Maze

The elevated plus-maze measures general exploratory tendencies and anxiety in mice [Cole and Rodgers 1993]. The cross-shaped maze contains two arms enclosed by walls and two arms open on all sides. General exploration is estimated by entries into the enclosed arms and anxiety levels by entries into the open arms, as well as open/total entry and duration ratios. Normal mice prefer to explore enclosed as opposed to open areas, indicating fear of open spaces. This agoraphobic response can be reduced after anxiolytic injections of benzodiazepines [Cole and Rodgers 1993]. These measures are relevant to Alzheimer symptoms, marked by elevated anxiety or else disinhibitory tendencies such as inappropriate hilarity and indifference [Chung and Cummings 2000; Daffner et al., 1992; Frisoni et al., 1999; Hart et al., 2003; Petry et al., 1988; Senanarong et al., 2004].

A loss of behavioral inhibition was revealed in *PS1*/A246E transgenic mice relative to human wild-type *PS1* transgenic mice, on the basis of higher open arms entries and duration as well as open/total entry and duration ratios in the former [Lalonde et al., 2003]. However, *PS1*/I213T KI mice did not differ from wild-type in open-arm exploration, although the former had fewer enclosed arm entries [Lalonde and Strazielle 2005]. Likewise, *PS1*/L235P [Huang et al., 2003] and APP_{695}SWE+*PS1*/M146L [Arendash et al., 2001a] mice did not differ from non-transgenic controls. On the contrary, anxiety levels were increased in APP_{695}SWEch+*PS1*/A246E mice, as indicated by lower open arm durations [Puoliväli et al., 2002].

The elevated plus-maze shows some promise in unraveling PS1-mediated effects, though the results depend on the mouse model. Some single *PS1* mutants appear to have low anxiety levels, whereas one bigenic mutant harboring Aβ plaques had the opposite pattern.

4.4. Emergence from a Small Compartment

The emergence test is a second commonly used method for evaluating anxiety in mice [Holmes 2001]. The animals are placed inside a small compartment contained within a larger one and the time taken before emerging from it is measured. This test has been shown to be sensitive to null mutations for corticotropin releasing factor [Smith et al., 1998], 5HT [Tecott et al., 1998], and dopamine [Dulawa et al., 1999] receptors in mice. However, *PS1*/I213T KI mice did not emerge sooner than controls [Lalonde and Strazielle 2005].

4.5. General Comments

The effects of transgene insertions of mutated *PS1* on motor activity and anxiety have only begun to be investigated. These experiments offer the potential of delineating the impact of amyloid pathology irrespective of the presence of Aβ plaques. The elevated plus-maze may be the most sensitive test relative to *PS1* mutations. But it remains to be determined to what extent the results are generalizable to other mutants and to other anxiety tests.

5. Motor Coordination in Presenilin Transgenic Mice

5.1. Rationale and Methods of Analysis

Posture and gait disturbances are prevalent even in early stage Alzheimer's disease [Pettersson et al., 2002]. This impairment may be due at least in part to vermal Purkinje cell loss [Sjöbeck et al., 2001]. The principal measure used for motor coordination tests in mice is the latency before falling from balance beams and wires [Lalonde and Strazielle 1999]. The distance travelled on a stationary beam is an alternative measure used for identifying those mice with postural deficits. If the stationary beam is open on one side, latencies before escaping can be used as yet another measure [Sadowski et al., 2004]. In addition, mice suspended on the horizontal wire of a coat-hanger can be evaluated for movement time before reaching either extremity and before climbing on a diagonal bar [Lalonde and Strazielle 1999]. In the rotorod test, the mice are trained to locomote forward on a beam rotating under a constant-speed or accelerated condition [Lalonde and Strazielle 1999]. Contrary to the stationary beam, the mice are forced to maintain their balance by adjusting their movements on a moving surface.

5.2. Stationary Beam

*APP_{695}*SWE+*PS1*/M146L mice were reported to fall sooner from the stationary beam than non-transgenic controls [Arendash et al., 2001a,b], indicating a vulnerability to postural deficits in a similar manner to Alzheimer patients [Pettersson et al., 2002]. However, the same *APP_{695}*SWE+*PS1*/M146L mice did not differ from controls in terms of latencies before escaping [Sadowski et al., 2004]. For single *PS1* mutants, the results vary depending on the specific model used. *PS1*/A246E mice fell more often than human wild-type *PS1* mice from two types of stationary beam [Lalonde et al., 2003]. But *PS1*/I213T KI mice did not fall sooner or more often than their controls [Lalonde and Strazielle 2005]. Moreover, *PS1*/M146L/*PDGFβ2* mice did not differ from non-transgenic controls in terms of latencies before escaping from the stationary beam [Sadowski et al., 2004].

5.3. Suspended Wire and Coat-Hanger

By comparison to non-transgenic controls, *APP_{695}*SWE+*PS1*/M146L mice fell more quickly from a horizontal wire [Arendash et al., 2001a]. However, by contrast to their impaired performance on the stationary beam, *PS1*/A246E mice did not fall more often and had normal movement times relative to human wild-type *PS1* mice in the coat-hanger test [Lalonde et al., 2003]. Likewise, *PS1*/I213T KI mice performed normally in this test relative to non-transgenic controls [Lalonde and Strazielle 2005].

5.4. Rotorod

PS1/A246E mice were impaired relative to human wild-type *PS1* on the constant-speed rotorod, but this occurred on only a few trials [Lalonde et al., 2003]. The performance of the same *PS1*/A246E mice was unimpaired relative to non-transgenic controls on the accelerating rotorod [Dineley et al., 2002]. Likewise, *PS1*/I213T KI mice performed normally on the accelerating rotorod [Lalonde and Strazielle 2005]. Although *PS1*/L286V transgenic mice performed worse than human wild-type *PS1* on the accelerated rotorod, they were no worse than non-transgenic controls [Vaucher et al., 2002].

5.5. General Comments

Although ignored by most investigators of Alzheimer's disease, motor coordination tests furnish the opportunity of delineating the neurobehavioral consequence of amyloid pathology in a wide sense and is all the more relevant in view of the gait disturbances observed in Alzheimer patients [Pettersson et al., 2002]. Moreover, there is a potential of improving motor deficits with experimental treatments in a parallel manner to cognitive defects. The data thus far indicate that the deficits of motor coordination found in some mutated *PS1* models are subtle.

6. Spatial Learning in Presenilin Transgenic Mice

6.1. Morris Water Maze

One of the earliest features of Alzheimer's dementia is a difficulty in acquiring new information [Jacobs 1995]. Their anterograde amnesia features attempts at remembering line drawings [Graham et al., 2004]. In addition to memory problems, constructional apraxia is manifested, as seen by difficulties in copying drawings and reproducing geometric figures with building blocks [Graham et al., 2004; Helmes and Ostbye 2002; Henderson et al., 1989; Nielson et al., 1996]. Moroever, Alzheimer patients got lost more often than controls in a maze [Rainville et al., 2002a], a hospital ward [Monacelli et al., 2003], and a motored vehicle [Uc et al., 2004]. The underlying reason behind these spatial deficits may be an impairment in flow perception [Tetewsky and Duffy 1999], visual search [Parasuraman et al., 2000; Tales et al., 2004], abstract reasoning [Helmes and Ostbye 2002], and executive functions [Collette et al., 1999; Graham et al., 2004; Rainville et al., 2002b].

One of the most often employed test measuring spatial function is the Morris water maze, in which mice are trained to escape from a pool by reaching a platform [Morris et al., 1982]. In the hidden platform condition, the mice must reach the goal on the basis of spatial landmarks (place learning). In the visible platform subtask, the mice may swim directly to the goal on the basis of local cues (visuomotor guidance). Since the location of the hidden platform is invariant from one trial to the next, this test reflects reference memory. To maximize mouse performances, special care must be taken to avoid water-induced hypothermia by using a suitably long intertrial interval. For water temperatures ranging between 20° and 24°C, body temperatures returned to normal values after their swim with a 13 min but not a 30 s intertrial interval [Iivonen et al., 2003]. After the acquisition phase, a probe trial is often conducted, when the submerged platform is removed from the pool, causing the mice to float or to swim in circles in the vicinity of where the platform used to be. Spatial learning in the Morris water maze has been shown to be sensitive to lesions in rat cortex, including extrastriate [Espinoza et al., 1999], parietal [DiMattia and Kesner 1988; Hoh et al., 2003], prefrontal [Hoh et al., 2003], posterior cingulate [Vann and Aggleton 2002], and entorhinal/perirhinal [Nagahara et al., 1995] regions, as well as large [Brun et al., 2002; Morris et al., 1982] and CA1-restricted [Stubley-Weatherly et al., 1996] parts of the hippocampus.

Under certain circumstances, transgenic mice with presenilin mutations are impaired during acquisition of place learning in the Morris water maze (Table 3). Indeed, place learning was deficient in *PS1*/M146L/*PDGFβ2* mice relative to non-transgenic controls at 8 though not at 22 months of age [Sadowski et al., 2004]. Likewise, the acquisition rate of *PS1*/I213T KI mice was slowed down relative to wild-type in the same task [Lalonde and Strazielle 2005]. The same results were found in *PS2*/VG transgenic mice relative to non-transgenic controls [Hwang et al., 2002]. Pooled data that included two *PS1*/M146L/*Prp* lines and the *PS1*/L286V line indicate normal acquisition but impaired probe trial performance [Janus et al., 2000]. These spatial defects may be mediated by increased $A\beta_{42}$ concentrations, as intracerebroventricular injections of soluble $A\beta_{25\text{-}35}$ in rats impaired the acquisition of spatial learning in the Morris water maze [Sun and Alkon 2002]. Moroever, intracerebroventricular injections of soluble Aβ oligomers in rats impaired 2-lever spatial alternation learning [Cleary et al., 2005]. However, negative results regarding *PS1* mutations have also been reported. For example, *PS1*/M146L/*PDGFβ2* mice had a normal learning curve during acquisition of place learning [Holcomb et al., 1999]. Likewise, *PS1*/L235P [Huang et al., 2003] and *PS1*/A246E [Lalonde et al., 2003] mice had normal learning rates in the same task.

Table 3. Performance of *PS1* transgenic (Tg) or knockin (KI) mutants and a *PS2* Tg mutant relative to normal controls or to human *PS1* wild-type for the *PS1*/A246E Tg in hidden and visible platform versions as well as the probe trial in the Morris water maze

Name	Hidden	Probe	Visible	Reference
PS1/M146L/*PDGFβ2* Tg	normal	normal	normal	[Holcomb et al., 1999]
PS1/M146L/*PDGFβ2* Tg	↓	NT	NT	[Sadowski et al., 2004]
PS1/M146L/*Prp* 2 lines+ *PS1*/L286V/*Prp* Tg pooled	normal	↓	normal	[Janus et al., 2000]
PS1/A246E Tg	normal	normal	normal	[Lalonde et al., 2003]
PS1/L235P Tg	normal	normal	NT	[Huang et al., 2003]
PS1/I213T KI	↓	normal	normal	[Lalonde and Strazielle 2005]
PS2/N141I Tg	↓	NT	NT	[Hwang et al., 2002]

↓=deficit, normal=not impaired vs controls; NT=not tested

One of the explanations for the variable results is the aging factor, as spatial learning in the Morris task deteriorates as a function of normal aging in mice [Lamberty and Gower 1992], possibly causing a floor effect. This may explain the age-related effect in the Sadowski et al., . [2004] study with *PS1*/M146L/*PDGFβ2* mice. Moroever, some normal strains show poor learning, for example because of poor visual acuity, thereby mitigating against the possibility of a transgene effect. A third possibility is that some *PS1* mutations are more likely than others to cause spatial impairments.

6.2. Radial Arm Maze

In the standard version of the radial arm maze task, hungry mice are trained to obtain food pellets at the end of each maze arm radiating from a central region without returning to the same one. The standard task has been adapted to a water maze configuration. In the 6-arm radial water maze task, the location of the escape platform remained constant during 4-trial daily sessions but was switched to a different maze arm from one day to the next [Arendash et al., 2001a; Gordon et al., 2001; Trinchese et al., 2004]. Thus, this version demands a win-stay strategy within days and a win-shift strategy across days. The acquisition of the radial arm maze task was slowed down in APP_{695}SWE+*PS1*/M146L mice relative to non-transgenic controls at 3 months of age [Trinchese et al., 2004]. Although the acquisition deficit was not reproduced at 5 or 15 months of age, APP_{695}SWE+*PS1*/M146L mice had poorer retention tested on trial 5 after a 30-min delay at 15 but not at 5 months [Arendash et al., 2001a; Gordon et al., 2001].

6.3. Barnes Circular Maze

The Barnes circular maze consists of a multiple hole array at the peripheral part of a board. As a consequence of photophobia, nocturnal animals are motivated to escape from the bright light illuminating the maze by entering into a dark hole. This task is dependent on spatial working memory, as an error is accumulated whenever the mice re-enter the same hole. Young APP_{695}SWE+*PS1*/M146L mice (7-month-old) acquired the original task normally, but were impaired in the reversal phase of the circular maze task [Pompl et al., 1999]. The acquisition of the task was also normal in older (15 month-old) APP_{695}SWE+*PS1*/M146L mice [Arendash et al., 2001a] as well as the single APP_{695}SWE mutant up to 19 months of age [Arendash et al., 2004; King and Arendash, 2002]. To our knowledge, single *PS1* mutants have not been evaluated in this task.

6.4. Spatial Discrimination in a T-Maze

By contrast to the circular maze task requiring a win-shift strategy, spatial discrimination learning requires a win-stay strategy, as the correct turning response (right or left) in this maze remains invariant throughout training. In the only experiment of this type, APP_{695}SWEch+*PS1*/A246E mice were not impaired in spatial discrimination learning [Liu et al., 2002]. Indeed, a paradoxical result was obtained, in that the bigenic mice learned faster than the non-transgenic controls. This result may be explained by the growth-promoting action of APP, overcoming the potentially deleterious influence of Aβ plaques. To our knowledge, single *PS1* mutants have not been evaluated in this task.

6.5. General Comments

Most evaluations of spatial learning have been conducted with Aβ plaque-containing *APP+PS1* bigenic models. Nevertheless, the findings available to date show that non-plaque containing single *PS1* mutants may be vulnerable to spatial learning deficits in the Morris water maze. There is interest in expanding this analysis by evaluating the mutants in other maze tasks, including radial arm, circular, and T- mazes.

7. Non-Spatial Learning in Presenilin Transgenic Mice

7.1. Object Recognition

Non-spatial memory may be evaluated by a form recognition test, in which the length of time mice explore a novel as opposed to a previously presented object is measured. In one version [Huang et al., 2003; Vaucher et al., 2002], the mice first explore two objects and then one of them is replaced by a new one. In the second version [Dewachter et al., 2002; Howlett et al., 2004], the mice first explore a single object and then this object is presented with a new one. In either version, retention is evaluated by the novel/total time preference ratio, dependent on the recognition of visual and tactile features of familiar objects.

The sensitity of the novel object recognition task to Alzheimer-like pathology was demonstrated in Aβ plaque-containing APP_{695}LD transgenic mice with the London mutation [Dewachter et al., 2002] and APP_{695}SWE+*PS1*/M146L/*thy1* (TgTASTPM) transgenic mice [Howlett et al., 2004]. Despite the absence of Aβ plaques, novel object recognition was also impaired in *PS1*/L235P transgenic mice [Huang et al., 2003]. However, object recognition was not impaired in a different model (*PS1*/L286V) relative either to human wild-type *PS1* or to non-transgenic mice at two retention intervals [Vaucher et al., 2002].

7.2. Shock Avoidance and Fear Conditioning

In contextual fear conditioning, the mice are given footshocks in a compartment and then removed from it by the experimenter. The mice are then placed back in the same compartment and the time spent in an immobile position (freezing) is measured. In cued fear conditioning, the mice are given footshocks during presentation of an auditory tone (or other cue). Afterwards, the mice are placed in the same compartment or in a different one and the immobility time during presentation of the auditory tone is measured [Dineley et al., 2002]. In active avoidance learning, the mice are warned by a conditioned stimulus (light or tone) to escape from a compartment before the onset of footshocks and to enter an adjacent safe one. The number of successful avoidances is measured.

By comparison to non-transgenic controls, APP_{695}SWE+*PS1*/A246E mice were deficient during contextual fear conditioning [Dineley et al., 2002]. In contrast, there was no intergroup difference in freezing to the cued stimulus. However, neither contextual nor cued fear conditioning tasks was altered in *PS1*/A246E transgenic mice relative to non-transgenic controls. The same result was found with *PS1*/M146L KI mice relative to wild-type, although the former group had poorer contextual learning when one wild-type allele was removed by interbreeding with a *PS1* knockout [Wang et al., 2004].

To our knowledge, active avoidance learning has only been evaluated in a double mutant. The acquisition of active avoidance responses was poorer in APP_{751}SWE+*PS2*/VG mice than non-transgenic controls [Richards et al., 2003]. But the single *PS2*/VG mutant was not evaluated.

7.3. General Comments

Object recognition and avoidance tasks offer the possibility of examining the effects of presenilin transgenesis on non-spatial learning. Although few experiments of this type have been conducted, preliminary results indicate that *PS1* mutants may be deficient at least in recognition memory. The effects of this manipulation on fear conditioning should be explored more extensively.

8. Neurochemical Characteristics of Presenilin Transgenic Mice

8.1. Calcium Homeostasis and Neurodegeneration

The effects of mutated *PS1* on calcium homeostasis and cell death have been examined *in vitro* and to a lesser extent *in vivo*. Intra-hippocampal kainate injections caused more severe neuronal degeneration in *PS1*/M146L KI mice than normal controls [Guo et al., 1999a]. In hippocampal cell cultures, the same mutants were more vulnerable to glutamate-induced toxicity. One of the possible reasons for these effects is the alteration of calcium homeostasis in response to excitotoxicity. Despite equivalent basal Ca^{2+} concentrations, glutamate-induced Ca^{2+} increases inside neurons were greater in *PS1*/M146L KI mice than those of controls, as well as cellular peroxides, mitochondrial reactive oxygen species (ROS), and lipid peroxidation. Moroever, dietary restriction decreased lipid peroxidation and protected these neurons from kainate-induced excitotoxicity [Zhu et al., 1999]. Under basal conditions of hippocampal cell cultures, growth and survival were normal in *PS1*/M146L KI mice [Guo et al., 1999b]. But when exposed to Aβ, *PS1*/M146L KI neurons were more susceptible to degeneration, presumably because of the increased ROS and lipid peroxidation detected in these mice, as the amyloid-induced neurotoxicity was counteracted by an antioxidant. After glucose deprivation, neuronal apoptosis was more pronounced in *PS1*/M146L KI hippocampal cells than that of controls [Xie et al., 2001]. The same model was evaluated for neural responses in prefrontal cortex slices to inositol triphosphate (IP_3), a secondary messenger binding to receptors on the endoplasmic reticulum membrane to release sequestered Ca^{2+} into the cytosol [Stutzmann et al., 2004]. The IP_3-evoked but not the spike-evoked transient rise in intracellular Ca^{2+} levels was enhanced in *PS1*/M146L KI mice. Anomalies in calcium homeostasis are not limited to neurons, as fibroblasts isolated from *PS1*/M146L KI neonatal tails had a higher transient rise in intracellular Ca^{2+} levels than controls after receptor activation of the phosphoinositide-calcium pathway induced by bradykinin [Leissring et al., 2000]. Changes in calcium homeostasis have been observed as well in transgenic models. Intracellular Ca^{2+} levels in lymphocytes obtained from spleens of transgenic mice with quintuple *PS1* mutations were higher than those of non-transgenic controls [Eckert et al., 2001]. The lymphocytes of these *PS1* mutants also had elevated ROS levels and were more liable to degeneration *in vitro*. In brain, the activity of two antioxidant enzymes (glutathione reductase and superoxide dismutase) was diminished in the same mutants [Leutner et al., 2000].

Despite normal basal levels, KCl- and Aβ-induced Ca^{2+} level increases in synaptosomes from *PS1*/M146L/*PDGFβ2* and *PS1*/M146V transgenic mice were elevated in comparison to non-transgenic mice [Begley et al., 1999]. Moreover, cells from hippocampal slices of *PS1*/M146L/*PDGFβ2* transgenic mice responded with higher Ca^{2+} levels to K^+-induced depolarization than those of non-transgenic controls [Barrow et al., 2000].

The effects of *PS1* mutations on the unfolded protein response (UPR), as defined by reduced expression or phosphorylation of chaperone proteins facilitating protein folding, have been studied in response to endoplasmic reticulum stress. *PS1* mutations caused disturbances in the UPR of fibroblasts cultured from *PS1*/I213T KI mouse embryos [Yasuda et al., 2002] and in SK-N-SH neuroblastoma cells transfected with complementary DNA containing *PS1*/A246E or *PS1*/ΔE9 mutations [Katayama et al., 2001]. Under other circumstances, cells with *PS1* mutations were not more prone to neurodegeneration or UPR deficiencies. For example, primary neocortical neurons from *PS1*/P264L KI mice were not more vulnerable than controls to apoptosis, as assessed by Hoechst 3342 staining of chromatin condensation, after exposure to three pro-apoptotic agents including $A\beta_{42}$ [Siman et al., 2000]. Likewise, the UPR of primary neocortical neurons and of embryonic fibroblasts derived from *PS1*/P264L KI mice did not differ from that of controls in response to endoplasmic reticulum stressors [Siman et al., 2001].

In any event, the calcium-related anomalies may be responsible for the altered electrophysiologic properties of mutated *PS1* neurons, such as long-term potentiation (LTP), dependent on intracellular calcium entry and associated with spatial learning in the Morris water maze [Morris 1989]. In particular, hippocampal LTP was enhanced in *PS1*/ΔE9 [Zaman et al., 2000], *PS1*/A246E [Parent et al., 1999], and *PS1*/M146L/*PDGFβ2* [Barrow et al., 2000] transgenic mice as well as *PS1*/M146V transgenic rats [Pybus et al., 2003]. Conversely, hippocampal LTP was diminished in mice with the $PS1^{-/+}$ null mutation [Morton et al., 2002].

Although there is little evidence of widespread atrophy in *PS1* mutants, few experimenters have attempted to determine cell numbers in specific brain regions. *PS1*/H163R and *PS1*/L286V transgene insertions driven by the *PDGFβ2* promoter contained more neurons with detectable $A\beta_{42}$ but not $A\beta_{40}$ than non-transgenic controls [Chui et al., 1999]. Despite the lack of extracellular Aβ deposits, these two *PS1* mutants had lower neuronal counts in frontal cortex and hippocampus and higher counts of GFAP-positive astrocytes. Other anomalies which may have behavioral significance include deficient adult neurogenesis detectable in the dentate gyrus of the hippocampal formation in *PS1*/P117L transgenic mice [Wen et al., 2004].

8.2. Cholinergic System

The loss of the basal forebrain cholinergic system is one of the most significant aspects of neurodegeneration in Alzheimer brain and plays a central role in cognitive deterioration [Whitehouse et al., 1982]. The activities of the specific acetylcholine-producing enzyme, choline acetyltransferase (ChAT) [Coyle et al., 1983; DeKosky et al., 1992; Hardy et al., 1985] and of the non-specific catabolizing enzyme, acetylcholinesterase (AChE) [Coyle et al., 1983; Atack et al., 1983; Blusztajn and Berse 2000; Kuhl et al., 1999; Sberna et al., 1998; Shinotoh et al., 2000], have often been found to be decreased in neocortex and hippocampus of Alzheimer patients. In line with these clinical findings, *APP*$_{695}$SWE+*PS1*/M146L mice had reduced vesicular acetylcholine transporter (VAChT) density in frontal and parietal cortex [Wong et al., 1999]. But no such effect was observable in *PS1*/M146L mice. No cholinergic deficiency was found in the *APP*$_{695}$SWEch+*PS1*/A246E line, as estimated by ChAT activity and AChE-positive fiber density in hippocampus and parietal cortex [Liu et al., 2002]. ChAT activity was also unchanged in frontoparietal cortex, hippocampus, and dorsomedial striatum of *PS1*/L286V [Vaucher et al., 2002] and in hippocampus of *PS1*/L235P [Huang et al., 2003] mice. Thus, cholinergic deficiency appears less obvious in plaque-containing *APP+PS1* models than in human dementia and has not yet been demonstrated in any single *PS1* mutant.

8.3. Synapse Density

The loss in synapse number is a critical neuromorphologic feature of Alzheimer's disease. Powerful correlations have been reported between synapse number loss and cognitive deficits in Alzheimer patients [Samuel et al., 1994: Terry et al., 1991]. However, a synapse-specific protein (synaptophysin) does not seem vulnerable even in plaque-containing transgenic models. Indeed, synaptophysin immunoreactivity was not affected in the CA1 hippocampal subregion of APP$_{695}$SWE+PS1/M146L mice [Hu et al., 2003] and was even increased in neocortex and dentate gyrus of APP$_{695}$SWE mice, attributable to the growth-promoting action of APP [King and Arendash 2002]. Thus, mutated APP appears to counteract the possible debilitating effects of plaque accumulation on synapse number and structure. To our knowledge, single PS1 mutants have not been evaluated for synaptophysin immunoreactivity. In view of the important association between synapse number and cognition in patients, it would seem to be of interest to determine how early stage Aβ accumulation caused by PS1 mutations affects synapses.

8.4. General Comments

Consistent data have been reported as to the deficient calcium homeostasis seen in mice genetically modified for *PS1* mutations, causing increased sensitivity to excitotoxic-mediated cell death. However, unlike Alzheimer brain and despite their electrophysiologic anomalies, there is little evidence of neuronal loss in widespread brain regions of *PS1* mutants. Moreover, there is no evidence at all that *PS1* mutations cause either cholinergic deficiency or the loss of synapses, neurochemical and neuropathologic signs associated in an important way with Alzheimer's disease. Further research must be conducted in order to determine the neurobehavioral consequences of increased $A\beta_{42}$ concentrations secondary to mutated *PS1*. Such research is likely to have therapeutic value in eventual pharmacological trials of prodromal Alzheimer's disease, particularly in patients with presenilin mutations.

REFERENCES

Anderson, A.J.; Stoltzner, S.; Lai, F.; Su, J.; Nixon, R.A. Morphological and biochemical assessment of DNA damage and apoptosis in Down syndrome and Alzheimer's disease, and effect of postmortem tissue archival on TUNEL. *Neurobiol. Aging* 21: 511-524, 2000.

Arendash, G.W.; King, D.L.; Gordon, M.N.; Morgan, D.; Hatcher, J.M.; Hope, C.E.; Diamond, D.M. Progressive, age-related behavioral impairments in transgenic mice carrying both mutant amyloid precursor protein and presenilin-1 transgenes. *Brain Res.* 891: 42-53, 2001a.

Arendash, G.W.; Gordon, M.N.; Diamond, D.M.; Austin, L.A.; Hatcher, J.M.; DiCarlo, G.; Wilcock, D.; Morgan, D. Behavioral assessment of Alzheimer's transgenic mice following long-term Aβ vaccination: task specificity and correlations between Aβ and spatial memory. *DNA Cell Biol.* 20: 737-744, 2001b.

Arendash, G.W.; Lewis, J.; Leighty, R.E.; McGowan, E.; Cracchiolo, J.R.; Hutton, M.; Garcia, M.F. Multi-metric behavioral comparison of APPsw and P301L models for Alzheimer's disease: linkage of poorer cognitive performance to tau pathology in forebrain. *Brain Res.* 1012: 29-41, 2004.

Atack, J.R.; Perry, E.K.; Bonham, J.R.; Perry, R.H.; Tomlinson, B.E.; Blessed, G.; Fairbairn, A. Molecular forms of acetylcholinesterase in senile dementia of Alzheimer type: selective loss of the intermediate (10S) form. *Neurosci. Lett.* 40: 199-204, 1983.

Barrow, P.A; Empson, R.M.; Gladwell, S.J.; Anderson, C.M.; Killick, R.; Yu, X.; Jefferys, J.G.R.; Duff, K. Functional phenotype in transgenic mice expressing human mutant *presenilin-1*. *Neurobiol. Dis.* 7: 119-126, 2000.

Bartzokis, G.; Cummings, J.L.; Sultzer, D.; Henderson, V.W.; Nuechterlein, K.H.; Mintz, J. White matter structural integrity in healthy aging adults and patients with Alzheimer disease: a magnetic resonance imaging study. *Arch. Neurol.* 60: 393-398, 2003.

Begley, J.G.; Duan, W.; Chan, S.; Duff, K.; Mattson, M.P. Altered calcium homeostasis and mitochondrial dysfunction in cortical synaptic compartments of presenilin-1 mutant mice. *J. Neurochem.* 72: 1030-1039, 1999.

Blusztajn, J.K.; Berse, B. The cholinergic neuronal phenotype in Alzheimer's disease. *Met. Brain Dis.* 15: 45-64, 2000.

Broe, M.; Shephard, C.E.; Milward, E.A.; Halliday, G.M. Relationship between DNA fragmentation; morphological changes and neuronal loss in Alzheimer's disease and dementia with Lewy bodies. *Acta Neuropathol.* 101: 616-624, 2001.

Brun, V.H.; Otnaess, M.K.; Molden, S.; Steffenach, H.-A.; Witter, M.P.; Moser, M.-B.; Moser, E.I. Place cells and place recognition maintained by direct entorhinal-hippocampal circuitry. *Science* 296: 2243-2246, 2002.

Callen, D.J.A.; Black, S.E.; Gao, F.; Caldwell, C.B.; Szalai, J.P. Beyond the hippocampus: MRI volumetry confirms widespread limbic atrophy in AD. *Neurology* 57: 1669-1674, 2001.

Campion, D.; Brice, A.; Dumanchin, C.; Puel, M.; Baulac, M.; De la Sayette, V.; Hannequin, D.; Duyckaerts, C.; Michon, A.; Martin, C.; Moreau, V.; Penet, C.; Martinez, M.; Clerget-Darpoux, F.; Agid, Y.; Frébourg, T. A novel presenilin *1* mutation resulting in familial Alzheimer's disease with an onset age of 29 years. *NeuroReport* 7: 1582-1584, 1996.

Campion, D.; Dumanchin, C.; Hannequin, D.; Dubois, B.; Bellliard, S.; Puel, M.; Thomas-Anerion, C.; Michon, A.; Martin, C.; Charbonnier, F.; Raux, G.; Camuzat, A.; Penet, C.; Mesnage, V.; Martinez, M.; Clerget-Darpoux, F.; Frebourg, T. Early-onset autosomal dominant Alzheimer disease: prevalence, genetic heterogeneity, and mutation spectrum. *Am. J. Hum. Genet.* 65: 664-670, 1999.

Chen, J.-Y.; Stern, Y.; Sano, M.; Mayeux, R. Cumulative risks of developing extrapyramidal signs, psychosis, or myoclonus in the course of Alzheimer's disease. *Arch. Neurol.* 48: 1141-1143, 1991.

Chui, D.-H.; Tanahashi, H.; Ozawa, K.; Ikeda, S.; Checler, F.; Ueda, O.; Suzuki, H.; Araki, W.; Inoue, H.; Shirotani, K.; Takahashi, K.; Gallyas, F.; Tabira, T. Transgenic mice with Alzheimer presenilin 1 mutations show accelerated neurodegeneration without amyloid plaque formation. *Nature Med.* 5: 560-564, 1999.

Chung, J.A.; Cummings, J.L. Neurobehavioral and neuropsychiatric symptoms in Alzheimer's disease. *Neurol. Clin.* 18: 829-846, 2000.

Citron, M.; Westway, D.; Xia, W.; Carlson, G.; Diehl, T.; Levesque, G.; Johnson-Wood, K.; Lee, M.; Seubert, P.; Davis, A.; Kholodenko, D.; Motter, R.; Sherrington, R.; Perry, B.; Yao, H.; Strome, R.; Lieberburg, I.; Rommens, J.; Kim, S.; Schenk, D.; Fraser, P.; St George Hyslop, P.; Selkoe, D.J. Mutant presenilins of Alzheimer's disease increase production of 42-residue amyloid β-protein in both transfected cells and transgenic mice. *Nature Med.* 3: 67-72, 1997.

Cleary, J.P.; Walsh, D.M.; Hofmeister, J.J.; Shankar, G.M.; Kuskowski, M.A.; Selkoe, D.J.; Ashe, K.H. Natural oligomers of the amyloid-β protein specifically disrupt cognitive function. *Nature Neurosci.* 8: 79-84, 2005.

Cole, J.C.; Rodgers, R.J. An ethological analysis of the effects of chlordiazepoxide and bretazenil (Ro 16-6028) in the murine elevated plus-maze. *Behav. Pharmacol.* 4: 573-580, 1993.

Collette, F.; Van der Linden, M.; Salmon E. Executive dysfunction in Alzheimer's disease. *Cortex* 35: 57-72, 1999.

Colurso, G.J.; Nilson, J.E.; Vervoort, L.G. Quantitative assessment of DNA fragmentation and beta-amyloid deposition in insular cortex and midfrontal gyrus from patients with Alzheimer's disease. *Life Sci.* 73: 1795-1803, 2003.

Coyle, J.T.; Price, D.L.; DeLong, M.R. Alzheimer's disease: a disorder of cortical cholinergic innervation. *Science* 219: 1184-1190, 1983.

Daffner, K.R.; Scinto, L.F.M.; Weintraub, S.; Guinessey, J.E.; Mesulam, M.M. Diminished curiosity in patients with probable Alzheimer's disease as measured by exploratory eye movements. *Neurology* 42: 320-328, 1992.

De Jonghe, C.; Van Broeckhoven, C. ? Crusio W.; Gerlai R. (Eds) Handbook of Molecular-genetic Techniques for Brain and Behavior Research, Techniques in the Behavioral and Neural Sciences, Vol. 13, pp. 841-862, Elsevier, Amsterdam, 1999.

DeKosky, S.T.; Harbaugh, R.E.; Schmitt, F.A.; Bakay, R.A.E.; Chui, H.C.; Knopman, D.S.; Reeder, T.M.; Shetter, A.G.; Senter, H.J.; Markesbery, W.R. Cortical biopsy in Alzheimer's disease: diagnostic accuracy and neurochemical, neuropathological, and cognitive correlations. *Ann. Neurol.* 32: 625-632, 1992.

Dember, W.N.; Fowler, H. Spontaneous alternation behavior. *Psychol. Bull.* 55: 412-428. 1958.

Dewachter, I.; Reversé, D.; Caluwaerts, N.; Ris, L.; Kuipéri, C.; Van den Haute, C.; Spittaels, K.; Umans, L.; Serneels, L.; Thiry, E.; Moechars, D.; Mercken, M.; Godaux, E.; Van Leuven, F. Neuronal deficiency of presenilin 1 inhibits amyloid plaque formation and corrects hippocampal long-term potentiation but not a cognitive defect of amyloid precursor protein [V717I] transgenic mice. *J. Neurosci.* 22: 3445-3453, 2002.

DiMattia, B.D.; Kesner, R.P. Spatial cognitive maps: differential role of parietal cortex and hippocampal formation. *Behav. Neurosci.* 102: 471-480, 1988.

Dineley, K.T.; Xia, X.; Bui, D.; Sweatt, J.D.; Zheng, H. Accelerated plaque accumulation, associative learning deficits, and up-regulation of α7 nicotinic receptor protein in transgenic mice co-expressing mutant human presenilin 1 and amyloid precursor proteins. *J. Biol. Chem.* 277: 22768-22780, 2002.

Duff, K.; Eckman, C.; Zehr, C.; Yu, X.; Prada, C.-M.; Perez-tur, J.; Hutton, J.; Buee, L.; Harigaya, Y.; Yager, D.; Morgan, D.; Gordon, M.N.; Holcomb, L.; Refolo, L.; Zenk, B.; Hardy, J.; Younkin, S. Increased β-amyloid42(43) in brains of mice expressing mutant presenilin 1. *Nature* 383: 710-713, 1996.

Dulawa, S.C.; Grandy, D.K.; Low, M.J.; Paulus, M.P.; Geyer, M.A. Dopamine D4 receptor-knockout mice exhibit reduced exploration of novel stimuli. *J. Neurosci.* 19: 9550-9556, 1999.

Eckert, A.; Schindowski, K.; Leutner, S.; Luckhaus, C.; Touchet, N.; Czech, C.; Müller, W.E. Alzheimer's disease-like alterations in peripheral cells from presenilin-1 transgenic mice. *Neurobiol. Dis.* 8: 331-342, 2001.

Eikelenboom, P.; Bate, C.; Van Gool, W.A.; Hoozemans, J.J.M.; Rozemuller J.M.; Veerhuis R.; Williams A. Neuroinflammation in Alzheimer's disease and prion disease. *Glia* 40: 232-239, 2002.

Eslinger, P.J.; Damasio, A.R. Preserved motor learning in Alzheimer's disease: implications for anatomy and behavior. *J. Neurosci.* 6: 3006-3009, 1986.

Espinoza, S.; Pinto-Hamuy, T.; Passig, C.; Carreno, F.; Marchant, F.; Urzua, C. Deficit in the water-maze after lesions in the anteromedial extrastriate cortex in rats. *Physiol. Behav.* 66: 493-496, 1999.

Förstl, H.; Burns, A.; Levy, R.; Cairns, N.; Luthert, P.; Lantos, P. Neurologic signs in Alzheimer's disease: results of a prospective clinical and neuropathologic study. *Arch. Neurol.* 49: 1038-1042, 1992.

Frisoni, G.B.; Testa, C.; Zorzan, A.; Sabattoli, F.; Beltramello, A.; Laakso, M.P. Detection of grey matter loss in mild Alzheimer's disease with voxel based morphometry. *J. Neurol. Neurosurg. Psychiat.* 73: 657-664, 2002.

Frisoni, G.B.; Rozzini, L.; Gozetti, A.; Binetti, G.; Zanetti, O.; Bianchetti, A.; Trabucchi, M.; Cummings, J.L. Behavioral syndromes in Alzheimer's disease: description and correlates. *Dement. Geriatr. Cogn. Disord.* 10: 130-138, 1999.

German, D.C.; Eisch, A.J. Mouse models of Alzheimer's disease: insight into treatment. *Rev. Neurosci.* 15: 353-369, 2004.

Gordon, M.N.; King, D.L.; Diamond, D.M.; Jantzen, P.T.; Boyett, K.V.; Hope, C.E.; Hatcher, J.M.; DiCarlo, G.; Gottscall, W.P.E.; Morgan, D.; Arendash, G.W. Correlation between cognitive deficits and Aβ deposits in transgenic APP+PS1 mice. *Neurobiol. Aging* 22: 377-386, 2001.

Gouras, G.K; Tsai, J.; Naslund, J.; Vincent, B.; Edgar, M.; Checler, F.; Greenfield, J.P.; Haroutunian, V.; Buxbaum, J.D.; Xu, H.; Greengard, P.; Relkin, N.R. Intraneuronal Aβ42 accumulation in human brain. *Am. J. Pathol.* 156: 15-20, 2000.

Graham, N.L.; Emery, T.; Hodges, J.R. Distinctive cognitive profiles in Alzheimer's disease and subcortical vascular dementia. *J. Neurol. Neurosurg. Psychiatry* 75: 61-71; 2004.

Guo, Q.; Fu, W.; Sopher, B.L.; Miller, M.W.; Ware, C.B.; Martin, G.M.; Mattson, M.P. Increased vulnerability of hippocampal neurons to excitotoxic necrosis in presenilin-1 mutant knock-in mice. *Nature Med.* 5: 101-106, 1999a.

Guo, Q.; Sebastian, L.; Sopher, B.L.; Miller, M.W.; Ware, C.B.; Martin, G.M.; Mattson, M.P. Increased vulnerability of hippocampal neurons from presenilin-1 mutant knock-in mice to amyloid β-peptide toxicity: central roles of superoxide production and caspase activation. *J. Neurochem.* 72: 1019-1029, 1999b.

Hardy, J.; Selkoe, D.J. The amyloid hypothesis of Alzheimer's disease: progress and problems on the road to therapeutics. *Science* 297: 353-356, 2002.

Hardy, J.; Adolfsson, R.; Alafuzoff, I.; Bucht, G.; Marcusson, J.; Nyberg, P.; Perdahl, E.; Wester, P.; Winblad, B. Transmitter deficits in Alzheimer's disease. *Neurochem. Int.* 7: 545-563, 1985.

Hart, D.J.; Craig, D.; Compton, S.A.; Critchlow, S.; Kerrigan, B.M.; McIlroy, S.P.; Passmore, A.P. A retrospective study of the behavioural and psychological symptoms of mid and late phase Alzheimer's disease. *Int. J. Geriatr. Psychiatry* 18: 1037-1042, 2003.

Helmes, E.; Ostbye, T. Beyond memory impairment: cognitive changes in Alzheimer's disease. *Arch. Clin. Neuropsychol.* 17: 179-193, 2002.

Henderson, V.W.; Mack, W.; Williams, B.W. Spatial disorientation in Alzheimer's disease. *Arch. Neurol.* 46: 391-394, 1989.

Hoh, T.E.; Kolb, B.; Eppel, A.; Vanderwolf, C.H.; Cain, D.P. Role of the neocortex in the water maze task in the rat: a detailed behavioral and Golgi-Cox analysis. *Behav. Brain Res.* 138: 81-94, 2003.

Holcomb, L.A.; Gordon, M.N.; Jantzen, P.; Hsiao, K.; Duff, K.; Morgan, D. Behavioral changes in transgenic mice expressing both amyloid precursor protein and presenilin-1 mutations: lack of association with amyloid deposits. *Behav. Genet.* 29: 177-185, 1999.

Holcomb, L.; Gordon, M.N.; McGowan, E.; Yu, X.; Benkowic, S.; Jantzen, P.; Wright, K.; Saad, I.; Mueller, R.; Morgan, D.; Sanders, S.; Zehr, C.; O'Campo, K.; Hardy, J.; Prada, C.-M.; Eckman, C.; Younkin, S.; Hsiao, K.; Duff, K. Accelerated Alzheimer-type phenotype in transgenic mice carrying both mutant amyloid precursor protein and presenilin transgenes. *Nature Med.* 4: 97-100, 1998.

Holmes, A. Targeted gene mutation approaches to the study of anxiety-like behavior in mice. *Neurosci. Biobehav. Rev.* 25: 261-273, 2001.

Howlett, D.R.; Richardson, J.C.; Austin, A.; Parsons, A.A.; Bate, S.T.; Davies, D.C.; Gonzalez, M.I. Cognitive correlates of Aβ deposition in male and female mice bearing amyloid precursor protein and presenilin-1 mutant transgenes. *Brain Res.* 1017: 130-136, 2004.

Hu, L.; Wong, T.P.; Cote, S.L.; Bell, K.F.; Cuello, A.C. The impact of Aβ-plaques on cortical cholinergic and non-cholinergic presynaptic boutons in Alzheimer's disease-like transgenic mice. *Neuroscience* 121: 421-432, 2003.

Huang, X.G.; Yee, B.K.; Nag, S.; Chan, S.T.H.; Tang, F. Behavioral and neurochemical characterization of transgenic mice carrying the human presenilin-1 gene with or without the leucine-to-proline mutation at codon 235. *Exp. Neurol.* 183: 673-681, 2003.

Huff, F.J.; Boller, F.; Lucchelli, F.; Querriera, R.; Beyer, J.; Belle, S. The neurologic examination in patients with probable Alzheimer's disease. *Arch. Neurol.* 44: 929-932, 1987.

Hwang, D.Y.; Chae, K.R.; Kang, T.S.; Hwang, J.H.; Lim, C.H.; Kang, H.K.; Goo, J.S.; Lee, M.R.; Lim, H.J.; Min, S.H.; Cho, J.Y.; Hong, J.T.; Song, C.W.; Paik, S.G.; Cho, J.S.; Kim, Y.K. Alterations in behavior, amyloid β-42, caspase-3, and Cox-2 in mutant *PS2* transgenic mouse model of Alzheimer's disease. *FASEB J.* 16: 805-813, 2002.

Iivonen, H.; Nurminen, L.; Harri, M.; Tanila, H.; Puoliväli, J. Hypothermia in mice tested in Morris water maze. *Behav. Brain Res.* 141: 207-213, 2003.

Ishino, H.; Higashi, S.; Chuta, M.; Ohta, H. Juvenile Alzheimer's disease with myoclonus: amyloid plaques and grumose alteration in the cerebellum. *Clin. Neuropathol.* 3: 193-198, 1984.

Iwatsubo, T.; Odaka, A.; Suzuki, N.; Mizusawa, H.; Nukina, N.; Ihara, Y. Visualization of Aβ42(43) and Aβ40 in senile plaques and end-specific Aβ monoclonals: evidence that an initially deposited species is Aβ42(43). *Neuron* 13: 45-53, 1994.

Jack, C.R.; Petersen, R.C.; Xu, Y.; O'Brien, P.C.; Smith, G.E.; Ivnik, R.J.; Tangalos, EG; Kokmen, E. Rate of medial temporal lobe atrophy in typical aging and Alzheimer's disease. *Neurology* 51: 993-999, 1998.

Jacobs, D.H.; Adair, J.C.; Williamson, D.J.; Na, D.L.; Gold, M.; Foundas, A.L.; Shuren, J.E.; Cibula, J.E.; Heilman, K.M. Apraxia and motor-skill acquisition in Alzheimer's disease are dissociable. *Neuropsychologia* 37: 875-880, 1999.

Jacobs, D.M.; Sano, M.; Dooneief, G.; Marder, K.; Bell, K.L.; Stern, Y. Neuropsychological detection and characterization of preclinical Alzheimer's disease. *Neurology* 45: 957-962, 1995.

Janus, C.; D'Amelio, S.; Amitay, O.; Chishti, M.A.; Strome, R.; Fraser, P.; Carlson, G.A.; Roder, J.C.; St. George-Hyslop, P.; Westaway, D. Spatial learning in transgenic mice expressing presenilin 1 (*PS1*) transgenes. *Neurobiol. Aging* 21: 541-549, 2000.

Juottonen, K.; Laakso, M.P.; Inausti, R.; Lehtovirta, M.; Pitänen, A.; Partanen, K.; Soininen, H. Volumes of the entorhinal and perirhinal cortices in Alzheimer's disease. *Neurobiol. Aging* 19: 15-22, 1998.

Karlsson, T.; Börjesson, A.; Adolfsson, R.; Nilsson, L.-G. Successive memory test performance and priming in Alzheimer's disease: evidence from the word-fragment completion task. *Cortex* 38: 341-355, 2002.

Katayama, T.; Imaizumi, K.; Honda, A.; Yoneda, T.; Kudo, T.; Takeda, M.; Mori, K.; Rozmahel, R.; Fraser, P.; St. George-Hyslop, P.; Tohyama, M. Disturbed activation of endoplasmic reticulum stress transducers by familial Alzheimer's disease-linked presenilin-1 mutations. *J. Biol. Chem.* 276: 43446-43454, 2001.

King, D.L.; Arendash, G.W. Behavioral characterization of the Tg2576 transgenic model of Alzheimer's disease through 19 months. *Physiol. Behav.* 75: 627-642, 2002.

Kuhl, D.E.; Koeppe, R.A.; Minoshima, S.; Snyder, S.E.; Ficaro, E.P.; Foster, N.L.; Frey, K.A.; Kilbourn, M.R.; Frey, K.A.; Kilbourn, M.R. *In vivo* mapping of cerebral acetylcholinesterase activity in aging and Alzheimer's disease. *Neurology* 52: 691-699, 1999.

Laakso, M.P.; Soininen, H.; Partanen, K.; Lehtovirta, M.; Hallikainen, M.; Hänninen, T.; Helkala, E.-L.; Riekkinen, P.J. Jr. MRI of the hippocampus in Alzheimer's disease: sensitivity, specificity, and analysis of the incorrectly classified subjects. *Neurobiol. Aging* 19: 23-31, 1998.

Lalonde, R. The neurobiological basis of spontaneous alternation. *Neurosci. Biobehav. Rev.* 26: 91-104, 2002.

Lalonde, R.; Strazielle, C. Motor performance of spontaneous murine mutations with cerebellar atrophy. Crusio W.; Gerlai R. (Eds). Handbook of Molecular-genetic Techniques for Brain and Behavior Research, Techniques in the Behavioral and Neural Sciences, Vol. 13, pp. 627-37, Elsevier, Amsterdam, 1999.

Lalonde, R.; Strazielle, C. *PS1* Knockin mice with the Japanese I213T mutation: effects on exploratory activity, motor coordination, and spatial learning. *Behav. Brain Res.* in press, 2005.

Lalonde, R.; Qian, S.; Strazielle, C. Transgenic mice expressing the *PS1*-A246E mutation: effects on spatial learning, exploration, anxiety, and motor coordination. *Behav. Brain Res.* 138: 71-79, 2003.

Lalonde, R.; Dumont, M.; Staufenbiel, M.; Strazielle, C. Neurobehavioral characterization of APP23 transgenic mice with the SHIRPA primary screen. *Behav. Brain Res.* 157: 91-98, 2005.

Lalonde, R.; Dumont, M.; Staufenbiel, M.; Sturchler-Pierrat, C.; Strazielle, C. Spatial learning, exploration, anxiety, and motor coordination in female APP23 transgenic mice with the Swedish mutation. *Brain Res.* 956: 36-44, 2002.

Lamberty, Y.; Gower, A.J. Age-related changes in spontaneous behavior and learning in NMRI mice from middle to old age. *Physiol. Behav.* 51: 81-88, 1992.

Lassmann, H.; Bancher, C.; Breitschopf, H.; Wegiel, J.; Bobinki, M.; Jellinger, K. Cell death in Alzheimer's disease evaluated by DNA fragmentation *in situ*. *Acta Neuropathol.* 89: 35-41, 1995.

Leissring, M.A.; Akbari, Y.; Fanger, C.M.; Cahalan, M.D.; Mattson, M.P.; LaFerla; F.M. Capacitative calcium entry deficits and elevated luminal calcium content in mutant presenilin-1 knockin mice. *J. Cell Biol.* 149: 793-798, 2000.

Lemere, C.A.; Lopera, F.; Kosik, K.S.; Lendon, C.L.; Ossa, J.; Saido, T.C.; Yamaguchi, H.; Ruiz, A.; Martinez, A.; Madrigal, L.; Hincapie, L.; Arango, J.C.L.; Anthony, D.C.; Koo, E.H.; Goate, A.M.; Selkoe, D.J.; Arango; J.C.V. The E280A presenilin 1 Alzheimer mutation produced increased Aβ42 deposition and severe cerebellar pathology. *Nature Med.* 2: 1146-1150, 1996.

Leutner, S.; Czech, C.; Schindowski, K.; Touchet, N.; Eckert, A.; Müller, W.E. Reduced antioxidant enzyme activity in brains of mice transgenic for human presenilin-1 with single or multiple mutations. *Neurosci. Lett.* 292: 87-90, 2000.

Liu, L.; Ikonen, S.; Heikkinen, T.; Tapiola, T.; van Groen, T.; Tanila, H. The effects of long-term treatment with metrifonate, a cholinesterase inhibitor, on cholinergic activity, amyloid pathology, and cognitive function in APP and PS1 doubly transgenic mice. *Exp. Neurol.* 173: 196-204, 2002.

Lyness; S.A., Zarow, C.; Chui, H.C. Neuron loss in key cholinergic and aminergic nuclei in Alzheimer's disease: a meta-analysis. *Neurobiol. Aging* 24: 1-23, 2003.

McKhann, G.; Drachman, D.; Folstein, M.; Katzman, R.; Price, D.; Stadlan, E.M. Clinical diagnosis of Alzheimer's disease: report of the NINCDS-ADRA work group under the Department of Health and Human Services task force on Alzheimer's disease. *Neurology* 34: 939-944, 1984.

Mendez, M.F.; Underwood, K.L.; Zander, B.A.; Mastri, A.R.; Sung, J.H.; Frey, W.H. II. Risk factors in Alzheimer's disease: a clinicopathologic study. *Neurology* 42: 770-775, 1992.

Monacelli, A.M.; Cushman, L.A.; Kavcic, V.; Duffy, C.J. Spatial disorientation in Alzheimer's disease: the remembrance of things passed. *Neurology* 61: 1491-1497, 2003.

Moran, M.A.; Mufson, E.J.; Gomez-Ramos, P. Colocalization of cholinesterases with β amyloid protein in aged and Alzheimer's brain. *Acta Neuropathol.* 85: 362-369, 1993.

Morgan, D. Learning and memory deficits in APP transgenic mouse models of amyloid deposition. *Neurochem. Res.* 28: 1029-1034, 2003.

Morris, R.G.M. Synaptic plasticity and learning: selective impairment of learning rats and blockade of long-term potentiation in vivo by the N-methyl-D-aspartate receptor antagonist AP5. *J. Neurosci.* 9: 3040-3057, 1989.

Morris, R.G.M.; Garrud, P.; Rawlins, J.N.P.; O'Keefe, J. Place navigation impaired in rats with hippocampal lesions. *Nature* 292: 681-683, 1982.

Morton, R.A.; Kuenzi, F.M.; Fitzjohn, S.M.; Rosahl, T.W.; Smith, D.; Zheng, H.; Shearman, M.; Collingridge, G.L.; Seabrook, G.R. Impairment in hippocampal long-term potentiation in mice underexpressing the Alzheimer's disease related gene presenilin-1. *Neurosci. Lett.* 319: 37-40, 2002.

Mosconi, L.; Sorbi, S; Nacmias, B.; De Cristofaro, M.T.; Fayyaz, M.; Cellini, E.; Bagnoli, S.; Bracco, L.; Herholz, K.; Pupi, A. Brain metabolic differences between sporadic and familial Alzheimer's disease. *Neurology* 61: 1138-1140, 2003.

Nagahara, A.H.; Otto, T.; Gallagher, M. Entorhinal-perirhinal lesions impair performance of rats on two versions of place learning in the Morris water maze. *Behav. Neurosci.* 109: 3-9, 1995.

Nakano, Y.; Kondoh, G.; Kudo, T.; Imaizumi, K.; Kato, M.; Miyazaki, J.; Tohyama, M.; Takeda, J.; Takeda, M. Accumulation of murine amyloidβ42 in a gene-dosage-dependent manner in *PS1* 'knock-in' mice. *Eur. J. Neurosci.* 11: 2577-2581, 1999.

Neve, R.L.; McPhie, D.L.; Chen, Y. Alzheimer's disease: a dysfunction of the amyloid precursor protein. *Brain Res.* 886: 54-66, 2000.

Nielson, K.A.; Cummings, B.J.; Cotman, C.W. Constructional apraxia in Alzheimer's disease correlates with neuritic neuropathology in occipital cortex. *Brain Res.* 741: 284-293, 1996.

Pappert, E.J.; Goetz, C.J.; Stebbins, M.; Belden, M.; Carvey, P.M. 5-Hydroxytryptamine-induced myoclonus in guinea pigs: mediation through 5-$HT_{1/2}$ receptor subtypes. *Eur. J. Pharmacol.* 347: 51-56, 1998.

Parasuraman, R.; Greenwood, P.M.; Alexander, G.E.; Parasuraman, R.; Greenwood, P.M.; Alexander, G.E. Alzheimer disease constricts the dynamic range of spatial attention in visual search. *Neuropsychologia* 38: 1126-1135, 2000.

Parent, A.; Linden, D.J.; Sisodia, S.S.; Borchelt, D.R. Synaptic transmission and hippocampal long-term potentiation in transgenic mice expressing FAD-linked presenilin 1. *Neurobiol. Dis.* 6: 56-62, 1999.

Perl D.P. Neuropathology of Alzheimer's disease and related disorders. *Neurol. Clin.* 18: 847-864, 2000.

Perry, G.; Cras, P.; Siedlak, S.L.; Tabaton, M.; Kawai, M. Beta protein immunoreactivity is found in the majority of neurofibrillary tangles of Alzheimer's disease. *Am. J. Pathol.* 140: 283-290, 1992.

Petry, S.; Cummings, J.L.; Hill, M.A.; Shapira, J. Personality alterations in dementia of the Alzheimer type. *Arch. Neurol.* 45: 1187-1190, 1988.

Pettersson, A.F.; Engardt, M.; Wahlund, L.-O. Activity and balance in subjects with mild Alzheimer's disease. *Dement. Geriatr. Cogn. Dis.* 13: 213-216, 2002.

Pillot, T.; Drouet, B.; Queillé, S.; Labeur, C.; Vandekerckhove, J.; Rosseneu, M.; Pinçon-Raymond, M.; Chambaz, J. The nonfibrillar β-amyloid peptide induces apopoptic neuronal cell death: involvement of its C-terminal fusogenic domain. *J. Neurochem.* 73: 1626-1634, 1999.

Pompl, P.N.; Mullan, M.J.; Bjugstad, K.; Arendash, G.W. Adaptation of the circular platform spatial memory task for mice: use in detecting cognitive impairment in the APP_{SW} transgenic mouse model for Alzheimer's disease. *J. Neurosci. Meth.* 87: 87-95, 1999.

Puoliväli, J.; Wang, J.; Heikkinen, T.; Heikkilä, M.; Tapiola, T.; van Groen, T.; Tanila, H. Hippocampal Aβ42 levels correlate with spatial memory deficit in APP and PS1 double transgenic mice. *Neurobiol. Dis.* 9: 339-347, 2002.

Pybus, R.; Barnard, E.; Estibeiro, P.; Mullins, J.; MacLeod, N. Enhanced long-term potentiation in the hippocampus of rats expressing mutant presenilin-1 is age related. *Neurobiol. Dis.* 12: 212-224, 2003.

Qian, S.; Jiang, P.; Guan, X.-M.; Singh, G.; Trumbauer, M.E.; Yu, H.; Chen, H.Y.; Van der Ploeg, L.H.T.; Zheng, H. Mutant human presenilin-1 protects presenilin 1 null mouse against embryonic lethality and elevates Aβ1-42/43 production. *Neuron* 20: 611-617, 1998.

Raina, A.K.; Hochman, A.; Zhu, X.; Rottkamp, A.; Nunomura, A.; Siedlak, S.L.; Boux, H.; Castellani, R.; Perry, G.; Smith, M.A. Abortive apoptosis in Alzheimer's disease. *Acta Neuropathol.* 101: 305-310, 2001.

Rainville, C.; Marchand, N.; Passini, R. Performances of patients with a dementia of the Alzheimer type in the Standardized Road-Map test of Direction Sense. *Neuropsychologia* 40: 567-573, 2002a.

Rainville, C.; Amieva, H.; Lafont, S.; Dartigues, J.F.; Orgogozo, J.M.; Fabrigoule, C. Executive function deficits in patients with dementia of the Alzheimer's type: a study with a Tower of London task. *Arch. Clin. Neuropsychol.* 17: 513-530, 2002b.

Richards, J.G.; Higgins, G.A.; Ouagazzal, A.-M.; Ozmen, L.; Kew, J.N.C.; Bohrmann, B.; Malherbe, P.; Brockhaus, M.; Loetscher, H.; Czech, C.; Huber, G.; Bluethmann, H.; Jacobsen, H.; Kemp, J.A. PS2APP transgenic mice, coexpressing hPS2mut and hAPPswe, show age-related cognitive deficits associated with discrete brain amyloid deposition and inflammation. *J. Neurosci.* 23: 8989-9003, 2003.

Rogers, D.C.; Fisher, E.M.C.; Brown, S.D.M.; Peters, J.; Hunter, A.J.; Martin, J.E. Behavioral and functional analysis of mouse phenotype: SHIRPA, a proposed protocol for comprehensive phenotype assessment. *Mamm. Genome* 8: 711-713, 1997.

Rohan de Silva, H.A.; Jen, A.; Wickenden, C.; Jen, L.-S.; Wilkinson, S.L.; Patel, A.J. Cell-specific expression of small β-amyloid precursor protein isoform mRNAs and proteins in neurons and astrocytes. *Mol. Brain Res.* 47: 147-156, 1997.

Rombouts, S.A.R.B.; Barkhof, F.; Witter, M.P.; Scheltens, P. Unbiased whole-brain analysis of gray matter loss in Alzheimer's disease. *Neurosci. Lett.* 285: 231-233, 2000.

Sadowski, M.; Pankiewicz, J.; Scholtzova, H.; Ji, Y.; Quartermain, D.; Jensen, C.H.; Duff, K.; Nixon, R.A.; Gruen, R.J.; Wisniewski, T. Amyloid-β deposition is associated with decreased hippocampal glucose metabolism and spatial memory impairment in APP/PS1 mice. *J. Neuropathol. Exp. Neurol.* 63: 418-428, 2004.

Samuel, W.; Terry, R.D.; DeTeresa, R.; Butters, N.; Masliah, E. Clinical correlates of cortical and nucleus basalis pathology in Alzheimer dementia. *Arch. Neurol.* 51: 772-778, 1994.

Sant'Angelo, A.; Trinchese, F.; Arancio, O. Usefulness of behavioral and electrophysiological studies in transgenic models of Alzheimer disease. *Neurochem. Res.* 28: 1009-1015, 2003.

Sberna, G.; Saez-Valero, J.; Li, Q.-X.; Czech, C.; Beyruther, K.; Masters, C.L; McLean, C.A.; Small, D.H. Acetylcholinesterase is increased in the brains of transgenic mice expressing the C-terminal fragment (CT100) of the β-amyloid protein precursor of Alzheimer's disease. *J. Neurochem.* 71: 723-731, 1998.

Scahill, R.I.; Schott, J.M.; Stevens, J.M.; Rossor, M.N.; Fox, N.C. Mapping the evolution of regional atrophy in Alzheimer's disease: unbiased analysis of fluid-registered serial MRI. *Proc. Natl. Acad. Sci. U.S.A.* 99: 4703-4707, 2002.

Senanarong, V.; Cummings, J.L.; Fairbanks, L.; Mega, M.; Masterman, D.M.; O'Connor, S.M.; Strickland, T.L. Agitation in Alzheimer's disease is a manifestation of frontal lobe dysfunction. *Dement. Geriatr. Cogn. Disord.* 17: 14-20, 2004.

Shinotoh, H.; Namba, H.; Fukushi, K.; Nagatsuka, S.; Tanaka, N.; Aotsuka, A.; Ota, T.; Tanada, S.; Irie, T. Progressive loss of cortical acetylcholinesterase activity in association with cognitive decline in Alzheimer's disease: a positron emission tomography study. *Ann. Neurol.* 48: 194-200, 2000.

Siman, R.; Flood, D.G.; Thinakaran, G.; Neumar, R.W. Endoplasmic reticulum stress-induced cysteine protease activation in cortical neurons. *J. Biol. Chem.* 276: 44736-44743, 2001.

Siman, R.; Reaume, A.G.; Savage, M.J.; Trusko, S.; Lin, Y.-G.; Scott, R.W.; Flood, D.G. Presenilin-1 P264L knock-in mutation: differential effects on Aβ production; amyloid deposition, and neuronal vulnerability. *J. Neurosci.* 20: 8717-8726, 2000.

Sjöbeck, M.; Englund, E.. Alzheimer's disease and the cerebellum: a morphologic study on neuronal and glial changes. *Dement. Geriatr. Cogn. Disord.* 12: 211-218, 2001.

Smith, G.W.; Aubry, J.-M.; Dellu, F.; Contarino, A.; Bilezikjian, L.; Gold, L.H.; Chen, R.; Marchuk, Y.; Hauser, C.; Bentley, C.A.; Sawchenko, P.E.; Koob, G.F.; Vale, W.; Lee, K.-F. Corticotropin releasing factor receptor 1-deficient mice display decreased anxiety, impaired stress response, and aberrant neuroendocrine development. *Neuron* 20: 1093-1102, 1998.

Stubley-Weatherly, L.; Harding, J.W.; Wright, J.W. Effects of discrete kainic acid-induced hippocampal lesions on spatial and contextual learning and memory in rats. *Brain Res.* 716: 29-38, 1996.

Stutzmann, G.E.; Caccamo, A.; LaFerla, F.M.; Parker, I. Dysregulated IP_3 signaling in cortical neurons of knock-in mice expressing an Alzheimer's-linked mutation in presenilin1 results in exaggerated Ca^{2+} signals and altered membrane excitability. *J. Neurosci.* 24: 508-513, 2004.

Su, J.H.; Kesslak, J.P.; Head, E.; Cotman, C.W. Caspase-cleaved amyloid precursor protein and activated caspase-3 are co-localized in the granules of granulovacuolar degeneration in Alzheimer's disease and Down's syndrome brain. *Acta Neuropathol.* 104: 1-6, 2002.

Sun M.-K.; Alkon D.L. Impairment of hippocampal CA1 heterosynaptic transformation and spatial memory by β-amyloid$_{25\text{-}35}$. *J. Neurophysiol.* 87: 2441-2449, 2002.

Swearer, J.M.; O'Donnell, B.F.; Drachman, D.A.; Woodward, B.M. Neuropsychological features of familial Alzheimer's disease. *Ann. Neurol.* 32: 687-694, 1992.

Takahashi, R.H.; Milner, T.A.; Li, F.; Nam, E.E.; Edgar, M.A.; Yamaguchi, H.; Beal, M.F.; Xu, H.; Greengard, P.; Gouras, G.K. Intraneuronal Alzheimer Aβ42 accumulates in multivesicular bodies and is associated with synaptic pathology. *Am. J. Pathol.* 161: 1869-1879, 2002.

Tales, A.; Muir, J.; Jones, R.; Bayer, A.; Snowden, R.J. The effects of saliency and task difficulty on visual search performance in ageing and Alzheimer's disease. *Neuropsychologia* 42: 335-345, 2004.

Tanzi, R.E.; Bertram, L. New frontiers in Alzheimer's disease genetics. *Neuron* 32: 181-184, 2001.

Tecott, L.H.; Logue, S.F.; Wehner, J.M.; Kauer, J.A. Perturbed dentate gyrus function in 5-HT_{2c} receptor mutant mice. *Proc. Natl. Acad. Sci. U.S.A.* 95: 15026-15031, 1998.

Terry, R.D.; Masliah, E.; Salmon, D.P.; Butters, N.; DeTeresa, R.; Hill, R.; Hansen, L.A.; Katzman, R. Physical basis of cognitive alterations in Alzheimer's disease: synapse loss is the major correlate of cognitive impairment. *Ann. Neurol.* 30: 572-580, 1991.

Tetewsky, S.J.; Duffy, C.J. Visual loss and getting lost in Alzheimer's disease. *Neurology* 52: 958-965, 1999.

Thal, D.R.; Rüb, U.; Schultz, C.; Sassin, I; Ghebremedhin, E.; Tredici, K.D.; Braak, E.; Braak H. Sequence of Aβ-protein deposition in the human medial temporal lobe. *J. Neuropathol. Exp. Neurol.* 59: 733-748, 2000.

Trinchese, F.; Liu, S.; Battaglia, F.; Walter, S.; Mathews, P.M; Arancio, O. Progressive age-related development of Alzheimer-like pathology in APP/PS1 mice. *Ann. Neurol.* 55: 801-814, 2004.

Troncoso, J.C.; Sukhov, R.R.; Kawas, C.H.; Koliatsos, V.E. *In situ* labeling of dying cortical neurons in normal aging and in Alzheimer's disease: correlations with senile plaques and disease progression. *J. Neuropathol. Exp. Neurol.* 55: 1134-1142, 1996.

Turner, P.R.; O'Connor, K.; Tate, W.P.; Abraham, W.C. Roles of amyloid precursor protein and its fragments in regulating neural activity; plasticity and memory. *Prog. Neurobiol.* 70: 1-32, 2003.

Uc, E.Y.; Rizzo, M.; Anderson, S.W.; Shi, Q.; Dawson, J.D. Driver route-following and safety errors in early Alzheimer disease. *Neurology* 63: 832-837, 2004.

Vann, S.D.; Aggleton, J.P. Extensive cytotoxic lesions of the rat retrosplenial cortex reveal consistent deficits on tasks that tax allocentric spatial memory. *Behav. Neurosci.* 116: 85-94, 2002.

Vaucher, E.; Fluit, P.; Chishti, M.A.; Westaway, D.; Mount, H.T.J.; Kar, S. Object recognition memory and cholinergic parameters in mice expressing human presenilin 1 transgenes. *Exp. Neurol.* 175: 398-406, 2002.

Walsh, R.N.; Cummins, R.A. The open-field test: a critical review. *Psychol. Bull.* 83: 482-504, 1976.

Wang, R.; Dineley, K.T.; Sweatt J.D.; Zheng H. Presenilin 1 familial Alzheimer's disease mutation leads to defective associative learning and impaired adult neurogenesis. *Neuroscience* 126: 305-312, 2004.

Weintraub, S.; Mesulam, M.M. Four neuropsychological profiles in dementia. Spinnler H.; Boller F. (Eds) Handbook of Neuropsychology, Vol. 8, pp. 253-282, Elsevier, Amsterdam, 1993.

Wen, P.H.; Hof, P.R.; Chen, X.; Gluck, K.; Austin, G.; Younkin, S.G.; Younkin, L.H.; DeGasperi, R.; Gama Sosa, M.A.; Robakis, N.K.; Haroutunian, V.; Elder, G.A. The presenilin-1 familial Alzheimer disease mutant P117L impairs neurogenesis in the hippocampus of adult mice. *Exp. Neurol.* 188: 224-237, 2004.

Whitehouse, P.J.; Price, D.L.; Struble, R.G.; Clark, A.W.; Coyle, J.T.; Delong, M.R. Alzheimer's disease and senile dementia: loss of neurons in the basal forebrain. *Science* 215: 1237-1239, 1982.

Wong, T.P.; Debeir, T.; Duff, K.; Cuello, A.C. Reorganization of cholinergic terminals in the cerebral cortex and hippocampus in transgenic mice carrying mutated presenilin-1 and amyloid precursor protein transgenes. *J. Neurosci.* 19: 2706-2716, 1999.

Woodruff-Pak, D.S.; Romano, S.; Papka, M. Training to criterion in eyeblink classical conditioning in Alzheimer's disease, Down's syndrome with Alzheimer's disease, and healthy elderly. *Behav. Neurosci.* 110: 22-29, 1996.

Xie, J.; Chang, X.; Zhang, X.; Guo, Q. Aberrant induction of Par-4 is involved in apoptosis of hippocampal neurons in presenilin-1 M146V mutant knock-in mice. *Brain Res.* 915: 1-10, 2001.

Yamaguchi, H.; Maat-Schieman, M.L.C.; van Duinen, S.G.; Prins, F.A.; Neeskens, P.; Natté, R.; Roos, R.A.C. Amyloid β protein (Aβ) starts to deposit as plasma membrane-bound form in diffuse plaques of brains from hereditary cerebral hemorrhage with amyloid-Dutch type; Alzheimer disease and nondemented aged subjects. *J. Neuropathol. Exp. Neurol.* 59: 723-732, 2000.

Yasuda, Y.; Kudo, T.; Katayama, T.; Imaizumi, K.; Yatera, M.; Okochi, M.; Yamamori, H.; Matsumoto, N.; Kida, T.; Fukumori, A.; Okumura, M.; Tohyama, M.; Takeda, M. FAD-linked presenilin-1 mutants impede translation regulation under ER stress. *Biochem. Biophys. Res. Comm.* 296: 313-318, 2002.

Zaman, S.H.; Parent, A.; Laskey, A.; Lee, M.K.; Borchelt, D.R.; Sisodia, S.S.; Manilow, R. Enhanced synaptic potentiation in transgenic mice expressing presenilin 1 familial Alzheimer's disease mutation is normalized with a benzodiazepine. *Neurobiol. Dis.* 7: 54-63, 2000.

Zhu, H.; Guo, Q.; Mattson, M.P. Dietary restriction protects hippocampal neurons against the death-promoting action of presenilin-1 mutation. *Brain Res.* 842: 224-229, 1999.

In: Cognitive Science Compendium, Volume 1
Editor: Miao-Kun Sun
ISBN: 978-1-60456-597-3

Chapter 7

SYNAPTIC PLASTICITY IN ADDICTION

Yan Dong [1,*] ***and R. Suzanne Zukin*** [2]

[1] Program in Neuroscience, Department of Vet and Comparative Anatomy, Pharmacology and Physiology, Washington State University, Pullman, WA 99164
Merck Research Laboratories, P.O. Box 2000, R80M-213, Rahway, NJ 07065, USA
[2] Department of Neuroscience, Albert Einstein College of Medicine, Bronx, NY 10461, USA

ABSTRACT

Drug addiction, defined as compulsive drug use despite serious negative consequences, has been one of the major social problems facing modern societies. A growing body of evidence suggests that drug exposure induces a series of adaptive changes within the brain reward circuitry, some of which are extremely long-lasting and which may mediate maladaptive emotion/reward learning and memory, thus leading to addiction. Here, we review recent findings concerning drug-induced neuronal plasticity occurring at excitatory synapses in the brain areas that make up the reward circuitry. Given that the synapse plays a critical role in neuronal plasticity, drug-induced synaptic plasticity may critically mediate the formation of drug-related memories and thereby, addictive behaviors.

1. INTRODUCTION

Acute exposure to addictive drugs usually produces an intensely positive affective state, or "high", which may lead to chronic drug use. After repeated drug exposure, a subpopulation of people develop addiction, manifested as compulsive drug use despite serious negative consequences including, for example, illness, disrupted personal relationships and failures in major life roles (Hyman and Malenka, 2001). Over the past two decades considerable efforts

* Correspondence: TEL 509-335-5960; FAX 509-335-4650.
E-mail: yandong@vetmed.wsu.edu

have been devoted to elucidating the neuronal mechanisms of drug addiction. Given that addictive drugs are usually rewarding and motivating, early attention was intuitively focused on brain circuits of the co-called the "reward pathway" which, under normal conditions, mediates natural rewards (such as food, drink and sex) and reward-associated emotion/motivation (Wise and Bozarth, 1987). As originally defined, the reward pathway includes the midbrain ventral tegmental area (VTA) and the nucleus accumbens (NAc), which receives dopaminergic innervation from the VTA, and which provides reciprocal inhibitory (GABAergic) afferents to and from the VTA. Subsequent work has expanded the reward pathways by including a number of several related neuroanatomic regions, such as the prefrontal cortex, the hippocampal formation and the amygdala.

The brain reward pathway mediates reward/emotion/motivation learning and memory. Unlike the classroom learning and memory that we are consciously aware of, reward/emotion/motivation learning and memory can happen subconsciously, which is perceived but implicit. For example for individuals who previously have experienced frightening car accidents, a glance at a related cue (same brand of car or similar street corner) can instantly elicit an entire panoply of emotions including strong fear and increased heart rate, before the explicit memory is recalled. The fearful emotional state is learned and stored as a memory that can be wakened by associated cues to produce behavioral consequences such as running away from the emotion-eliciting site. The normophysiologic process mediated by the reward pathway can be usurped by addictive drugs to produce a state of addiction. Exposure to addictive drugs induces an extremely strong and long-lasting pathological emotional memory that, once awakened, manifests as a robust sense of craving.

An abundance of evidence suggests that the process by which addictive drugs produce persistent alterations in brain function may share common molecular and cellular mechanisms with experience-dependent neuronal plasticity at excitatory synapses of key brain regions such as the VTA and NAc (Kelley, 2004; Wolf et al., 2004). As such, one of the major goals of the field is to identify drug-induced neuronal plasticity in the brain reward pathway, with the hope that such efforts may result in the identification of critical maladaptive neuronal plasticity that leads to addiction. Given the leading role of the excitatory synapse in neuronal plasticity, in this review we focus on drug-induced plasticity occurring at excitatory synapses in the brain reward pathway.

2. Synaptic Plasticity

Synaptic plasticity, or more precisely, experience-dependent changes in the efficacy or strength of synaptic transmission, plays a critical role in formation of memories (Bear, 1996). The most intensely studied and well-known cellular models of synaptic plasticity are long-term potentiation (LTP), or the persistent increase in the strength of synaptic transmission, and long-term depression (LTD), or the persistent decrease in the strength of synaptic transmission (Malenka and Nicoll, 1999). There are several different forms of LTP and LTD with very different molecular mechanisms, among which the best characterized are LTP and LTD at hippocampal CA1 synapses.

Two relevant glutamate receptor types, the alpha-amino-3-hydroxy-5-methyl-4-isoxazole propionic acid (AMPA) receptor and the N-methyl-D-aspartate (NMDA) receptor, are expressed at most excitatory synapses of hippocampal CA1 neurons. Binding of presynaptically-released glutamate is sufficient for the activation of AMPA receptors and conduction of a depolarizing current, rendering the postsynaptic neuron more likely to fire action potentials. However, NMDA receptor activation requires not only the binding of glutamate but also strong depolarization of the postsynaptic membrane to relieve the magnesium blockade of the channel pore. Unlike AMPA receptors, which mainly flux sodium (Na^+) ions, activated NMDA receptors readily conduct calcium (Ca^{2+}) ions into the postsynaptic cell. It is this NMDA receptor-dependent calcium influx that acts, via multiple intracellular signaling pathways, to trigger both LTP and LTD, respectively (Malenka and Nicoll, 1999). More specifically, depending on the intracellular level and spatial compartmentalization, calcium ions can either activate either protein kinases (e.g. calcium/calmodulin-dependent protein kinase II, or CaMKII) or protein phosphotases (e.g. calcium/calmodulin-dependent protein phosphatase calcineurin and protein phosphatase 1, or PP1) (Malinow and Malenka, 2002). CaMKII appears to trigger the delivery of intracellular AMPA receptors to the postsynaptic membrane surface, producing LTP, whereas PPI and/or calcineurin can induce internalization of the AMPA receptors from the postsynaptic membrane, and thus produce LTD (Malenka, 2003). Thus the strength of the excitatory synapses will be persistently upregulated (LTP) when the number of postsynaptic AMPA receptors is increased and persistently downregulated (LTD) when the number of postsynaptic AMPA receptors is decreased.

AMPA receptors are tetramers composed of combinations of subunits including GluR1, GluR2, GluR3 and in the early developmental stage also Glu4 (Malinow and Malenka, 2002). Each subunit contains a unique intracellular C-terminal that selectively interacts with different binding partners and may activate different intracellular signaling cascades and, as such, is subject to differential regulation during LTP and LTD. For example, GluR1 appears to be essential for LTP, as hippocampal LTP is absent either in GluR1 knock out animals (Zamanillo et al., 1999) and or when the interaction between the GluR1 C-terminus and intracellular signaling cascades is interrupted (Shi et al., 1999). The molecular mechanisms of hippocampal LTD and LTD have been used as a template for synaptic plasticity study in other brain regions.

3. Synaptic Plasticity in the Brain Reward Pathway

As the backbone of the brain reward pathway, the mesocorticolimic dopamine (DA) system comprises the VTA, the NAc and the PFC (Tzschentke, 2001). This pathway mediates natural reinforcement as well as in the reinforcing properties of addictive drugs (Wise and Bozarth, 1987; Tzschentke, 2001). The VTA contains DA-releasing neurons, which project to the NAc and the PFC, and which receive excitatory projections from the PFC, the pedunculopontine nucleus and the bed nucleus of the stria terminalis (Tzschentke, 2001). The PFC sends glutamatergic projections to the VTA and the NAc. The major cell type of the NAc is the inhibitory GABAergic medium spiny neurons, which receive excitatory projections from the PFC, the hippocampus and the amygdala (Tzschentke, 2001). NAc cells are for the most part quiescent and their activity depends on excitatory input from cortical and subcortical limbic afferents (Nicola et al., 2000).

It is well documented that synaptic plasticity in the brain reward pathway share many core features in common with, but also distinctly different from that of hippocampal CA1 neurons. Excitatory synapses on neurons in the brain reward pathway can undergo both LTP and LTD. As in the hippocampus, LTP at excitatory synapses on VTA DA neurons requires activation of NMDA receptors (Bonci and Malenka, 1999), although it remains unclear as to whether CaMKII is the downstream effector (or target) of NMDA receptor-mediated calcium influx. LTD at excitatory synapses on VTA DA neurons, however, does not require activation of NMDA receptors but rather appears to be triggered by voltage-dependent calcium channels (Jones and Kauer, 1999; Thomas et al., 2000; Gutlerner et al., 2002). In addition, expression of LTD within VTA DA neurons involves regulated internalization of postsynaptic AMPA receptors, a process involving cAMP and activation of protein kinase A (PKA), but not activation of protein phosphatases 1/2A (Gutlerner et al., 2002).

Similar in VTA DA neurons, LTP can be induced in NAc neurons in an NMDA receptor-dependent manner and is likely expressed by upregulation of the postsynaptic AMPA receptor function/number (Kombian and Malenka, 1994). LTD in NAc neurons appears to be more complicated. Thus far there are two forms of LTD have been observed in NAc neurons: (1) NMDA receptor-dependent LTD, involving NMDA receptor mediated calcium influx and removal of postsynaptic AMPA receptors (Thomas et al., 2000); and (2) CB1 receptor-dependent LTD, in which endocannabinoind-mediated negative feedback causes a long-lasting depression of presynaptic glutamate release (Robbe et al., 2002b).

Current evidence suggests that both LTD and LTP can also be induced in the PFC neurons. LTP in PFC neurons requires activation of the NMDA receptor, whereas LTD appears to be NMDA receptor-independent and probably involves the CB1 receptor and metabotropic glutamate receptors (Huang et al., 2004). Since both the NAc and PFC receive dense DA projections, it is not surprising that DA plays a critical role in modulating the development and/or maintenance of synaptic plasticity in these two brain regions (Law-Tho et al., 1995; Otani et al., 1998; Gurden et al., 2000; Otani, 2003; Huang et al., 2004).

4. Drug-induced Synaptic Plasticity in the Brain Reward Pathway

Ventral Tegmental Area

There are two major cell types in the VTA, dopaminergic neurons and inhibitory GABA neurons, both of which project to target areas such as the NAc and PFC (Tzschentke, 2001). The neuronal activity of DA neurons is thought to encode information about the rewarding and motivationallly relevant properties of external stimuli and thus particularly important for the development of drug-induced addictive behaviors (Overton and Clark, 1997; Schultz, 2002). Glutamatergic afferents from the PFC and other limbic areas synapse on DA and GABA neurons in the VTA and importantly modulate the neuronal activity of VTA DA cells (Overton and Clark, 1997).

A critical question has been whether drugs of abuse induce long lasting synaptic plasticity in the brain reward pathway. Since LTP and LTD may involve insertion/internalization of AMPA receptors to/from the postsynaptic membrane, several early studies focused on the effects of drug of abuse on GluR1 protein levels within the VTA. A number of studies reported that repeated administration of cocaine, morphine (Fitzgerald et al., 1996) or ethanol (Ortiz et al., 1995) results in an increase in the level of the GluR1 subunit of AMPA glutamate receptors in the VTA. Other groups, however, found that there was no net change in GluR1 levels following cocaine or amphetamine administration (Lu et al., 2002). Since these studies used different schedules of drug administration, inconsistencies in their findings may reflect the fact that drug effects critically depend on the type of drug and the regimen of drug administration. It is worth noting that in most neurons there is a significant intracellular pool of AMPA receptors, and that expression studies of GluR1 transcript or protein levels alone may fail to reveal functionally important alterations in receptor trafficking. To directly assess the effects of drugs of abuse on VTA AMPA receptors, White and colleagues performed extracellular electrophysiological recording in living animals and observed that repeated cocaine or amphetamine exposure caused VTA DA neurons to be more responsive to iontophoresed AMPA (White et al., 1995). This important result can be interpreted as physiological evidence of an "LTP" like phenomena in VTA DA neurons following drug exposure. However, the direct physiological measurement in the above study is the AMPA receptor-evoked action potential bursting, not the synaptic AMPA receptor-mediate EPSCs *per se*. It remains unclear whether the increased responsiveness occurs as a consequence of regulated AMPA receptor trafficking, and/or as a result of modulation of voltage-dependent ion channels, which are also important in synaptic responses to excitatory stimulation.

To directly assess the effects of drugs of abuse on the excitatory synaptic strength at VTA DA neurons, a series of studies have been conducted to measure the ratio of AMPA receptor-mediated synaptic current to NMDA receptor mediated synaptic current. This approach is predicted on the premise that LTP and LTD occur as a result of modification of synaptic AMPA receptor function/number, while the NMDA receptor function/number during LTP and LTD is relatively stable. Thus LTP is accompanied by an increase in the AMPA/NMDA ratio, whereas LTD is accompanied by a decrease in the AMPA/NMDA ratio. Therefore if a stimulus (e.g. administration of a drug of abuse) increases or decreases the AMPA/NMDA ratio, it is likely that this treatment may induce an LTP- or LTD-like process, accordingly. This approach offers the important advantage that it is independent of the number of synapses activated and therefore independent of such variables as the positioning of the electrodes or anatomy of the tissue.

Using this approach, Ungless et al., . observed a substantial increase the AMPA/NMDA ratio in VTA DA neurons in acute brain slices prepared from animals receiving a single cocaine administration (Ungless et al., 2001). This cocaine-induced increase in the AMPA/NMDA ratio was abolished either in GluR1 (-/-) knockout mice or when NMDA receptors were selectively blocked by co-administration of the NMDA receptor antagonist MK801 (Ungless et al., 2001; Dong et al., 2004). Moreover, the ability to further enhance AMPA receptor mediated EPSCs (LTP induction) was diminished, whereas the ability to decrease the AMPA receptor-mediated EPSCs (LTD induction) was increased in VTA DA neurons from animals previously treated with cocaine, consistent with the idea that cocaine exposure, through an LTP-like process has already potentiated the AMPA receptor-mediated EPSCs. Taken together, these results indicate that *in vivo* cocaine administration induces a form of plasticity at excitatory synapses on midbrain DA neurons having in common many of the core features of hippocampal LTP.

Time course studies provide important clues regarding the behavioral correlates of this drug-induced synaptic adaptation. Cocaine-induced "LTP" in VTA DA neurons appears to have a fast onset and is relatively transient. It can be detected as soon as two hours after *in vivo* drug administration (Faleiro et al., 2004), and persists for at least five, but no more than ten, days (Ungless et al., 2001). The duration of the potentiation is also independent of the paradigm of cocaine treatment as the AMPA/NMDA ratio returns to the basal levels within ten days following the cessation of prolonged cocaine treatment (Borgland et al., 2004). Both the early onset and the relatively brief duration of drug-induced VTA "LTP" are consistent with the notion that this plasticity functions to prime the initial addictive processes, but are not the neuronal substrate for the long-term maintenance of the addicted state (Kalivas, 2004).

To further address whether the synaptic modification of VTA DA neurons is a mechanism specific to cocaine or, rather, is more generally applicable to all drugs of abuse, two subsequent studies (Saal et al., 2003; Dong et al., 2004) examined the effects of five addictive drugs: cocaine, amphetamine, morphine, nicotine and ethanol as well as two non-addictive psychoactive drugs Prozac (fluoxetine) and carbamazepine. The addictive drugs differ markedly in their *in vivo* triggering mechanisms. Cocaine and amphetamine acutely enhance DA transmission by blocking or reversing DA transporters. Morphine acutely activates opioid receptors; nicotine acutely activates nicotinic receptors. Ethanol acutely enhances GABAergic transmission. But they have one thing in common: they all cause addiction. The two control drugs are psychoactive and share some pharmacological effects with addictive drugs but do not cause addiction. For example, like cocaine and amphetamine, fluoxetine is a monoamine reuptake inhibitor, whereas carbamazepine like cocaine, blocks sodium channels. When administered *in vivo*, all five addictive drugs, but not the non-addictive drugs, increased the AMPA/NMDA ratio in VTA DA neurons (Saal et al., 2003). These findings suggest that this form of experience-dependent synaptic modification is specific for addictive drugs.

Stress is critically involved in addictive behaviors (Marinelli and Piazza, 2002; Shaham et al., 2003). Stress potentiates the acquisition of self-administration (Kabbaj et al., 2001), triggers reinstatement (Piazza and Le Moal, 1998), enhances conditioned place preference (CPP), and induces reinstatement of drug-seeking behaviors following extinction (Will et al., 1998; Sanchez and Sorg, 2001; Sanchez et al., 2003). Moreover, single exposure to stress increases the AMPA/NMDA ratio in VTA DA neurons in a manner similar to drugs of abuse (Saal et al., 2003). The stress- and cocaine-elicited effects on synaptic scaling exhibit mutual occlusion, and are both blocked by the NMDA receptor antagonist MK801. On the other hand, cocaine-, but not stress- induced synaptic potentiation in VTA DA neurons is blocked by DA D1-like antagonists, whereas the converse is true for glucocorticoid receptor blockade (Dong et al., 2004). Together, these findings indicate that whereas the *in vivo* triggering mechanisms for these forms of experience-dependent plasticity share some features in common, they differ in several important aspects (Dong et al., 2004).

The direct consequence of short-term LTP in VTA DA neurons following withdrawal from drug exposure is the enhancement of DA transmission. Enhanced DA transmission can be interpreted as an increase in the incentive salience attributed to drugs and drug-related cues, which may be experimentally manifested as behavioral sensitization (Robinson and Berridge, 2000). In parallel, VTA DA neurons may also function to establish and maintain the prediction of the reinforcing effect (e.g. reward) from the conditioned cues (Schultz, 2002) and are thus critically involved in conditioned behaviors related to addiction and reward.

Behavioral sensitization to drugs is an animal model for the intensification of drug craving, a core component of addiction. Behavioral sensitization arises during repeated drug administration and persists even after long periods of withdrawal (Robinson and Becker, 1986). This addiction-related behavior is highly conditional, as animals exhibiting higher levels of sensitization when exposed to the same environmental context in which their previous drug experience occurred (Badiani and Robinson, 2004). Another model for studying conditioned drug behavior is called conditioned place preference (CPP), in which animals show a strong preference for the place associated with drug experience.

Glutamatergic afferents on DA cells in the VTA are thought to be critical to several aspects of addictive behaviors, including behavioral sensitization and relapse to cocaine self-administration. In accordance with the general cognitive function of DA, it has been proposed that glutamatergic afferents to the NAc and cocaine-induced "LTP" in VTA DA neurons may play a critical role in cocaine induced behavioral sensitization, cocaine-conditioned sensitization, cocaine-induced CPP and relapse to cocaine self-administration. Indeed, this prevailing notion (Overton et al., 1999; Vanderschuren and Kalivas, 2000; Wolf et al., 2004) has been supported by many results. For example, intra VTA injection of NMDA receptor antagonist, which blocks drug-induced "LTP", blocks the induction of behavioral sensitization (Wolf, 1998) as well as CPP in response to cocaine administration (Kim et al., 1996; Harris and Aston-Jones, 2003). Animals with higher cocaine-induced AMPA/NMDA ratio in the VTA DA neurons exhibit a higher level of locomotor sensitization (Borgland et al., 2004). After discontinuing repeated psychostimulant administration, the VTA DA neurons remain superactive for several days, a time course consistent with that of cocaine-induced increases in the AMPA/NMDA ratio (White et al., 1995). This relatively transient superactivity of VTA DA neurons has been regarded as part of the mechanism that mediates "transfer" from the induction to the expression phase of locomotor sensitization (Wolf et al., 2004). In the case of CPP, it has been demonstrated that overexpression of GluR1 in the VTA causes a significant enhancement in morphine-induced CPP, while overexpression of GluR2, an AMPA receptor subunit not required for LTP, does not (Carlezon et al., 2001; Carlezon and Nestler, 2002). The role of VTA "LTP" in CPP is further suggested by the finding that the same dose of DA D1 receptor antagonist that blocks cocaine induced "LTP" in the VTA (Dong et al., 2004) blocks cocaine-induced CPP (Nazarian et al., 2004).

To directly examine whether the VTA "LTP" is required for cocaine-induced sensitization and CPP, we conducted a series of behavioral experiments using GluR1(-/-) knockout mice (Dong et al., 2004). These mice do not exhibit cocaine-induced potentiation of excitatory synaptic transmission (cocaine-induced "LTP") in VTA DA neurons. Surprisingly, the GluR1(-/-) knockout mice exhibit a robust behavioral sensitization to cocaine (Dong et al., 2004). This finding indicates that cocaine-induced LTP is not required for this form of behavioral plasticity. On the other hand, it remains unknown whether LTP in the VTA may contribute to the attribution of incentive value (motivational significance) to drug-related cues. While GluR1 deletion did not abolish sensitization to cocaine, two other behaviors were impaired in the knockout mice (Dong et al., 2004). The knockout mice did not exhibit cocaine-induced CPP. Moreover, the GluR1 knockout mice did not exhibit context-dependent enhancement of locomotor activity when placed in a conditioning apparatus in which they had received cocaine 24 hours previously (Dong et al., 2004). Cocaine-elicited LTP in VTA DA cells and CPP (Nazarian et al., 2004), as well as context-dependent enhancement of locomotor activity (Dong et al., 2004) were blocked by the same dose of a D1 receptor antagonist. Furthermore, both the cocaine-induced LTP in VTA DA cells and the cocaine-primed enhancement of morphine CPP (Kim et al., 2004) last five, but not ten, days (Borgland et al., 2004), and both can be blocked by an NMDA receptor antagonist. Taken collectively, these studies indicate that cocaine-induced LTP in VTA DA neurons may contribute to context-dependent associative learning and may contribute to attribution of incentive value (motivational significance) to drug-associated cues. More generally, since enhanced strength of excitatory synaptic strength in midbrain DA neurons will cause increased activities in DA cells (Overton and Clark, 1997; Grillner and Mercuri, 2002) and a consequent increase in DA release in target regions such as the NAc, experiences such as acute stress, which elicit LTP in the VTA may prime the animal such that subsequent experiences have greater incentive or motivational value.

Nucleus Accumbens

The NAc together with the VTA are the major components of the reward circuitry. Long-term changes in synaptic strength at excitatory synapses on NAc medium spiny neurons are critically important for many addictive behaviors. Early in situ hybridization and immunolabeling studies demonstrate that levels of GluR1 and GluR2 mRNA and protein are decreased in animals 14 days following withdrawal of repeated amphetamine administration (Lu et al., 1997; Lu et al., 1999; Lu and Wolf, 1999), whereas upregulation of GluR1 (but not GluR2) protein was also observed in the NAc of animals 30 days after withdrawal from repeated cocaine administration (Churchill et al., 1999). In the late 1990s White and his colleagues used extracellular recording combined with microiontophoretic techniques to measure glutamate-elicited action potential firing in NAc neurons and demonstrated that NAc neurons became hyposensitive to glutamate after withdrawal from chronic exposure to either cocaine or amphetamine (White et al., 1995). These results, while suggestive that drug exposure may elicit a postsynaptically- expressed LTD in the NAc, nevertheless have several possible caveats. For example, downregulation of mRNA levels of GluR1/2 does not necessarily correlate with downregulation of GluR1/2 protein, and may not be indicative of a downregulation of synaptic AMPA receptors. Furthermore, the reduced electrophysiological responses of NAc neurons are assessed by measuring action potential firing, which in addition to being mediated by metabotropic glutamate receptors, is also influenced by the activities of voltage-gated sodium, potassium and calcium channels, all of which may operate independently of glutamate receptor activities. Indeed, it has been demonstrated that exposure to cocaine persistently reduces both sodium (Zhang et al., 1998) and calcium currents (Zhang et al., 2002) in NAc neurons, changes that can attenuate neuronal excitability.

More recently, Thomas et al., . showed that repetitive cocaine administration produces a decrease in the AMPA/NMDA ratio in NAc neurons (Thomas et al., 2001), consistent with a reduction in AMPA receptor number/function in the postsynaptic membrane and LTD. Furthermore, cocaine pretreatment enhances the amplitude of AMPA receptor miniature EPSCs, reduces the ability to generate LTD (Thomas et al., 2001) and increases the ability to generate LTP (Yao et al., 2004) in NAc neurons, implying that synaptic strength has been previously depressed by chronic drug exposure. Interestingly, exposure to amphetamine, another psychostimulant that shares similar molecular mechanismss of action with cocaine, does not alter NAc LTP (Li and Kauer, 2004).

Although the exact molecular mechanisms underlying drug-induced LTD in NAc neurons are as yet unknown, the available evidence suggests that DA may be critically involved in the induction phase of LTD. Unlike neurons of the dorsal striatum, DA is not required in the induction in vitro of LTD and LTP in NAc neurons (Pennartz et al., 1993; Thomas et al., 2000). Rather, DA, the selective DA D1 receptor agonist SKF82958 and the indirect agonist amphetamine all block tetanus-induced LTP (Li and Kauer, 2004). Moreover, cocaine exposure elevates the sensitivity of the D1 receptor in the NAc to endogenous and exogenous agonists (Beurrier and Malenka, 2002). It is tempting to speculate that excitatory synaptic strength in NAc neurons is constitutively modulated by endogenous factors such as DA and is finely balanced between potentiation and depression. Chronic drug administration enhances DA tone and thereby suppresses potentiation, thus shifting the equilibrium towards depression. This model predicts that any elevation of DA tone, even in the absence of drug administration, should also produce "LTD" in NAc neurons. Indeed, a recent finding indicates that excitatory synapses appear to be pre-depressed in the NAc slices from animals lacking DA transporters, norepinephrine transporters, or vesicular monoamine transporters, and that LTP of a greater magnitude could be induced (Yao et al., 2004). Clearly, more experiments are required to elucidate the role of DA in drug-induced NAc "LTD". An important unanswered question is whether direct microinjection of selective DA D1 receptor agonists can mimic, or whether antagonists can block, cocaine-induced "LTD" in NAc neurons.

Although the detailed molecular processes underlying the expression of drug-induced NAc LTD are also unclear, regulated glutamate receptor trafficking represents an intriguing model supported by a growing body of evidence. Recent studies show that repeated cocaine exposure reduces the levels of PSD-95 (Yao et al., 2004), a postsynaptic scaffolding protein that may control the AMPA receptor trafficking. Overexpression of PSD-95 increases the number of synaptic AMPA receptor and basal AMPA receptor mediated EPSC, occludes LTP and enhances LTD (Schnell et al., 2002; Beique and Andrade, 2003; Stein et al., 2003), whereas reduction of PSD-95 level decreases the number of synaptic AMPA receptors and the AMPA receptor mediated EPSCs (Nakagawa et al., 2004). Consistent with these findings, targeted disruption of PSD-95 abolishes LTD and facilitates the induction of LTP (Migaud et al., 1998; Komiyama et al., 2002). It has also been shown that controlling the availability and stability of PSD-95 by either ubiquitination or palmitoylation may serve as a mechanism to regulate the expression and maintenance of surface AMPA receptors (El-Husseini Ael et al., 2002; Colledge et al., 2003). It is conceivable that PSD-95 may play a similar role in the NAc, and thus diminished PSD-95 levels could contribute to cocaine-induced decreases in synaptic strength. It would be interesting to know whether cocaine exposure also affects the level of SAP97, the ubiquitination and/or palmitoylation of PSD-95, and whether cocaine-induced NAc "LTD" exists in PSD-95 null animals.

While the above observations indicate that drug exposure generates postynaptically-expressed LTD in NAc neurons, they do not exclude several other possible means by which drugs of abuse could alter synaptic strength. Evidence suggests that drug exposure may also modulate excitatory synaptic strength via presynaptic mechanisms. For example, DA presynaptically modulates excitatory synaptic transmission (Nicola et al., 2000). This presynaptic modulation is significantly enhanced by cocaine exposure (Beurrier and Malenka, 2002), and repeated cocaine administration decreases glutamate immunolabeling within the nerve terminals of the NAc shell (Keys et al., 1998; Meshul et al., 1998), suggesting enhanced presynaptic neurotransmitter release. In addition to evidence that addictive drugs can enhance presynaptic release, presynaptically expressed LTD, observed in both the NAc and dorsal striatum (Gerdeman et al., 2002; Robbe et al., 2002a), is blocked by the administration of a single dose of THC (Mato et al., 2004) or cocaine (Fourgeaud et al., 2004), suggesting that drugs of abuse can also modify the synaptic strength via presynaptic mechanisms.

Long-term changes in excitatory synaptic transmission, such as drug-induced "LTD", have been proposed to be critically involved in addiction-related behaviors (Wolf et al., 2004). This notion is supported by several behavioral studies. For example, intra-NAc infusion of NMDA receptor antagonists, which block several forms of synaptic plasticity, prevent the expression of behavioral sensitization (Li et al., 1997), the establishment of reinforcement learning (Kelley et al., 1997) and the acquisition (but not performance) of cue-elicited reinstatement of drug intake (McFarland and Kalivas, 2001; Park et al., 2002). Unfortunately so far there is no study directly describing the behavioral significance of cocaine-induced "LTD" in the NAc, but emerging evidence has started to provide clues that this form of plasticity may critically contribute to cocaine reward and behavioral sensitization. As shown by Kelz et al., . (Kelz et al., 1999), overexpression of GluR1 in the NAc, the opposite to cocaine-induced "LTD", produces cocaine aversion. By analogy, NAc LTD may contribute to the enhancement of the incentive value of cocaine following repeated exposure. Consistent with this concept, blockade of calcineurin activity by the selective inhibitor FK506, a treatment that may interfere with the LTD process, results in a significantly attenuated drug-induced behavioral sensitization (Tsukamoto et al., 2001). Clearly, additional studies are required to more directly elucidate the functional significance of NAc LTD as it relates to addiction. For example, drug-induced behavioral modification of animals in which NAc LTD is prevented (e.g. by PP1 peptide, (Morishita et al., 2001), or in which excitatory synaptic strength is otherwise decreased would prove particularly illuminating.

Other Brain Regions

The prefrontal cortex, which provides the major excitatory inputs to the brain reward pathway, functions to mediate working memory, impulsive behavior and reward-related learning (Jentsch and Taylor 1999). There have been no studies directly addressing the effect of repeated administration of psychostimulants on excitatory synaptic transmission within the PFC although one study shows that the PFC neurons in amphetamine-pretreated rats exhibited enhanced glutamate responsiveness to glutamate (Peterson et al., 2000), implying that a possible LTP like process may occur in the PFC following drug exposure. On the other hand, it has been consistently demonstrated that cocaine exposure increases the intrinsic membrane excitability of PFC pyramidal neurons by affecting ion channel activity (Dong et al., 2005; Nasif et al., 2005a; Nasif et al., 2005b). Upregulation of the intrinsic membrane excitability of PFC neurons presumably increases presynaptic neurotransmitter release at the excitatory synapses, a process that may serve as a trigger for drug-induced synaptic plasticity in both the VTA and NAc. In addition, drug-induced LTP in the VTA can presumably increase DA tones in the PFC, which should significantly increase the membrane excitability via D1 receptor-mediated modulation of ionic conductances (Yang and Seamans, 1996; Yang et al., 1996; Dong and White, 2003; Dong et al., 2004). It appears that drug-induced physiological adaptations in the PFC, NAc and VTA comprise a positive feedback loop through which drugs of abuse persistently activate the brain reward pathway.

In the ventral lateral bed nucleus of the stria terminalis excitatory synapses may also generate "LTP" in response to cocaine administration (Dumont et al., 2005). Interestingly this LTP could be induced only when the drug is actively consumed by animals, indicating that the triggering process may require co-activation of other brain regions (e.g. PFC) that participate in goal-directed behaviors. It would be interesting to determine the molecular mechanism of this novel "LTP", as well as the functional and behavioral correlates.

5. Summary

"Addictive behavior does not result from repeated drug exposure *per se*, but is a result of learning" (Tzschentke and Schmidt, 2003). The concept that addiction is a maladaptive form of reward/emotion/motivation learning and memory (Kelley, 2004) has been recently more readily embraced by the field. This article has predominately focused on drug-induced synaptic plasticity, an important type of plasticity likely to mediate many forms of learning and memory. However, there are many other forms of neuronal plasticity that are likely to also be important for learning and memory, and may also critically contribute to the addiction process. Thus, drug exposure can also cause persistent changes in the intrinsic membrane properties of neurons within the brain reward pathway, which in turn may contribute to certain addiction-related behaviors, such as the sensitivity to cocaine. While the mechanisms underlying drug-induced structural and functional neuronal adaptations continue to be explored, it should be borne in mind that the link between drug-induced neuronal plasticity and maladaptive reward/emotion memory remains to be firmly established. In large part, a successful outcome will depend upon the continued development of appropriate animal models and behavioral assay with which to assess the affective and emotional states. Indeed, an approach in which our advancing appreciation of the molecular basis of drug-induced neural adaptation is integrated with an understanding of its functional and behavioral correlates remains the best hope for eventually unlocking the secrets to drug addiction.

Acknowledgments

We thank Drs. Nick Hastings, Yanhua Huang, and Tung Fong for critical suggestions for the manuscript.

References

Badiani A, Robinson TE (2004) Drug-induced neurobehavioral plasticity: the role of environmental context. *Behav Pharmacol* 15:327-339.

Bear MF (1996) A synaptic basis for memory storage in the cerebral cortex. *Proc Natl Acad Sci U S A* 93:13453-13459.

Beique JC, Andrade R (2003) PSD-95 regulates synaptic transmission and plasticity in rat cerebral cortex. *J Physiol* 546:859-867.

Beurrier C, Malenka RC (2002) Enhanced inhibition of synaptic transmission by dopamine in the nucleus accumbens during behavioral sensitization to cocaine. *J Neurosci* 22:5817-5822.

Bonci A, Malenka RC (1999) Properties and plasticity of excitatory synapses on dopaminergic and GABAergic cells in the ventral tegmental area. *J Neurosci* 19:3723-3730.

Borgland SL, Malenka RC, Bonci A (2004) Acute and chronic cocaine-induced potentiation of synaptic strength in the ventral tegmental area: electrophysiological and behavioral correlates in individual rats. *J Neurosci* 24:7482-7490.

Carlezon WA, Jr., Nestler EJ (2002) Elevated levels of GluR1 in the midbrain: a trigger for sensitization to drugs of abuse? *Trends Neurosci* 25:610-615.

Carlezon WA, Jr., Todtenkopf MS, McPhie DL, Pimentel P, Pliakas AM, Stellar JR, Trzcinska M (2001) Repeated exposure to rewarding brain stimulation downregulates GluR1 expression in the ventral tegmental area. *Neuropsychopharmacology* 25:234-241.

Churchill L, Swanson CJ, Urbina M, Kalivas PW (1999) Repeated cocaine alters glutamate receptor subunit levels in the nucleus accumbens and ventral tegmental area of rats that develop behavioral sensitization. *J Neurochem* 72:2397-2403.

Colledge M, Snyder EM, Crozier RA, Soderling JA, Jin Y, Langeberg LK, Lu H, Bear MF, Scott JD (2003) Ubiquitination regulates PSD-95 degradation and AMPA receptor surface expression. *Neuron* 40:595-607.

Dong Y, White FJ (2003) Dopamine D1-class receptors selectively modulate a slowly inactivating potassium current in rat medial prefrontal cortex pyramidal neurons. *J Neurosci* 23:2686-2695.

Dong Y, Saal D, Thomas M, Faust R, Bonci A, Robinson T, Malenka RC (2004) Cocaine-induced potentiation of synaptic strength in dopamine neurons: behavioral correlates in GluRA(-/-) mice. *Proc Natl Acad Sci U S A* 101:14282-14287.

Dong Y, Nasif FJ, Tsui JJ, Ju WY, Cooper DC, Hu XT, Malenka RC, White FJ (2005) Cocaine-induced plasticity of intrinsic membrane properties in prefrontal cortex pyramidal neurons: adaptations in potassium currents. *J Neurosci* 25:936-940.

Dumont EC, Mark GP, Mader S, Williams JT (2005) Self-administration enhances excitatory synaptic transmission in the bed nucleus of the stria terminalis. *Nat Neurosci* 8:413-414.

El-Husseini Ael D, Schnell E, Dakoji S, Sweeney N, Zhou Q, Prange O, Gauthier-Campbell C, Aguilera-Moreno A, Nicoll RA, Bredt DS (2002) Synaptic strength regulated by palmitate cycling on PSD-95. *Cell* 108:849-863.

Faleiro LJ, Jones S, Kauer JA (2004) Rapid synaptic plasticity of glutamatergic synapses on dopamine neurons in the ventral tegmental area in response to acute amphetamine injection. *Neuropsychopharmacology* 29:2115-2125.

Fitzgerald LW, Ortiz J, Hamedani AG, Nestler EJ (1996) Drugs of abuse and stress increase the expression of GluR1 and NMDAR1 glutamate receptor subunits in the rat ventral tegmental area: common adaptations among cross-sensitizing agents. *J Neurosci* 16:274-282.

Fourgeaud L, Mato S, Bouchet D, Hemar A, Worley PF, Manzoni OJ (2004) A single in vivo exposure to cocaine abolishes endocannabinoid-mediated long-term depression in the nucleus accumbens. *J Neurosci* 24:6939-6945.

Gerdeman GL, Ronesi J, Lovinger DM (2002) Postsynaptic endocannabinoid release is critical to long-term depression in the striatum. *Nat Neurosci* 5:446-451.

Grillner P, Mercuri NB (2002) Intrinsic membrane properties and synaptic inputs regulating the firing activity of the dopamine neurons. *Behav Brain Res* 130:149-169.

Gurden H, Takita M, Jay TM (2000) Essential role of D1 but not D2 receptors in the NMDA receptor-dependent long-term potentiation at hippocampal-prefrontal cortex synapses in vivo. *J Neurosci* 20:RC106.

Gutlerner JL, Penick EC, Snyder EM, Kauer JA (2002) Novel protein kinase A-dependent long-term depression of excitatory synapses. Neuron 36:921-931.

Harris GC, Aston-Jones G (2003) Critical role for ventral tegmental glutamate in preference for a cocaine-conditioned environment. *Neuropsycho-pharmacology* 28:73-76.

Huang YY, Simpson E, Kellendonk C, Kandel ER (2004) Genetic evidence for the bidirectional modulation of synaptic plasticity in the prefrontal cortex by D1 receptors. *Proc Natl Acad Sci U S A* 101:3236-3241.

Hyman SE, Malenka RC (2001) Addiction and the brain: the neurobiology of compulsion and its persistence. *Nat Rev Neurosci* 2:695-703.

Jones S, Kauer JA (1999) Amphetamine depresses excitatory synaptic transmission via serotonin receptors in the ventral tegmental area. *J Neurosci* 19:9780-9787.

Kabbaj M, Norton CS, Kollack-Walker S, Watson SJ, Robinson TE, Akil H (2001) Social defeat alters the acquisition of cocaine self-administration in rats: role of individual differences in cocaine-taking behavior. *Psychopharmacology* (Berl) 158:382-387.

Kalivas PW (2004) Glutamate systems in cocaine addiction. *Curr Opin Pharmacol* 4:23-29.

Kelley AE (2004) Memory and addiction: shared neural circuitry and molecular mechanisms. *Neuron* 44:161-179.

Kelley AE, Smith-Roe SL, Holahan MR (1997) Response-reinforcement learning is dependent on N-methyl-D-aspartate receptor activation in the nucleus accumbens core. *Proc Natl Acad Sci U S A* 94:12174-12179.

Kelz MB, Chen J, Carlezon WA, Jr., Whisler K, Gilden L, Beckmann AM, Steffen C, Zhang YJ, Marotti L, Self DW, Tkatch T, Baranauskas G, Surmeier DJ, Neve RL, Duman RS, Picciotto MR, Nestler EJ (1999) Expression of the transcription factor deltaFosB in the brain controls sensitivity to cocaine. *Nature* 401:272-276.

Keys AS, Mark GP, Emre N, Meshul CK (1998) Reduced glutamate immunolabeling in the nucleus accumbens following extended withdrawal from self-administered cocaine. *Synapse* 30:393-401.

Kim HS, Park WK, Jang CG, Oh S (1996) Inhibition by MK-801 of cocaine-induced sensitization, conditioned place preference, and dopamine-receptor supersensitivity in mice. *Brain Res Bull* 40:201-207.

Kim JA, Pollak KA, Hjelmstad GO, Fields HL (2004) A single cocaine exposure enhances both opioid reward and aversion through a ventral tegmental area-dependent mechanism. *Proc Natl Acad Sci U S A* 101:5664-5669.

Kombian SB, Malenka RC (1994) Simultaneous LTP of non-NMDA- and LTD of NMDA-receptor-mediated responses in the nucleus accumbens. *Nature* 368:242-246.

Komiyama NH, Watabe AM, Carlisle HJ, Porter K, Charlesworth P, Monti J, Strathdee DJ, O'Carroll CM, Martin SJ, Morris RG, O'Dell TJ, Grant SG (2002) SynGAP regulates ERK/MAPK signaling, synaptic plasticity, and learning in the complex with postsynaptic density 95 and NMDA receptor. *J Neurosci* 22:9721-9732.

Law-Tho D, Desce JM, Crepel F (1995) Dopamine favours the emergence of long-term depression versus long-term potentiation in slices of rat prefrontal cortex. *Neurosci Lett* 188:125-128.

Li Y, Kauer JA (2004) Repeated exposure to amphetamine disrupts dopaminergic modulation of excitatory synaptic plasticity and neurotransmission in nucleus accumbens. *Synapse* 51:1-10.

Li Y, Vartanian AJ, White FJ, Xue CJ, Wolf ME (1997) Effects of the AMPA receptor antagonist NBQX on the development and expression of behavioral sensitization to cocaine and amphetamine. *Psychopharmacology* (Berl) 134:266-276.

Lu W, Wolf ME (1999) Repeated amphetamine administration alters AMPA receptor subunit expression in rat nucleus accumbens and medial prefrontal cortex. *Synapse* 32:119-131.

Lu W, Monteggia LM, Wolf ME (1999) Withdrawal from repeated amphetamine administration reduces NMDAR1 expression in the rat substantia nigra, nucleus accumbens and medial prefrontal cortex. *Eur J Neurosci* 11:3167-3177.

Lu W, Monteggia LM, Wolf ME (2002) Repeated administration of amphetamine or cocaine does not alter AMPA receptor subunit expression in the rat midbrain. *Neuropsychopharmacology* 26:1-13.

Lu W, Chen H, Xue CJ, Wolf ME (1997) Repeated amphetamine administration alters the expression of mRNA for AMPA receptor subunits in rat nucleus accumbens and prefrontal cortex. *Synapse* 26:269-280.

Malenka RC (2003) Synaptic plasticity and AMPA receptor trafficking. *Ann N Y Acad Sci* 1003:1-11.

Malenka RC, Nicoll RA (1999) Long-term potentiation--a decade of progress? *Science* 285:1870-1874.

Malinow R, Malenka RC (2002) AMPA receptor trafficking and synaptic plasticity. *Annu Rev Neurosci* 25:103-126.

Marinelli M, Piazza PV (2002) Interaction between glucocorticoid hormones, stress and psychostimulant drugs. *Eur J Neurosci* 16:387-394.

Mato S, Chevaleyre V, Robbe D, Pazos A, Castillo PE, Manzoni OJ (2004) A single in-vivo exposure to delta 9THC blocks endocannabinoid-mediated synaptic plasticity. *Nat Neurosci* 7:585-586.

McFarland K, Kalivas PW (2001) The circuitry mediating cocaine-induced reinstatement of drug-seeking behavior. *J Neurosci* 21:8655-8663.

Meshul CK, Noguchi K, Emre N, Ellison G (1998) Cocaine-induced changes in glutamate and GABA immunolabeling within rat habenula and nucleus accumbens. *Synapse* 30:211-220.

Migaud M, Charlesworth P, Dempster M, Webster LC, Watabe AM, Makhinson M, He Y, Ramsay MF, Morris RG, Morrison JH, O'Dell TJ, Grant SG (1998) Enhanced long-term potentiation and impaired learning in mice with mutant postsynaptic density-95 protein. *Nature* 396:433-439.

Morishita W, Connor JH, Xia H, Quinlan EM, Shenolikar S, Malenka RC (2001) Regulation of synaptic strength by protein phosphatase 1. *Neuron* 32:1133-1148.

Nakagawa T, Futai K, Lashuel HA, Lo I, Okamoto K, Walz T, Hayashi Y, Sheng M (2004) Quaternary structure, protein dynamics, and synaptic function of SAP97 controlled by L27 domain interactions. *Neuron* 44:453-467.

Nasif FJ, Hu XT, White FJ (2005a) Repeated cocaine administration increases voltage-sensitive calcium currents in response to membrane depolarization in medial prefrontal cortex pyramidal neurons. *J Neurosci* 25:3674-3679.

Nasif FJ, Sidiropoulou K, Hu XT, White FJ (2005b) Repeated cocaine administration increases membrane excitability of pyramidal neurons in the rat medial prefrontal cortex. *J Pharmacol Exp Ther* 312:1305-1313.

Nazarian A, Russo SJ, Festa ED, Kraish M, Quinones-Jenab V (2004) The role of D1 and D2 receptors in the cocaine conditioned place preference of male and female rats. *Brain Res Bull* 63:295-299.

Nicola SM, Surmeier J, Malenka RC (2000) Dopaminergic modulation of neuronal excitability in the striatum and nucleus accumbens. *Annu Rev Neurosci* 23:185-215.

Ortiz J, Fitzgerald LW, Charlton M, Lane S, Trevisan L, Guitart X, Shoemaker W, Duman RS, Nestler EJ (1995) Biochemical actions of chronic ethanol exposure in the mesolimbic dopamine system. *Synapse* 21:289-298.

Otani S (2003) Prefrontal cortex function, quasi-physiological stimuli, and synaptic plasticity. *J Physiol Paris* 97:423-430.

Otani S, Blond O, Desce JM, Crepel F (1998) Dopamine facilitates long-term depression of glutamatergic transmission in rat prefrontal cortex. *Neuroscience* 85:669-676.

Overton PG, Clark D (1997) Burst firing in midbrain dopaminergic neurons. *Brain Res Brain Res Rev* 25:312-334.

Overton PG, Richards CD, Berry MS, Clark D (1999) Long-term potentiation at excitatory amino acid synapses on midbrain dopamine neurons. *Neuroreport* 10:221-226.

Park WK, Bari AA, Jey AR, Anderson SM, Spealman RD, Rowlett JK, Pierce RC (2002) Cocaine administered into the medial prefrontal cortex reinstates cocaine-seeking behavior by increasing AMPA receptor-mediated glutamate transmission in the nucleus accumbens. *J Neurosci* 22:2916-2925.

Pennartz CM, Ameerun RF, Groenewegen HJ, Lopes da Silva FH (1993) Synaptic plasticity in an in vitro slice preparation of the rat nucleus accumbens. *Eur J Neurosci* 5:107-117.

Peterson JD, Wolf ME, White FJ (2000) Altered responsiveness of medial prefrontal cortex neurons to glutamate and dopamine after withdrawal from repeated amphetamine treatment. *Synapse* 36:342-344.

Piazza PV, Le Moal M (1998) The role of stress in drug self-administration. *Trends Pharmacol Sci* 19:67-74.

Robbe D, Bockaert J, Manzoni OJ (2002a) Metabotropic glutamate receptor 2/3-dependent long-term depression in the nucleus accumbens is blocked in morphine withdrawn mice. *Eur J Neurosci* 16:2231-2235.

Robbe D, Kopf M, Remaury A, Bockaert J, Manzoni OJ (2002b) Endogenous cannabinoids mediate long-term synaptic depression in the nucleus accumbens. *Proc Natl Acad Sci U S A* 99:8384-8388.

Robinson TE, Becker JB (1986) Enduring changes in brain and behavior produced by chronic amphetamine administration: a review and evaluation of animal models of amphetamine psychosis. *Brain Res* 396:157-198.

Robinson TE, Berridge KC (2000) The psychology and neurobiology of addiction: an incentive-sensitization view. *Addiction* 95 Suppl 2:S91-117.

Saal D, Dong Y, Bonci A, Malenka RC (2003) Drugs of abuse and stress trigger a common synaptic adaptation in dopamine neurons. *Neuron* 37:577-582.

Sanchez CJ, Sorg BA (2001) Conditioned fear stimuli reinstate cocaine-induced conditioned place preference. *Brain Res* 908:86-92.
Sanchez CJ, Bailie TM, Wu WR, Li N, Sorg BA (2003) Manipulation of dopamine d1-like receptor activation in the rat medial prefrontal cortex alters stress- and cocaine-induced reinstatement of conditioned place preference behavior. *Neuroscience* 119:497-505.
Schnell E, Sizemore M, Karimzadegan S, Chen L, Bredt DS, Nicoll RA (2002) Direct interactions between PSD-95 and stargazin control synaptic AMPA receptor number. *Proc Natl Acad Sci U S A* 99:13902-13907.
Schultz W (2002) Getting formal with dopamine and reward. *Neuron* 36:241-263.
Shaham Y, Shalev U, Lu L, De Wit H, Stewart J (2003) The reinstatement model of drug relapse: history, methodology and major findings. *Psychopharmacology* (Berl) 168:3-20.
Shi SH, Hayashi Y, Petralia RS, Zaman SH, Wenthold RJ, Svoboda K, Malinow R (1999) Rapid spine delivery and redistribution of AMPA receptors after synaptic NMDA receptor activation. *Science* 284:1811-1816.
Stein V, House DR, Bredt DS, Nicoll RA (2003) Postsynaptic density-95 mimics and occludes hippocampal long-term potentiation and enhances long-term depression. *J Neurosci* 23:5503-5506.
Thomas MJ, Malenka RC, Bonci A (2000) Modulation of long-term depression by dopamine in the mesolimbic system. *J Neurosci* 20:5581-5586.
Thomas MJ, Beurrier C, Bonci A, Malenka RC (2001) Long-term depression in the nucleus accumbens: a neural correlate of behavioral sensitization to cocaine. *Nat Neurosci* 4:1217-1223.
Tsukamoto T, Iyo M, Tani K, Sekine Y, Hashimoto K, Ohashi Y, Suzuki K, Iwata Y, Mori N (2001) The effects of FK506, a specific calcineurin inhibitor, on methamphetamine-induced behavioral change and its sensitization in rats. *Psychopharmacology* (Berl) 158:107-113.
Tzschentke TM (2001) Pharmacology and behavioral pharmacology of the mesocortical dopamine system. *Prog Neurobiol* 63:241-320.
Tzschentke TM, Schmidt WJ (2003) Glutamatergic mechanisms in addiction. *Mol Psychiatry* 8:373-382.
Ungless MA, Whistler JL, Malenka RC, Bonci A (2001) Single cocaine exposure in vivo induces long-term potentiation in dopamine neurons. *Nature* 411:583-587.
Vanderschuren LJ, Kalivas PW (2000) Alterations in dopaminergic and glutamatergic transmission in the induction and expression of behavioral sensitization: a critical review of preclinical studies. *Psychopharmacology* (Berl) 151:99-120.
White FJ, Hu XT, Zhang XF, Wolf ME (1995) Repeated administration of cocaine or amphetamine alters neuronal responses to glutamate in the mesoaccumbens dopamine system. *J Pharmacol Exp Ther* 273:445-454.
Will MJ, Watkins LR, Maier SF (1998) Uncontrollable stress potentiates morphine's rewarding properties. *Pharmacol Biochem Behav* 60:655-664.
Wise RA, Bozarth MA (1987) A psychomotor stimulant theory of addiction. *Psychol Rev* 94:469-492.
Wolf ME (1998) The role of excitatory amino acids in behavioral sensitization to psychomotor stimulants. *Prog Neurobiol* 54:679-720.
Wolf ME, Sun X, Mangiavacchi S, Chao SZ (2004) Psychomotor stimulants and neuronal plasticity. *Neuropharmacology* 47 Suppl 1:61-79.

Yang CR, Seamans JK (1996) Dopamine D1 receptor actions in layers V-VI rat prefrontal cortex neurons in vitro: modulation of dendritic-somatic signal integration. *J Neurosci* 16:1922-1935.

Yang CR, Seamans JK, Gorelova N (1996) Electrophysiological and morphological properties of layers V-VI principal pyramidal cells in rat prefrontal cortex in vitro. *J Neurosci* 16:1904-1921.

Yao WD, Gainetdinov RR, Arbuckle MI, Sotnikova TD, Cyr M, Beaulieu JM, Torres GE, Grant SG, Caron MG (2004) Identification of PSD-95 as a regulator of dopamine-mediated synaptic and behavioral plasticity. *Neuron* 41:625-638.

Zamanillo D, Sprengel R, Hvalby O, Jensen V, Burnashev N, Rozov A, Kaiser KM, Koster HJ, Borchardt T, Worley P, Lubke J, Frotscher M, Kelly PH, Sommer B, Andersen P, Seeburg PH, Sakmann B (1999) Importance of AMPA receptors for hippocampal synaptic plasticity but not for spatial learning. *Science* 284:1805-1811.

Zhang XF, Hu XT, White FJ (1998) Whole-cell plasticity in cocaine withdrawal: reduced sodium currents in nucleus accumbens neurons. *J Neurosci* 18:488-498.

Zhang XF, Cooper DC, White FJ (2002) Repeated cocaine treatment decreases whole-cell calcium current in rat nucleus accumbens neurons. *J Pharmacol Exp Ther* 301:1119-1125.

In: Cognitive Science Compendium, Volume 1
Editor: Miao-Kun Sun
ISBN: 978-1-60456-597-3

Chapter 8

CORTICAL PLASTICITY FOR LANGUAGE PROCESSING IN THE HUMAN BRAIN

Kuniyoshi L. Sakai [1,2,*] ***and Makiko Muto*** [1,2]

[1] Department of Cognitive and Behavioral Science, Graduate School of Arts and Sciences, The University of Tokyo, Komaba, 3-8-1 Komaba, Meguro-ku, Tokyo 153-8902, Japan

[2] CREST, Japan Science and Technology Corporation, Kawaguchi-shi 332-0012, Japan

ABSTRACT

A second language (L2) can be mastered at any time in life, though the L2 ability becomes rarely comparable to first language (L1) if it is acquired beyond the hypothesized critical period around puberty. The neural substrates of plasticity for acquiring languages are largely unknown. We used functional imaging to examine whether training in the conjugation of English verbs from present to past tense alters brain activations in 13-year-old twins. A novel experimental design contrasted past tense verb identification and verb matching, which were tested in either English (L2) or Japanese (L1). After two-month classroom training in the past tense, the left dorsal inferior frontal gyrus (IFG) exhibited significantly correlated activation increases for L2 within each pair of twins, and the increases were positively correlated with individual performance improvements. Moreover, the cortical plasticity for L2 acquisition was guided toward the L1 specialization of the left dorsal IFG, in spite of notable differences between L1 and L2 in the students' linguistic knowledge and in their performance in making past tense forms. These findings suggest a cortical mechanism underlying L2 acquisition of grammatical knowledge, which critically depends on shared genetic and environmental factors for each twin in a surprisingly predictive manner. Accumulating evidence from recent imaging and stimulation studies has suggested that the left IFG is specialized in grammatical processing, which is thus referred to as the "grammar center".

* Corresponding author. Email: sakai@mind.c.u-tokyo.ac.jp

On the other hand, reading skills of letters and words are usually acquired at school age but can still be mastered in adulthood. To identify which brain regions in adults show plasticity for learning letters, Hangul letters were experimentally associated with either speech sounds (HS condition) or non-speech sounds (HN condition) in imaging sessions over two consecutive days. Selective activations under the HS condition were found in regions including the left posterior inferior temporal gyrus (PITG) and the parieto-occipital cortex (PO), as compared with activations under a condition for familiar letters and speech sounds, and with those under the HN condition. The left PITG showed a selective activation increase under the HS condition over two days, the degree of which predicted individual performance improvement. Therefore, the left PITG can be referred to as the "letter center". Furthermore, functional connectivity between the left PITG and the left PO was selectively enhanced under the HS condition. These results demonstrate that a new link between orthography and phonology is formed by the plasticity of a functional system involving the left PITG in association with the left PO. Our approach to evaluate learning procedures in terms of not only indirect behavioral changes but direct functional changes takes a first step toward a new era in the brain science of education.

Keywords: Language acquisition; Syntax; Universal grammar (UG); Letter learning; Functional magnetic resonance imaging (fMRI)

1. The Plasticity of the "Grammar Center"

The specialization of language processing in human cognitive systems is one of the central issues in neuroscience, and it has been highly debated from a number of perspectives based on lesion studies of Broca's aphasia and functional imaging studies of the prefrontal cortex (Caplan et al., 1999; Friederici, 2002; Sakai et al., 2003). The narrowest definition of Broca's area is the left pars opercularis [F3op, Brodmann's area (BA) 44] and the left pars triangularis (F3t, BA 45), a part of the third frontal convolution (F3) or the left inferior frontal gyrus (IFG). However, the syndrome referred to as permanent Broca's aphasia arises from a considerably larger brain lesion that includes the insula and subjacent white matter (Mohr, 1976), and the region medial to Broca's area (the left precentral gyrus of the insula) has been implicated in the motor planning of speech (Dronkers, 1996). Recent functional magnetic resonance imaging (fMRI) studies have provided accumulating evidence that the lateral surrounding region of Broca's area, including the left IFG and the left lateral premotor cortex (LPMC), is selectively related to grammatical processing (Stromswold et al., 1996; Dapretto and Bookheimer, 1999; Kang et al., 1999; Embick et al., 2000; Friederici et al., 2000; Suzuki and Sakai, 2003), independent of domain-general factors such as task difficulty and short-term memory (Hashimoto and Sakai, 2002), and of unreal language rule usage (Musso et al., 2003). Furthermore, the causal link between the activation of the left IFG and grammatical decisions has been recently proved by a transcranial magnetic stimulation (TMS) study (Sakai et al., 2002). Therefore, the left prefrontal regions can be suggested as the most likely candidate of the "grammar center" (Figure 1).

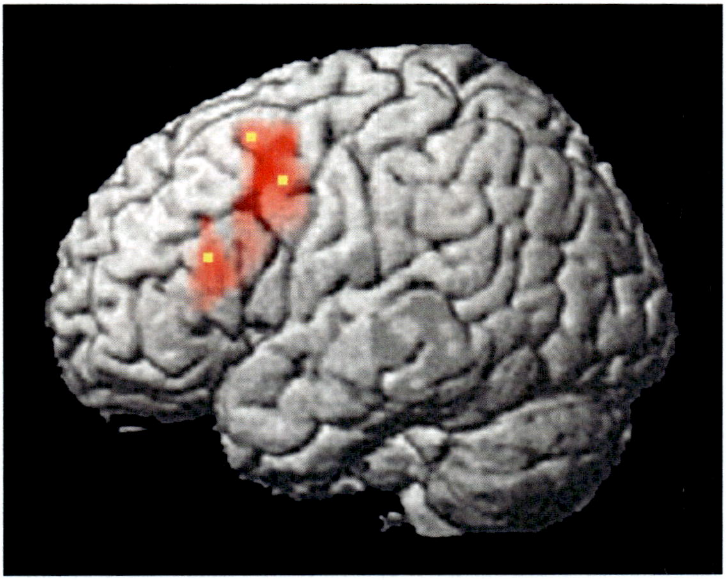

Figure 1. The grammar center in the left prefrontal cortex. Regions (red) identified by directly comparing syntactic decision tasks and a short-term memory task for words are projected together onto the left hemisphere of a surface-rendered standard brain.

The next primary challenge is how the function of the grammar center can be modified during the course of acquiring new languages. The native or first language (L1) is normally acquired in the first several years of life, during which children rapidly expand their linguistic knowledge (Gopnik et al., 1999; Boysson-Bardies, 1999). In contrast, a second language (L2) can be mastered at any time in life, though the L2 ability becomes rarely comparable to L1 if it is acquired beyond the hypothesized critical period around puberty (Lenneberg, 1967; Johnson and Newport, 1989). Whether L2 relies on the same dedicated mechanism of L1, which may critically involve the maturation of the grammar center, is thus a matter of debate (Epstein et al., 1996). It has been recently reported that L1 and L2 are represented differentially in cortical areas during discourse production or listening tasks (Kim et al., 1997; Dehaene et al., 1997). However, other imaging studies have reported common neural substrates of L1 and L2 during word generation tasks (Klein et al., 1995; Chee et al., 1999). The critical experiments thus would be to clarify brain plasticity that represents new acquisition of L2, i.e., the dynamic processes of L2 acquisition and associated cortical changes, rather than static states that reflect a certain level of L2 performance.

We have examined whether the learning of English past tense verbs as L2 knowledge alters brain activations for the 13-year-old students (native Japanese speakers) studying English for the first time at a secondary education school in Japan (Sakai et al., 2004). We targeted twins as subjects (six monozygotic and one dizygotic twin pairs), because it is intriguing to ask whether shared factors of twins actually influence their language abilities and neural substrates for Japanese (L1) and English (L2). For two months, the students participated in intensive training in English verbs as part of their standard classroom education. To directly evaluate the brain's changes in activation due to this training, the twins completed two sets of fMRI sessions, one before the training (Day 1) and one after the training (Day 2). Figure 2A illustrates the experimental paradigm with four tasks used in fMRI sessions: an English verb-matching (EM) task, an English past tense (EP) task, a Japanese verb-matching (JM) task, and a Japanese past tense (JP) task. General cognitive factors such as word recognition and response selection were controlled by the EM and JM tasks, which were directly compared with the EP and JP tasks, respectively. A particular challenge in this study was to assess the effect of an educational method used in classroom lessons directly in terms of brain activation.

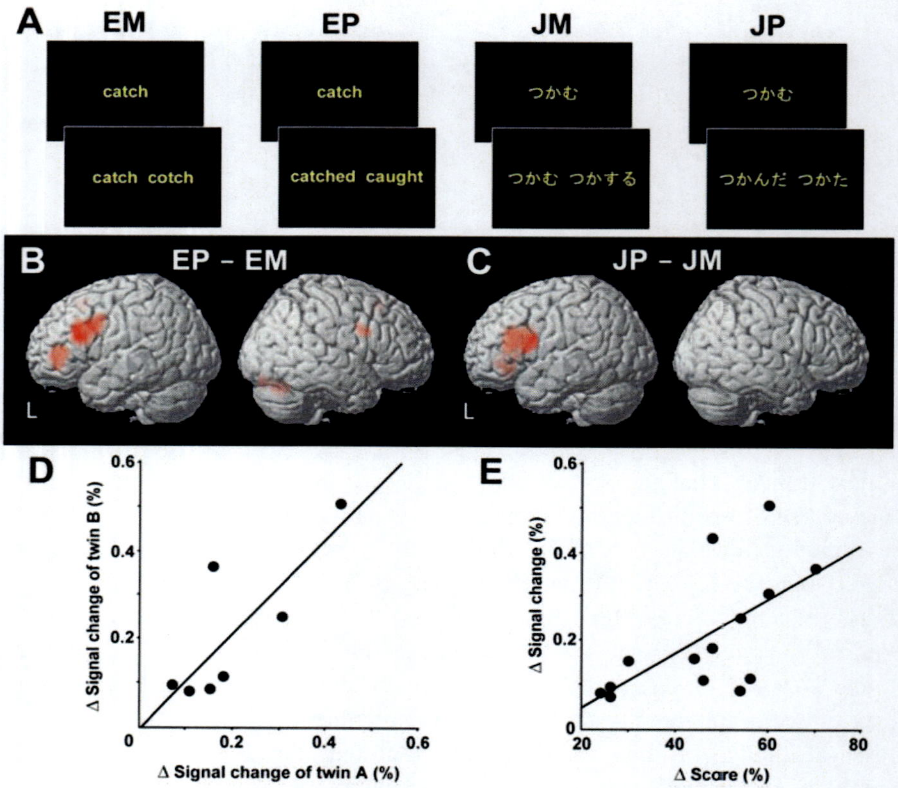

Figure 2. Correlated functional changes of the grammar center in twins induced by classroom education of second language. (A) The four language tasks used to test the plasticity of the grammar center. (*Left to right:* English verb-matching (EM) task, English past tense (EP) task, Japanese verb-matching (JM) task, and Japanese past tense (JP) task. For JP, one example (*tsukamu*; gross: catch) is shown, and the one on the left (*tsukannda*; gross: caught) is correct. Japanese stimuli were presented in both the *hiragana* and *kanji* writing systems and were matched with English words of the same meaning. (B) EP – EM for Day 2. The most prominent activation change was observed in the left (L) inferior frontal gyrus (IFG). (C) JP – JM for Day 2. Note the consistent activation in the left IFG. (D) The activation increases (ΔSignal change: Day 2 – Day 1) of the left dorsal IFG in EP – EM, plotted for each pair of twins (identified as twin A and twin B). (E) Significant correlation between L2 performance improvements and activation increases in the left dorsal IFG. For all individual participants, the activation increases (ΔSignal change: Day 2 – Day 1) of the left IFG in EP – EM are plotted against improvements in examination scores (ΔScore: Day 2 – Day 1). A regression line is shown.

When the EP task was contrasted with the EM task (EP – EM) in a random effects analysis for Day 2, activations were found primarily in the left IFG (Figure 2B), which were absent in EP - EM for Day 1. Moreover, the JP – JM contrast for Day 2 revealed the left IFG activation (Figure 2C), which was stable and remained unchanged between Days 1 and 2. It is thus striking to note that the left IFG activated for the English past tense verbs exactly match the regions activated for the Japanese past tense verbs, which is in agreement with the universal nature of grammatical processing. These results suggest that the cortical plasticity for L2 acquisition is guided toward the L1 specialization of the left dorsal IFG, at least at the age of 13, in spite of notable differences between L1 and L2 in the students' linguistic knowledge and in their performance in making past tense forms.

We next examined whether or not shared factors of twins influence the functional changes observed here. The activation increases at one local maximum (coordinates in the standard brain: –45, 21, 30) of the left dorsal IFG across Days 1 and 2 showed a highly significant correlation within each pair of twins ($r = 0.80$) (Figure 2D). The permutation procedure of randomly assigning each individual to a new pair resulted in the normal distribution with mean $r = 0.0$ and $SD = 0.39$, confirming this statistical significance at $P = 0.02$. The activation changes of the same region in JP – JM showed no positive correlation within each pair of twins ($r = -0.46$), indicating that the significant correlation is selective to L2 acquisition. These results suggest that the functional changes specifically observed in the left IFG are susceptible to shared genetic and environmental factors for each twin in a surprisingly predictive manner. Furthermore, the activation increases in the left IFG predicted the extent to which each individual subject improved his/her knowledge of the past tense, as shown by a significant positive correlation ($r = 0.63$) between the improvements in examination scores and the activation increases in EP – EM across Days 1 and 2 (Figure 1E). Therefore, we conclude that the amount by which activation increases in the left dorsal IFG is a good indicator of individual improvement in acquiring L2 past tense knowledge. The present fMRI study thus successfully elucidates one type of cerebral mechanisms involving the grammar center, indicating how new linguistic knowledge is acquired and represented in individual brains. To our knowledge, the present study is the first direct demonstration that classroom education, if properly executed, can change the function of the prefrontal cortex.

2. The Plasticity of the "Letter Center"

Reading letters and words consists of multiple steps including orthographic, phonological, and lexico-semantic processing, which are usually acquired at school age but can still be mastered in adulthood. Previous imaging studies for adult subjects have reported that the left fusiform gyrus (FG) showed more activation for visually presented consonant letter strings than for pseudofonts or faces (Price et al., 1996; Puce et al., 1996; Tarkiainen et al., 2002), indicating that this region is selectively involved in letter recognition. Moreover, recent functional imaging studies have clarified that the left FG showed larger activation for both pronounceable words and pseudowords than for unpronounceable consonant letter strings (Fiez et al., 1999; Brunswick et al., 1999; Cohen et al., 2002), suggesting that the left FG discriminates, on the basis of orthographic regularities, between legal and illegal strings. Several studies have also reported that the activation for letter strings extended into the left posterior inferior temporal gyrus (PITG) (Paulesu et al., 2000; Xu et al., 2001; Mechelli et al., 2003). While the left FG/PITG is thus referred to as the "visual word form area" (McCandliss et al., 2003), other researchers have clarified that this region is also activated during various multimodal tasks that do not engage visual word form processing (Price and Devlin, 2003). It is thus crucial to determine which cognitive factors are responsible for activating the left FG/PITG. In this paper, we tentatively call the left FG/PITG the "letter center".

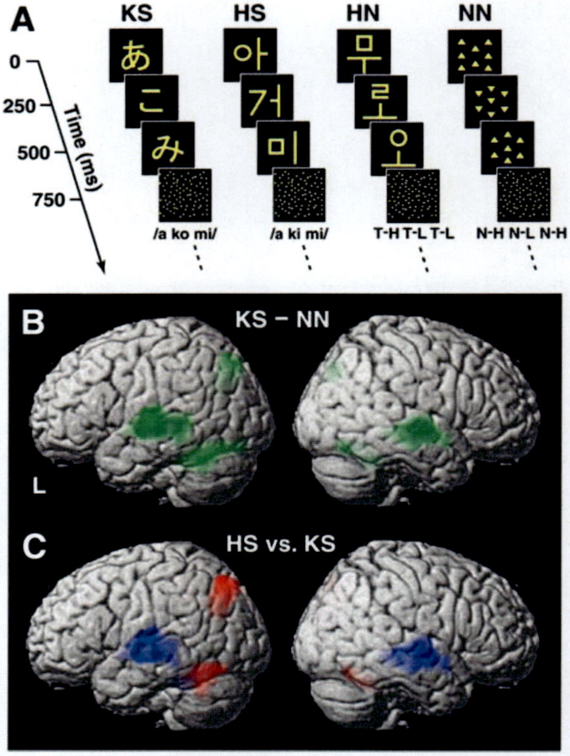

Figure 3. An audiovisual matching paradigm for testing a new link between orthography and phonology. (A) The four experimental conditions: kana and speech (KS), Hangul and speech (HS), Hangul and non-speech (HN), and non-letter and non-speech (NN). Under one of these conditions, a string of three visual stimuli (250 ms each) was presented at the center of the screen for each trial of 4 s. The subjects were asked to judge whether or not the visual string matched a string of auditory stimuli (total of 550 ms) that were presented either before or after the visual stimuli. A random dot pattern was presented between visual strings. A red cross for fixation was always shown at the center of the screen, but it was omitted from the figure. For the KS and NN examples the visual and auditory strings matched, and they did not match for the HS and HN examples. (B) Regions identified by the contrast KS – NN (green), using the data in Days 1 and 2. Activations of the superior and middle temporal gyri (STG/MTG) are bilateral, while left-dominant activation is observed over the fusiform gyrus (FG), the posterior inferior temporal gyrus (PITG), and the parieto-occipital cortex (PO). (C) Regions identified by HS – KS (red) and KS – HS (blue). Whereas the bilateral STG/MTG and the left FG are selectively activated under the KS condition, the left PITG and the left PO show selective activations under the HS condition.

A fundamental question is whether the left FG/PITG or other regions show plasticity for learning to associate letters with speech sounds, which is one critical subprocess of reading. In a recent fMRI study, we examined whether forming a new link between orthography and other linguistic information is necessary for the cortical plasticity in those regions (Hashimoto and Sakai, 2004). Cortical activations were monitored while the adult subjects were trained with Hangul letters for the first time. The subjects were Japanese who had learned kana letters by the age of seven, but they had no knowledge about Hangul letters before the experiments. Kana and Hangul letters are exclusively used in the Japanese and Korean languages, respectively. For both kana and Hangul letters used in the present study, there was a one-to-one correspondence between each letter and a speech sound. Here we focused on prelexical processing and used nonsense words for all the linguistic stimuli. The subjects participated in fMRI sessions over two consecutive days (Days 1 and 2, approximately 24 hours apart). The scanning sessions served as both testing and repeated exposures to the Hangul letters, and there was no separate training session except the initial exposure to the letters.

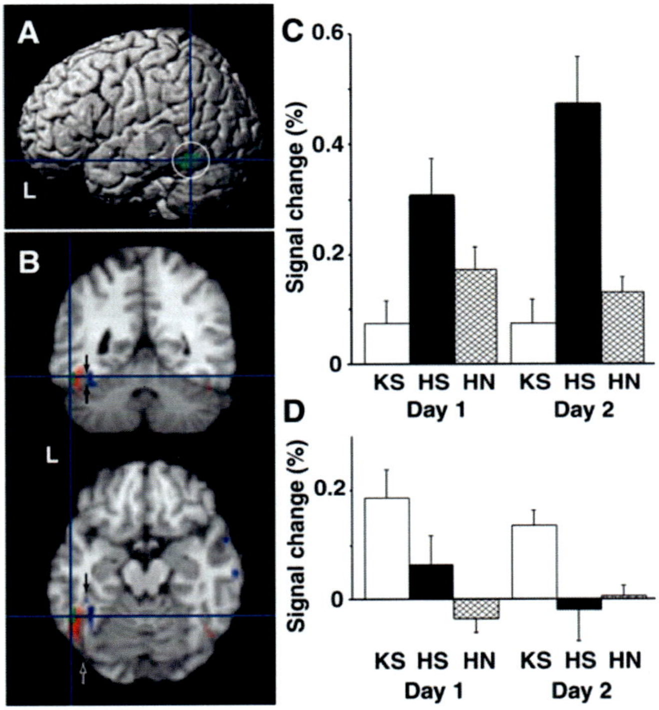

Figure 4. Direct visualization of cortical plasticity of the letter center for forming a new link between orthography and phonology. (A) A region exhibiting an activation increase induced by learning Hangul letters with speech sounds. The interaction of task by day in HS – NN for Day 2 – Day 1 resulted in significant activation in the left PITG (green), projected onto a surface-rendered standard brain. The location of its local maximum is shown by blue crosshairs (y = –51, z = –18). (B) The separation within the left FG/PITG. On the coronal (y = –51) and axial (z = –18) slices of a standard brain, the regions identified by KS – HS and HS – KS are shown in blue and red, respectively. Arrows indicate the lateral occipitotemporal sulcus that corresponds to the border between the PITG (lateral) and the FG (medial). Note that the region (green) with the interaction of task by day shown in (A) is localized within the left PITG. (C) Averaged activation data among subjects at the local maximum (–54, –51, –18) of the left PITG. Mean percent signal changes for KS (open bars), HS (filled bars), and HN_I (crosshatched bars) on Days 1 and 2 are shown with reference to the corresponding NN condition. Error bars denote standard errors among the subjects. Note the selective activation increase under the HS condition from Day 1 to Day 2. (D) Averaged activation data at the local maximum (–36, –42, –24) of the left FG. In contrast to the left PITG, the left FG shows larger activation for KS than for HS or HN, with no significant activation change between Days 1 and 2.

The fMRI sessions consisted of four audiovisual matching tasks characterized as the following conditions: kana and speech (KS), Hangul and speech (HS), Hangul and non-speech (HN), and non-letter and non-speech (NN) (Figure 3A). Under the KS condition, the subjects were presented with a string of three kana letters for each trial of 4 s, and they were asked to judge whether or not it matched a string of three-syllable speech sounds. Individual letters were serially presented to ensure that the subjects attended equally to each letter. Because each letter was immediately masked by the next letter or a masking stimulus for the last letter, the subjects had to recognize each individual letter within the brief presentation time (250 ms). Under the HS condition, the subjects were asked to judge whether a string of three Hangul letters matched a string of three-syllable speech sounds. Under the HN condition, we used another set of eight Hangul letters, and the subjects were trained to associate these complex Hangul letters with non-speech sounds. This condition lacks phonology; thus, no link between orthography and phonology is trained throughout the experiments. Under the NN condition, in order to control for general cognitive factors related to low-level cross-modal matching and association, we further replaced letters with geometric patterns that are associated with non-speech sounds. We targeted HS as a major learning condition of interest for matching speech sounds with newly learned letters, whereas KS and HN served as control conditions for speech sounds and Hangul letters, respectively.

By employing this novel audiovisual matching paradigm to examine cross-modal learning effects on cortical plasticity, we obtained three striking results. First, we found that the regions recruited for integrating newly learned orthographic and phonological information, the left PITG and the parieto-occipital cortex (PO), are indeed separable from the specific regions of the superior and middle temporal gyri (STG/MTG), necessary for processing associative information including already acquired orthography (Figs. 3B & 3C). Moreover, the left PITG and the left PO activations, which are selectively enhanced under the HS condition in comparison with those under the KS and HN conditions, are not due to auditory factors alone, because the same set of speech sounds were presented under the KS and HS conditions. We also excluded the possibility that they reflect visual exposure to new and complex letters alone, given that the complexity of the visual stimuli used under the HS and HN conditions was basically equivalent. Therefore, we conclude that the left PITG and the left PO subserve the formation of a new link between letter and speech representations. Second, the left PITG, but not the left FG, showed a selective activation increase under the HS condition over two days (Figure 4), the degree of which was predictive of individual performance improvement. Behavioral results also indicate that learning effects of HS and HN on Day 1 were clearly manifested on Day 2. Moreover, it is intriguing that the functional separation between the two adjacent regions of the left PITG and the left FG followed the anatomical boundary of the lateral occipitotemporal sulcus (Figure 4B). These results suggest that the left PITG plays a primary and critical role in the cortical plasticity involved in learning new letters. Third, we demonstrated that functional connectivity between the left PITG and the left PO was selectively enhanced under the HS condition, but not under the KS or HN condition. The present fMRI study thus establishes that the human adult brain is still adaptable to new associations between orthographic and phonological stimuli, and that this type of learning is achieved in only two days by the rapid plasticity of a functional system critically involving the left PITG in association with the left PO.

The ability of learning to read is selectively impaired in patients diagnosed with developmental dyslexia (Shaywitz, 1996; Habib, 2000). With regard to the neurological basis of this impairment, previous imaging studies have revealed reduced activation in the left FG/PITG of dyslexic adults relative to that in normal controls when reading words or pseudowords (Shaywitz et al., 1998; Brunswick et al., 1999; Paulesu et al., 2001). A magnetoencephalography (MEG) study revealed cortical responses selective to letter strings in the left inferior occipitotemporal cortex in normal adults; these responses were not detectable in dyslexics (Helenius et al., 1999). The findings in the present study are consistent with those in these studies, in that they suggest that the lesions in the left FG/PITG may impair linking orthography to phonology. Furthermore, we have clearly established differential roles for the left FG and the left PITG; the left FG processes information about already acquired orthography, whereas the left PITG integrates newly learned orthography and phonology. We suggest that in the case of the literate adult reader, a selective lesion of the left FG or the left PITG might lead to double dissociation of deficits in reading letters, such that a left FG lesion selectively impairs already acquired orthography, whereas a left PITG impairment selectively impairs the formation of a new link between orthography and phonology. Taken together, the present results provide the initial step towards understanding the neural mechanisms that underlie the scope and limits of learning in school-age children and adulthood. Research in this direction could contribute to the neuroscience of education, and be utilized to develop effective training programs for language learners and for patients with language disabilities.

Acknowledgments

We would like to thank Natsuki Saeki for her technical assistance. This work was supported by a Core Research of Evolutional Science and Technology (CREST) grant from the Japan Science and Technology Corporation.

References

Boysson-Bardies, B. *How Language Comes to Children: From Birth to Two Years*, The MIT Press, Cambridge, MA, 1999.

Brunswick, N.; McCrory, E.; Price, C.J.; Frith, C.D.; Frith, U. Explicit and implicit processing of words and pseudowords by adult developmental dyslexics - A search for Wernicke's Wortschatz? *Brain* 122: 1901-1917, 1999.

Caplan, D.; Alpert, N.; Waters, G. PET studies of syntactic processing with auditory sentence presentation. *Neuroimage* 9: 343-351, 1999.

Chee, M.W.L.; Tan, E.W.L.; Thiel, T. Mandarin and English single word processing studied with functional magnetic resonance imaging. *J. Neurosci.* 19: 3050-3056, 1999.

Cohen, L.; Lehéricy, S.; Chochon, F.; Lemer, C.; Rivaud, S.; Dehaene, S. Language-specific tuning of visual cortex? Functional properties of the visual word form area. *Brain* 125: 1054-1069, 2002.

Dapretto, M.; Bookheimer, S.Y. Form and content: Dissociating syntax and semantics in sentence comprehension. *Neuron* 24: 427-432, 1999.

Dehaene, S.; Dupoux, E.; Mehler, J.; Cohen, L.; Paulesu, E.; Perani, D.; Van de Moortele, P.-F.; Lehéricy, S.; Le Bihan, D. Anatomical variability in the cortical representation of first and second language. *NeuroReport* 8: 3809-3815, 1997.

Dronkers, N.F. A new brain region for coordinating speech articulation. *Nature* 384: 159-161, 1996.

Embick, D.; Marantz, A.; Miyashita, Y.; O'Neil, W.; Sakai, K.L. A syntactic specialization for Broca's area. *Proc. Natl. Acad. Sci. USA* 97: 6150-6154, 2000.

Epstein, S.D.; Flynn, S.; Martohardjono, G. Second language acquisition: Theoretical and experimental issues in contemporary research. *Behav. Brain Sci.* 19: 677-714, 1996.

Fiez, J.A.; Balota, D.A.; Raichle, M.E.; Petersen, S.E. Effects of lexicality, frequency, and spelling-to-sound consistency on the functional anatomy of reading. *Neuron* 24: 205-218, 1999.

Friederici, A.D. Towards a neural basis of auditory sentence processing. *Trends Cognit. Sci.* 6: 78-84, 2002.

Friederici, A.D.; Opitz, B.; von Cramon, D.Y. Segregating semantic and syntactic aspects of processing in the human brain: An fMRI investigation of different word types. *Cereb. Cortex* 10: 698-705, 2000.

Gopnik, A.; Meltzoff, A.N.; Kuhl, P.K. *The Scientist in the Crib: Mind, Brains, and How Children Learn,* William Morrow and Co., Inc., New York, 1999.

Habib, M. The neurological basis of developmental dyslexia - An overview and working hypothesis. *Brain* 123: 2373-2399, 2000.

Hashimoto, R.; Sakai, K.L. Specialization in the left prefrontal cortex for sentence comprehension. *Neuron* 35: 589-597, 2002.

Hashimoto, R.; Sakai, K.L. Learning letters in adulthood: Direct visualization of cortical plasticity for forming a new link between orthography and phonology. *Neuron* 42: 311-322, 2004.

Helenius, P.; Tarkiainen, A.; Cornelissen, P.; Hansen, P.C.; Salmelin, R. Dissociation of normal feature analysis and deficient processing of letter-strings in dyslexic adults. *Cereb. Cortex* 9: 476-483, 1999.

Johnson, J.S.; Newport, E.L. Critical period effects in second language learning: The influence of maturational state on the acquisition of English as a second language. *Cognit. Psychol.* 21: 60-99, 1989.

Kang, A.M.; Constable, R.T.; Gore, J.C.; Avrutin, S. An event-related fMRI study of implicit phrase-level syntactic and semantic processing. *Neuroimage* 10: 555-561, 1999.

Kim, K.H.S.; Relkin, N.R.; Lee, K.M.; Hirsch, J. Distinct cortical areas associated with native and second languages. *Nature* 388: 171-174, 1997.

Klein, D.; Milner, B.; Zatorre, R.J.; Meyer, E.; Evans, A.C. The neural substrates underlying word generation: A bilingual functional-imaging study. *Proc. Natl. Acad. Sci. USA* 92: 2899-2903, 1995.

Lenneberg, E.H. Biological Foundations of Language, John Wiley & Sons, Inc., New York, 1967.

McCandliss, B.D.; Cohen, L.; Dehaene, S. The visual word form area: Expertise for reading in the fusiform gyrus. *Trends Cognit. Sci.* 7: 293-299, 2003.

Mechelli, A.; Gorno-Tempini, M.L.; Price, C.J. Neuroimaging studies of word and pseudoword reading: Consistencies, inconsistencies, and limitations. *J. Cognit. Neurosci.* 15: 260-271, 2003.

Mohr, J.P. Broca's area and Broca's aphasia. Whitaker, H.; Whitaker, H.A. (Eds.), *Studies in Neurolinguistics*, pp. 201-235, Academic Press, New York, 1976.

Musso, M.; Moro, A.; Glauche, V.; Rijntjes, M.; Reichenbach, J.; Büchel, C.; Weiller, C. Broca's area and the language instinct. *Nat. Neurosci.* 6: 774-781, 2003.

Paulesu, E.; Démonet, J.F.; Fazio, F.; McCrory, E.; Chanoine, V.; Brunswick, N.; Cappa, S.F.; Cossu, G.; Habib, M.; Frith, C.D. et al., .Dyslexia: Cultural diversity and biological unity. *Science* 291: 2165-2167, 2001.

Paulesu, E.; McCrory, E.; Fazio, F.; Menoncello, L.; Brunswick, N.; Cappa, S.F.; Cotelli, M.; Cossu, G.; Corte, F.; Lorusso, M. et al., . A cultural effect on brain function. *Nat. Neurosci.* 3: 91-96, 2000.

Price, C.J.; Devlin, J.T. The myth of visual word form area. *Neuroimage* 19: 473-481, 2003.

Price, C.J.; Wise, R.J.S.; Frackowiak, R.S.J. Demonstrating the implicit processing of visually presented words and pseudowords. *Cereb. Cortex* 6: 62-70, 1996.

Puce, A.; Allison, T.; Asgari, M.; Gore, J.C.; McCarthy, G. Differential sensitivity of human visual cortex to faces, letterstrings, and textures: A functional magnetic resonance imaging study. *J. Neurosci.* 16: 5205-5215, 1996.

Sakai, K.L.; Homae, F.; Hashimoto, R. Sentence processing is uniquely human. *Neurosci. Res.* 46: 273-279, 2003.

Sakai, K.L.; Miura, K.; Narafu, N.; Muraishi, M. Correlated functional changes of the prefrontal cortex in twins induced by classroom education of second language. *Cereb. Cortex* 14: 1233-1239, 2004.

Sakai, K.L.; Noguchi, Y.; Takeuchi, T.; Watanabe, E. Selective priming of syntactic processing by event-related transcranial magnetic stimulation of Broca's area. *Neuron* 35: 1177-1182, 2002.

Shaywitz, S.E. Dyslexia. *Sci. Am.* 275 (5): 78-84, 1996.

Shaywitz, S.E.; Shaywitz, B.A.; Pugh, K.R.; Fulbright, R.K.; Constable, R.T.; Mencl, W.E.; Shankweiler, D.P.; Liberman, A.M.; Skudlarski, P.; Fletcher, J.M. et al., . Functional disruption in the organization of the brain for reading in dyslexia. *Proc. Natl. Acad. Sci. USA* 95: 2636-2641, 1998.

Stromswold, K.; Caplan, D.; Alpert, N.; Rauch, S. Localization of syntactic comprehension by positron emission tomography. *Brain Lang.* 52: 452-473, 1996.

Suzuki, K.; Sakai, K.L. An event-related fMRI study of explicit syntactic processing of normal/anomalous sentences in contrast to implicit syntactic processing. *Cereb. Cortex* 13: 517-526, 2003.

Tarkiainen, A.; Cornelissen, P.L.; Salmelin, R. Dynamics of visual feature analysis and object-level processing in face versus letter-string perception. *Brain* 125: 1125-1136, 2002.

Xu, B.; Grafman, J.; Gaillard, W.D.; Ishii, K.; Vega-Bermudez, F.; Pietrini, P.; Reeves-Tyer, P.; DiCamillo, P.; Theodore, W. Conjoint and extended neural networks for the computation of speech codes: The neural basis of selective impairment in reading words and pseudowords. *Cereb. Cortex* 11: 267-277, 2001

.

In: Cognitive Science Compendium, Volume 1
Editor: Miao-Kun Sun
ISBN: 978-1-60456-597-3

Chapter 9

MEMORY RECONSOLIDATION: A MOLECULAR PERSPECTIVE

Laura S. J. von Hertzen and K. Peter Giese*
Wolfson Institute for Biomedical Research, University College London, Gower Street, London, WC1E 6BT, UK

ABSTRACT

A memory undergoes a consolidation process in order to become stable. This new memory is de-stabilized when it is subsequently retrieved, but may undergo a reconsolidation phase, which allows the memory to be maintained. In recent years, memory reconsolidation has been the subject of much debate as well as excitement in the neuroscience community; not only because the phenomenon has extended the scope of the consolidation theory, but also because its discovery promises clinical applications, such as the treatment of post-traumatic stress disorders or drug addiction. Here we review recent advances in the study of the molecular processes underlying memory reconsolidation. Such studies aimed to provide mechanistic insights into memory reconsolidation, and to establish the relationship between memory consolidation and reconsolidation. The current findings lead to the hypothesis that memory reconsolidation is a partial recapitulation of memory consolidation.

INTRODUCTION

Fresh memories need to be stabilized in order to become long lasting through a process generally referred to as memory consolidation. Memory consolidation begins directly after training, and interference with the process impairs long-term but not short-term memory (for review, see Silva and Giese, 1994; Dudai, 2004). During memory consolidation de novo transcription and protein synthesis occur, and post-genomic studies suggest that a large

* Corresponding author: p.giese@ucl.ac.uk

number of proteins are synthesized (e.g., Levenson et al., 2004). It was long believed that memories undergo consolidation only once and that subsequently they remain permanent and unmodifiable. However, this so-called consolidation theory has been challenged by the finding that the retrieval of a consolidated memory can return the memory to a labile state, inducing a further consolidation process termed memory reconsolidation (Nader et al., 2000; Sara, 2000). During memory reconsolidation de novo transcription occurs (von Hertzen and Giese, 2005a), and inhibition of transcription or protein synthesis within a short time window after memory retrieval blocks memory reconsolidation, resulting in the loss of the previously consolidated memory (Nader et al., 2000; Taubenfeld et al., 2001; Anokhin et al., 2002; Debiec et al., 2002; Kida et al., 2002; Milekic and Alberini, 2002).

Much current research aims to understand the connection between memory consolidation and reconsolidation. One way of establishing this relationship is to undertake a molecular comparison of the two processes, because such a study can reveal whether memory reconsolidation is a recapitulation of memory consolidation, or whether the two processes actually differ. Understanding the nature of memory reconsolidation promises not only to advance the knowledge of long-term memory formation, but it also has potential clinical implications, including the development of physiopathological and therapeutic treatments for drug addiction and post-traumatic stress disorder using the concept of memory erasure.

Comparing Molecular Processes Underlying Memory Consolidation and Memory Reconsolidation

During memory consolidation and memory reconsolidation de novo transcription and protein synthesis occur. For the molecular comparison of the two processes one can apply various techniques to monitor changes in gene expression, such as real-time polymerase chain reaction or in situ hybridizations (e.g., von Hertzen and Giese, 2005a). For the analysis of gene expression during memory consolidation animals are trained in a memory task, and killed at different time points. Control groups (see below) are included in the analysis of gene expression to ensure that the expression changes are specific for memory consolidation. For the analysis of gene expression during memory reconsolidation animals are trained in a memory task, the memory is reactivated after memory consolidation is thought to be completed, and the animals are killed at different time points. Again control groups are used to ensure the expression changes are specific for memory reconsolidation. These gene expression studies reveal which molecular processes are engaged during memory consolidation and memory reconsolidation. Additionally, functional studies can be performed, which determine whether particular expression changes are essential for memory consolidation and/or memory reconsolidation.

A big confound in the comparative study of memory reconsolidation has been the use of behavioural tasks with different cognitive demands and anatomical substrates, which might explain inconsistencies observed in the field. Ideally, the mechanisms underlying memory consolidation and memory reconsolidation should be studied in single trial learning paradigms, as these tasks allow a clear temporal dissociation between the training phase and the memory consolidation phase, an advantage which cannot be exploited in multi-trial learning tasks. Fear conditioning, in which the animal learns to associate a neutral stimulus with a mild foot shock, has been frequently used in the study of memory reconsolidation (e.g., von Hertzen and Giese, 2005a). But other single trial learning tasks such as inhibitory avoidance (Taubenfeld et al., 2001), object recognition (Bozon et al., 2003), and odor discrimination have also been used (Tronel et al., 2002).

Another important factor to take into consideration is that memory consolidation and memory reconsolidation may engage distinct anatomical systems. For a particular behavioural task, one can properly compare memory consolidation and memory reconsolidation at the molecular level, if both processes involve the brain structure under investigation. For example, in studies of passive avoidance memory Alberini and colleagues showed that protein synthesis in the hippocampus is required for consolidation but not reconsolidation (Taubenfeld et al., 2001), whereas protein synthesis in the amygdala is required for reconsolidation but not consolidation (Tronel et al., 2005). Accordingly, the investigators showed that the transcription factor C/EBPβ is required in the hippocampus for consolidation but not reconsolidation of passive avoidance memory (Taubenfeld et al., 2001). It would be wrong to interpret these results as a demonstration that the hippocampus supports distinct molecular mechanisms for consolidation and reconsolidation of passive avoidance memory. Such a claim would be truly only possible if the hippocampus was required for both consolidation and reconsolidation in this behavioural paradigm. The hippocampus is essential for both the consolidation and reconsolidation of contextual fear memory (Debiec et al., 2001). Therefore, the molecular differences between memory phases in the hippocampus are informative when the animals have been trained in contextual fear protocols (Lee et al., 2004; von Hertzen and Giese, 2005a).

Finally, the use of good experimental controls is also crucial. For example, the duration of the memory reactivation session determines whether or not it induces memory reconsolidation (Suzuki et al., 2004). When comparing molecular processes occurring after training to those observed after memory reactivation, one has to ensure that the memory reactivation triggers reconsolidation and not, for example, memory extinction, which would result in reduced task performance after memory reactivation (Suzuki et al., 2004). Furthermore, one must ensure that the changes observed are specific to memory consolidation or memory reconsolidation, and are not triggered, for example, by environmental factors alone. This can be accomplished by Pavlovian contextual fear conditioning, where one can neatly dissociate gene expression changes induced specifically by the conditioned stimulus (CS; context) or unconditioned stimulus (US; shock) from those induced by the CS-US association or memory reactivation (von Hertzen and Giese, 2005a). Such controls cannot be implemented for single trial learning tasks that use instinctive behaviour, such as novel object recognition. For such tasks as these, one cannot exclude the possibility that the gene expression changes observed after training are induced by the environment rather than by the initial learning.

Do Memory Consolidation and Memory Reconsolidation Share Common Mechanisms?

To understand the nature of memory reconsolidation, recent studies have focused on molecular correlates of memory reconsolidation versus memory consolidation. An increasing number of such comparative studies have shown that the two phenomena share common molecular processes.

Comparisons of post-translational modifications occurring during memory consolidation and reconsolidation have shown that mitogen-activated protein kinase is phosphorylated in the hippocampus after both contextual fear conditioning and reactivation of the contextual fear memory (Kelly et al., 2003). Mitogen-activated protein kinase is also phosphorylated in the amydgala after both auditory fear conditioning and reactivation of the tone fear memory (Duvarci et al., 2005). Similar results have been reported for the transcription factor NF-KappaB, which is activated by phosphorylation in the hippocampus after both contextual fear conditioning and contextual fear memory reactivation (Merlo et al., 2005). However, these studies on post-translational events must be interpreted with caution; the modification of a protein by behavioural training need not be involved in the stabilising processes of memory consolidation and memory reconsolidation. Indeed, recently Chen and colleagues have shown that phosphorylation of phosphatidyinositol 3 kinase after contextual fear memory reactivation is required for memory retrieval, but not memory reconsolidation (Chen et al., 2005). Thus, phosphorylation events after memory retrieval might not contribute to memory reconsolidation, though they might subserve quasi-instantaneous processes such as recall. It is open to question whether post-translational modifications are involved at all in stabilising memory.

In contrast, the relatively slow processes of memory consolidation and memory reconsolidation point to cellular events with slower kinetics, such as novel gene expression. Transcriptional analyses have shown that the genes encoding the serum/glucocorticoid kinase 1 (SGK-1) and 3 (SGK-3) as well as the transcription factors zif268 and c-fos are all upregulated in the hippocampus after contextual fear conditioning, as well as after re-exposure to the training context (Hall et al., 2000, 2001; Stanciu et al., 2000; Strekalova et al., 2003; von Hertzen and Giese, 2005a). The same carefully controlled experiments have shown, however, that three of these genes –SGK1, zif268 and c-fos– in addition to being regulated after conditioning and memory reactivation, are also regulated by the shock alone. As these genes may be fear and not memory regulated, they can therefore not be used to compare memory consolidation and reconsolidation; indeed, the upregulation observed after context–shock memory reactivation could be attributable to fear and not memory reactivation. Only genes that are specifically regulated by the learned CS-US association during consolidation are suitable for this molecular analysis. Interestingly, we have recently reported that during consolidation of contextual fear memory, SGK3 mRNA is upregulated specifically by the CS-US association, and that this up-regulation is recapitulated during contextual memory reconsolidation. As SGK-3 is not triggered by fear or other environmental factors during memory consolidation, the upregulation observed after re-exposure must be specific to memory reactivation. These data therefore show for the first time that memory consolidation and reconsolidation share at least some common processes at the transcriptional level (von Hertzen and Giese, 2005a).

Comparisons at the functional level have also shown similarities between memory consolidation and memory reconsolidation. Both the mitogen-activated protein kinase and the transcription factor zif268 are required for the consolidation and reconsolidation of recognition memory (Kelly et al., 2003, Bozon et al., 2003).

Thus, there is increasing evidence, both at the functional and molecular level, showing that memory consolidation and memory reconsolidation share at least some molecular processes.

Are There Molecular Processes Specific to Memory Consolidation or Memory Reconsolidation?

The increasing amount of data showing that memory consolidation and memory reconsolidation share common molecular processes suggests that the two phenomena are actually very similar in nature. However, this does not mean that they are identical. Indeed, while the evidence for common molecular mechanisms increases, so does that for distinctive features. Studies that have used contextual fear conditioning and analyzed molecular processes in the hippocampus, have revealed differences between memory consolidation and memory reconsolidation. Importantly, the hippocampus is required for both consolidation and reconsolidation of contextual fear memory (Debiec et all., 2001) so that these molecular distinctions within the tissue are valid.

Two studies have provided evidence for the existence of memory consolidation-specific processes; processes that occur during memory consolidation but not reconsolidation. The hippocampal expression of the nerve growth factor inducible gene B (NGFI-B; also called nur77) is up-regulated during consolidation but not reconsolidation of contextual fear memory (von Hertzen and Giese, 2005a, b). Additionally, an antisense knock-down approach has shown that the brain-derived neurotrophic factor (BDNF) is required in the hippocampus for consolidation but not reconsolidation of contextual fear memory (Lee et al., 2004). Consistent with this finding, BDNF is up-regulated in an association-specific manner during consolidation of contextual fear memory (Hall et al., 2000). The up-regulation of both NGFI-B and BDNF during memory consolidation occurs predominantly in area CA1 of the hippocampus (Hall et al., 2000; von Hertzen and Giese, 2005a), thus area CA1 is likely to play an important role in consolidation-specific processes. Although BDNF expression has not been studied during reconsolidation, and thus it remains to be determined whether, as suggested by the antisense studies, the expression of BDNF is consolidation-specific, these two studies nonetheless provide convincing evidence that there are molecular processes specific to memory consolidation.

It has been argued that the occurrence of consolidation-specific processes is not surprising, and indeed should exist, because the animal experiences a CS-US pairing during training, whereas only the CS is presented during memory reactivation (Nader et al., 2005). However, these considerations neglect the fact that memory reactivation induces innate fear which is likely to be comparable to the fear triggered by the US. Thus, differences in gene expression observed between memory consolidation and reconsolidation are not at all reducible to differences in the external stimuli present. This idea is supported by our finding that the expression of the association-specific transcript SGK-3 is up-regulated during both the consolidation and reconsolidation of contextual fear memory (von Hertzen and Giese, 2005a). Since in principle all molecular processes could behave like the SGK-3 expression, the existence of molecular processes specific to memory consolidation shows that there is a mechanistic difference between memory consolidation and memory reconsolidation.

A recent antisense study has tended to confirm that in addition to consolidation-specific processes, reconsolidation-specific processes also exist. Knock-down of the transcription factor zif268 in the hippocampus was shown to block reconsolidation but not consolidation of contextual fear memory (Lee et al., 2004). This finding is nevertheless surprising because other functional studies have shown that zif268 is required for the consolidation of various forms of memory (Jones et al., 2000; Bozon et al., 2003; Malkani et al., 2004). Furthermore, zif268 expression is upregulated in the hippocampus during both consolidation and reconsolidation (Hall et al., 2000, 2001), and it is not clear whether the expression change during reconsolidation is specific to memory reactivation. A problem is that repeated exposure to the context without conditioning leads to an increase in hippocampal zif268 expression. Because this zif268 expression change is smaller than the expression change after reactivation of contextual fear memory it has been argued that there would be memory reconsolidation-specific zif268 expression. However, a latent inhibition control group showed that fear can up-regulate hippocampal zif268 expression. Thus, the innate fear during reactivation of contextual fear memory may have potentiated the increase in zif268 expression, and a memory reconsolidation-specific zif268 expression change might not exist. Therefore, it is at this stage premature to conclude that reconsolidation specific processes exist. However, Lee et al., . nevertheless provide further evidence that there are differences between memory consolidation and memory reconsolidation.

Memory Reconsolidation as a Partial Recapitulation of Memory Consolidation

The data from comparative studies on the molecular mechanisms underlying memory consolidation and memory reconsolidation have led us to the hypothesis that reconsolidation is a partial recapitulation of consolidation (von Hertzen and Giese, 2005a). This hypothesis is based on the findings that I) memory consolidation and memory reconsolidation share common molecular processes, such as the up-regulation of SGK-3 expression, and II) memory consolidation-specific processes exist in addition. This theory therefore proposes that a larger number of molecular processes are engaged during memory consolidation than during memory reconsolidation. This would be consistent with the idea that establishing a new memory is likely to require more substantial alteration at the cellular level than maintaining an already existing memory. The reconsolidated memory might have a sparser representation that is actively maintained by fewer genes.

It is very likely that there is functional redundancy between some of the molecular processes occurring during memory formation, because microarray studies imply that some functionally related genes might be regulated in the same way. If this assumption is correct, then, according to our hypothesis, there would be less functional redundancy during memory reconsolidation than during memory consolidation. Such a difference in functional redundancy would explain why contextual memory reconsolidation but not consolidation is affected by a knock-down of zif268 in the hippocampus.

In conclusion, memory reconsolidation appears to be a partial recapitulation of consolidation, differing from consolidation merely in that it only engages a subset of the processes required for memory consolidation. Such partial rather than total recapitulation may have evolved as a more economic and reliable mechanism for an organism to maintain a memory.

Acknowledgments

We are grateful to Jeff Vernon for comments on an earlier draft of this manuscript. This work was supported by the Wellcome Trust.

References

Anokhin, K.V., Tiunova, A.A., Rose, S.P. Reminder effects - reconsolidation or retrieval deficit? Pharmacological dissection with protein synthesis inhibitors following reminder for a passiveavoidance task in young chicks. *Eur. J. Neurosci.* 15: 1759-1765, 2002.

Bozon, B., Davis, S., Laroche, S. A requirement for the immediate early gene zif268 in reconsolidation of recognition memory after retrieval. Neuron 40: 695-701, 2003.

Chen, X., Garelick, M.G., Wang, H., Lil, V., Athos, J., Storm, D.R. PI3 kinase signaling is required for retrieval and extinction of contextual memory. *Nat. Neurosci.* 8: 925-931, 2005.

Debiec, J., LeDoux, J.E., Nader, K. Cellular and systems reconsolidation in the hippocampus. *Neuron* 36: 527-538, 2002.

Dudai, Y. The neurobiology of consolidations, or, how stable is the engram? *Annu. Rev. Psychol.* 55: 51-86, 2004.

Duvarci, S., Nader, K., LeDoux, J.E. Activation of extracellular signal-regulated kinase-mitogenactivated protein kinase cascade in the amygdala is required for memory reconsolidation of auditory fear conditioning. *Eur. J. Neurosci.* 21: 283-289, 2005.

Hall, J., Thomas, K.L., Everitt, B.J. Rapid and selective induction of BDNF expression in the hippocampus during contextual learning. *Nat. Neurosci.* 3: 533-535, 2000.

Hall, J., Thomas, K.L., Everitt, B.J. Cellular imaging of zif268 expression in the hippocampus and amygdala during contextual and cued fear memory retrieval: selective activation of hippocampal CA1 neurons during the recall of contextual memories. *J. Neurosci.* 21: 2186-2193, 2001.

Jones, M.W., Errington, M.L., French, P.J., Fine, A., Bliss, T.V.P., Garel, S., Charnay, P., Bozon, B., Laroche, S., Davis, S. A requirement for the immediate early gene zif268 in the expression of late LTP and long-term memories. *Nat. Neurosci.* 4: 289-296, 2001.

Kelly, A., Laroche, S., Davis, S. Activation of mitogen-activated protein kinase/extracellular signalregulated kinase in hippocampal circuitry is required for consolidation and reconsolidation of recognition memory. *J. Neurosci.* 23: 5354-5360, 2003.

Kida, S., Josselyn, S.A., de Ortiz, S.P., Kogan, J.H., Chevere, I., Masushige, S., Silva, A.J. CREB required for the stability of new and reactivated fear memories. *Nat. Neurosci.* 5: 348-355, 2002.

Lee, J.L., Everitt, B.J., Thomas, K.L. Independent cellular processes for hippocampal memory consolidation and reconsolidation. *Science* 304: 839-843, 2004.

Levenson, J.M., Choi, S., Lee, S.Y., Cao, Y.A., Ahn, H.J., Worley, K.C., Pizzi, M., Liou, H.C., Sweatt, J.D. A bioinformatics analysis of memory consolidation reveals involvement of the transcription factor c-rel. *J. Neurosci.* 24: 3933-3943, 2004.

Malkani, S., Wallace, K.J., Donley, M.P., Rosen, J.B. An egr-1 (zif268) antisense oligodeoxynucleotide infused into the amygdala disrupts fear conditioning. *Learn. Mem.* 11: 617-624, 2004.

Merlo, E., Freudenthal, R., Maldonado, H., Romano, A. Activation of the transcription factor NF-kappaB by retrieval is required for long-term memory reconsolidation. *Learn. Mem.* 12: 23-29, 2005.

Milekic, M.H., Alberini, C.M. Temporally graded requirement for protein synthesis following memory reactivation. *Neuron* 36: 521-525, 2002.

Nader, K., Schafe, G.E., LeDoux, J.E. Fear memories require protein synthesis in the amygdala for reconsolidation after retrieval. *Nature* 406: 722-726, 2000.

Nader, K., Hardt, O., Wang, S.H. Response to Alberini: right answer, wrong question. *Trends Neurosci.* 28: 346-347, 2005.

Sara, S.J. Retrieval and reconsolidation: toward a neurobiology of remembering. *Learn. Mem.* 7: 73-84, 2000.

Silva, A.J., Giese, K.P. Plastic genes are in! Curr. Opin. Neurobiol. 4: 413-420, 1994.

Stanciu, M., Radulovic, J., Spiess, J. Phosphorylated cAMP response element binding protein in the mouse brain after fear conditioning: relationship to Fos production. *Mol. Brain Res.* 94: 15-24, 2001.

Strekalova, T., Zorner, B., Zacher, C., Sadovska, G., Herdegen, T., Gass, P. Memory retrieval after contextual fear conditioning induces c-Fos and JunB expression in CA1 hippocampus. *Genes Brain Behav.* 2: 3-10, 2003.

Suzuki, A., Josselyn, S.A., Frankland, P.W., Masushige, S., Silva, A.J., Kida, S. Memory reconsolidation and extinction have distinct temporal and biochemical signatures. *J. Neurosci.* 24: 4787-4795, 2004.

Taubenfeld, S.M., Milekic, M.H., Monti, B., Alberini, C.M. The consolidation of new but not reactivated memory requires hippocampal C/EBPß. *Nat. Neurosci.* 4: 813- 818, 2001.

Tronel, S., Sara, S.J. Mapping of olfactory memory circuits: region-specific c-fos activation after odor-reward associative learning or after its retrieval. *Learn. Mem.* 9: 105-111, 2002.

von Hertzen, L.S.J., Giese, K.P. Memory reconsolidation engages only a subset of immediate-early genes induced during consolidation. J. Neurosci. 25: 1935-1942, 2005a.

von Hertzen, L.S.J., Giese, K.P. α-isoform of Ca^{2+}/calmodulin-dependent kinase II autophosphorylation is required for memory consolidation-specific transcription. *Neuroreport* 16: 1411-1414, 2005b.

In: Cognitive Science Compendium, Volume 1
Editor: Miao-Kun Sun
ISBN: 978-1-60456-597-3

Chapter 10

NEUROPROTECTIVE EFFECTS OF INSULIN-LIKE GROWTH FACTOR I AGAINST NEUROTOXIC AMYLOID

David Aguado-Llera[1,2], Eduardo Arilla-Ferreiro[2] and Vicente Barrios[1,*]

[1] Department of Endocrinology, Hospital Infantil Universitario Niño Jesús, Universidad Autónoma, E-28009 Madrid

[2] Neurobiochemistry Unit, Universidad de Alcalá, E-28871 Alcalá de Henares, Spain

ABSTRACT

Alzheimer's disease (AD) is a neurodegenerative disorder anatomically characterized by the appearance of neurofibrillary tangles and extracellular deposits of amyloid peptides (Aβ). Although the primary alteration is not established, accumulative data indicate that deposition of Aβ in nerve terminals is the primary pathogenic event in this neurodegenerative process. Evidence indicates that glucose utilization is impaired in this disorder, suggesting that disruption of insulin and probably insulin-like growth factor I (IGF-I) signaling may be involved in the pathogenesis of AD. In this context, AD appears in the later periods of life when there is a progressive loss of sensitivity to IGF-I, as well as a decline in its circulating levels. In fact, decreased levels of IGF-I have been reported in familial forms of AD. However, high IGF-I and insulin levels are reported in the late onset form of this disease, reflecting a possible state of IGF-I resistance. Furthermore, animal models of familial AD suggest a disruption of IGF-I input that could be involved in Aβ accumulation. IGF-I is a physiologically important neurotrophic factor for both neurons and glial cells that is widely expressed in brain tissue early in development and, although reduced, this expression is maintained throughout adulthood.

* Corresponding author: Vicente Barrios, Ph.D., Department of Endocrinology, Hospital Infantil Universitario Niño Jesús, Avda. Menéndez Pelayo, 65, 28009 Madrid (Spain), Telephone: +34-91-5035900, Fax: +34-91-5035939, E-mail: vbarrios@telefonica.net

This growth factor possesses neurotrophic activities in areas severely affected in AD and its potential neuroprotective role could occur at least two levels. First, the activation of IGF-I receptors protects neurons against cell death induced by Aβ and IGF-I normalizes intracellular signaling pathways affected by Aβ. In this regard, IGF-I activates the phosphatidylinositol 3 kinase (PI3K)/Akt pathway, which is reduced in patients with a mutation of the Aβ precursor protein, and the Aβ induced increase in c-Jun kinase (JNK) is also blocked by IGF-I administration. In addition, given its capacity to modulate insulin actions, IGF-I may protect against the decrease in neuronal insulin sensitivity described in AD. Furthermore, IGF-I is reported to regulate brain Aβ levels by enhancing clearance of neurotoxic fragments, possibly by increasing the transport of Aβ carrier proteins such as albumin and transthyretin into the brain. Given that preliminary clinical trials show that treatment of insulin resistance in AD is beneficial and IGF-I infusion reduces brain Aβ deposits in mice over-expressing mutant Aβ, it follows that IGF-I could be a useful therapeutic agent for the treatment of AD.

Keywords: Alzheimer's disease, β-amyloid, insulin-like growth factor I, neuroprotection

1. The Role of IGF-I in Normal Brain Function

Insulin-like growth factor I (IGF-I) mediates many of the effects of growth hormone (GH) in the control of somatic growth. This growth factor, localized in many tissues and body fluids, exerts trophic effects acting as an endocrine and paracrine/autocrine factor (Jones and Clemmons, 1995). IGF-I, its membrane receptor and the IGF-binding proteins (IGFBPs), are all widely found in brain tissue during development and their expression is maintained in different brain areas during adulthood (Torres-Alemán, 1999). Emerging evidence indicates that IGF-I has diverse and relevant functions in the brain, including metabolic, neuroendocrine, neurotrophic and neuromodulatory actions (Torres-Alemán, 2005). Thus, during development of the nervous system IGF-I promotes neuron differentiation, growth and survival (Lobie et al., 2000), whereas in the adult brain it performs a wide-spectrum of trophic and reparative actions.

Many of the functions of IGF-I overlap with those of insulin. In fact, both hormones belong to the same protein family (Mattson, 2002). One common function of insulin and IGF-I is that they both modulate brain function and energy balance (Bondy and Cheng, 2002). IGF-I and insulin bind to their own receptors with the highest affinity, but cross-reactivity occurs at higher hormone concentrations. Although these two factors have virtually identical signaling cascades, and many of their functions are common, they are not redundant. In fact, their distinct biological roles could be due to cell-specific expression patterns of receptors during development (Bondy and Cheng, 2004).

The actions of IGF-I are mediated by the IGF receptors and its biological activity is modulated through IGFBPs. The IGF-I receptor is a heterotetrameric membrane glycoprotein, that consists of two α- and two β-subunits and that binds IGF-I with higher affinity than IGF-II or insulin. After ligand binding to the α-subunits and activation of this tyrosine kinase receptor a variety of target molecules are phosphorylated, including its major substrate the insulin receptor substrate (IRS) proteins (Fig. 1). In turn, IRS phosphorylation activates growth-factor receptor-bound protein 2 (Grb2), the p85 regulatory subunit of phosphatidylinositol-3 kinase (PI3 kinase), phosphotyrosine phosphatases and transcription factors leading to modifications in calcium mobilization (Sun et. al, 1993; LeRoith et al., 1995), and subsequently changes in growth, metabolic functions and cell viability (Saltiel and Pessin, 2002). Binding of IRS to Grb2 activates mitogen-activated protein kinase (MAPK), which contributes to mitogenesis, neurite development and modulation of gene expression (De la Monte and Wands, 2005). The association of IRS to p85 inhibits apoptosis through activation of PI3/Akt kinase (Dudek et al., 1997) and inhibition of glycogen synthase kinase-3β (GSK-3β) (Pap and Cooper, 1998). Akt kinase restrains apoptosis by phosphorylation of GSK-3β and Bad. Inactivation of Bad through this mechanism releases anti-apoptotic Bcl proteins and inhibits caspase activation, resulting in reduced neuronal death (Condorelli et al., 2001).

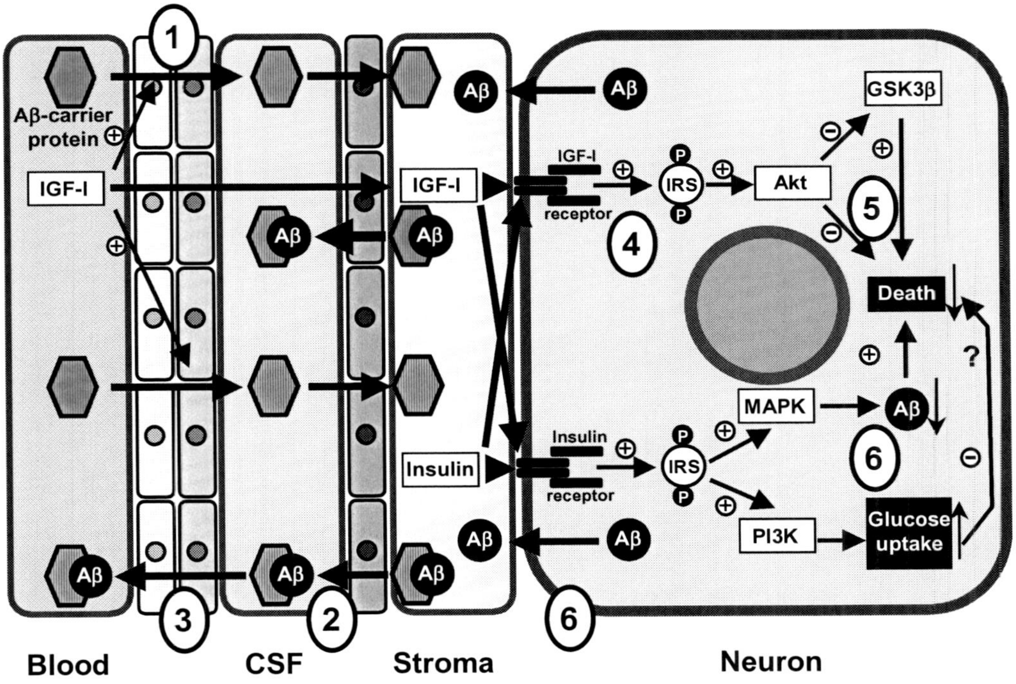

Figure 1. Roles for insulin-like growth factor I (IGF-I) against β-amyloid (Aβ). (1) Serum IGF-I stimulates the passage of Aβ-carrier proteins from the blood into the cerebrospinal fluid (CSF). (2) Aβ-carrier proteins accumulate in the brain stroma where they bind Aβ and transfer it into the CSF. (3) The complexes Aβ-Aβ carrier proteins are transported to the blood. (4) IGF-I binds to IGF-I and/or insulin receptors and induces phosphorylation of insulin receptor substrate (IRS). (5) IRS phosphorylation activates Akt that restrains apoptosis directly or indirectly through inhibition of glycogen synthetase kinase 3β (GSK3β), an enzyme that disrupts trafficking of Aβ and increases cell death. (6) IGF-I/insulin activation of mitogen-activated protein kinase (MAPK) promotes intracellular trafficking of Aβ to the plasma cell membrane. Phosphatidylinositol 3 kinase (PI3K) activation enhances glucose uptake and could increase cell survival.

2. Alzheimer's Disease and Cerebral Amyloidosis

Old age is a risk factor for important neurological impairments such as Alzheimer's disease (AD) and with the increased life span in modern societies the incidence of these conditions has also increased. This neurodegenerative disorder affects nearly 2% of the population in industrialized countries and the risk increases in subjects beyond the age of 70 (Mattson, 2004). Alzheimer's disease is characterized by plaques that represent extracellular deposition of amyloid β-peptide (Aβ), intraneuronal cytoskeletal abnormalities and increased neuronal death. An abnormally high level of Aβ in brain seems to have a central role in the pathogenesis of AD (Hardy and Selkoe, 2002), activating an inflammatory response that contributes to the increased cell death and the cognitive alterations that typify this disease (Walsh and Selkoe, 2004). Brain regions rich in these plaques are severely atrophic, in particular those involved in learning and memory processes such as the temporal and frontal lobes. The reduction in brain size in patients with AD is the result of degeneration of synapses and death of neurons (Mattson et al., 2001).

Aβ is a 39-43 amino acid peptide that originates from the proteolytic cleavage of the Aβ precursor protein (APP) (Selkoe, 1994), although fragments containing 25-35 amino acids have also been described (Kubo et al., 2002). Aberrant proteolytic processing of the Aβ precursor protein (APP) may promote neuronal degeneration by enhancing the concentrations of specific Aβ forms and diminishing the levels of the neuroprotective form of APP alpha (Kerr and Small, 2005). APP gene mutations are responsible for early onset AD (Goate et al., 1991), while in the more common sporadic forms of AD the initiating causes are unclear, but most likely involve increased levels of oxidative stress and impaired energy metabolism (Blass et al., 2002).

Elevated levels of Aβ are present in AD brains and its deposition inside the senile plaques is a hallmark of AD and appears to be a primary cause of the cognitive dysfunctions that occur. Indeed, Aβ-induced neuronal death could be responsible for the cognitive deterioration (Saura et al., 2005). In this context, Aβ administration increases cell death and leads to a significant reduction in p-Akt, a mechanism known to be involved in neuronal protection (Dudek et al., 1997). Indeed, this intracellular pathway has also been shown to be down-regulated in AD patients with altered forms of APP (Ryder et al., 2004). Another molecular mechanism by which Aβ induces toxicity has also been reported (Morishima et al., 2001). In this paradigm Aβ increases c-Jun N-terminal kinase (JNK) that then phosphorylates and activates the transcription factor c-Jun, which in turn stimulates the transcription of several genes including the death inducer Fas ligand. The binding of Fas to its receptor induces different signals leading to caspase activation and cell death. However, in spite of multiple data suggesting that Aβ deposition and neuronal loss are correlated, the direct link between Aβ production by neurons and cell death has only recently been clearly established.

3. IGF-I AND AMYLOIDOSIS

3.1. IGF-I in Alzheimer's Disease

We have previously shown that different neurodegenerative conditions have reduced levels of IGF-I in plasma and brain, both in humans (Torres-Alemán et al., 1996; Torres-Alemán et al., 1998) and in animal models (Busiguina et al., 2000), suggesting that IGF-I may be implicated in the ethiopathogenesis of these disorders. The pathogenic findings in AD suggest that many of the brain alterations may be related to impaired insulin/IGF-I signaling. Initial data in AD patients showed high serum IGF-I and IGFBP concentrations in the more common sporadic forms of AD (Tham et al., 1993) and low IGF-I levels in familial AD (Mustafa et al., 1999). Early studies seemed to indicate that IGF-I binding sites were augmented in cortical areas of the AD brain (Crews et al., 1992), but subsequent studies showed similar IGF-I binding in control and AD brains (Jafferali et al., 2000). In spite of these contradictory data, the concept of IGF-I resistance in AD should still be considered. First, despite normal or increased IGF-I/insulin receptor expression, tyrosine kinase activity in the brain is reduced in AD (Frolich et al., 1999). Second, the compensatory increase in IGF-I levels in late-onset AD patients (Tham et al., 1993) could indicate a loss in sensitivity to IGF-I, as does the progressive decrease in IGF-I sensitivity during aging (Willis et al., 1997), which occurs at the same time of AD appearance (Mrak et al., 1997). Third, as Aβ can bind to the insulin receptor (Xie et al., 2002), it is possible that it modulates IGF-I signaling by also competing for IGF-I binding sites. Four, Aβ could modulate IGF-I actions indirectly through cytokines, such as TNF-α, that induce IGF-I resistance. In this regard Aβ may lead to an inflammatory response by activation of mechanisms involving glial cells (Venters et al., 1999). Lastly, extensive abnormalities in IGF-I signaling mechanisms in AD brains have been reported including reduced levels of IRS mRNA, IRS-associated PI3 kinase and the activated form of Akt (p-Akt), as well as increased GSK-3β activity (Steen et al., 2005). Overall, disrupted IGF-I signaling may be considered as one of the primary events in sporadic forms of AD, which is the most common form of this disease.

3.2. IGF-I modulates Neurotoxicity and Brain Levels of Aβ

The protective functions of IGF-I against Aβ may occur through different mechanisms. Some neuroprotective actions of IGF-I could be regarded as independent of Aβ modulation and involve counteraction of amyloid effects. Preliminary *in vitro* studies demonstrated that activation of the IGF-I receptor in rat hippocampal neurons protects and rescues neurons against cell death induced by Aβ (Doré et al., 1997). Interestingly, IGF-I activates intracellular signaling pathways decreased by Aβ accumulation (Fig. 1). We have shown that subcutaneous IGF-I administration increases the levels of p-Akt in the temporal cortex of Aβ-treated rats (Aguado-Llera et al., 2005) and p-Akt has been shown to be down-regulated in AD patients with APP mutations (Ryder et al., 2004). Another mechanism by which IGF-I could suppress Aβ neurotoxicity is through the JNK pathway (Wei et al., 2002). JNK, which is triggered in response to stress signals, is increased in response to Aβ and IGF-I protects against Aβ-induced cell death by blocking JNK activation. In addition, IGF-I may exert its neuroprotective properties through the tripeptide Gly-Pro-Glu (GPE), which is naturally cleaved from the N-terminal sequence of IGF-I (Sara et al., 1989). Both GPE and IGF-I molecules have a beneficial effect in restoring at least one neurotransmitter system severely affected by Aβ insult (Aguado-Llera et al., 2004; Aguado-Llera et al., 2005). However, IGF-I actions are not specific to Aβ-induced cell death, as a similar effect has been described after ethanol exposure. Indeed, neuronal rescue from toxic effects of ethanol is also mediated by IGF-I activation of Akt (De la Monte et al., 2000).

Another mechanism by which IGF-I may exert its neuroprotective effects is through modulation of Aβ levels. At a cellular level, IGF-I imitates insulin in its capacity to increase Aβ release from neurons, but it is not clear whether IGF-I and insulin act through the same receptor to modulate Aβ release (Gasparini et al., 2001). IGF-I activation of extracellular signal-regulated kinase (ERK) MAPK promotes physiological trafficking of Aβ to the plasma cell membrane (Fig. 1). Furthermore, oxidative stress caused by increased GSK-3β disrupts intracellular trafficking of Aβ and IGF-I reduces GSK-3β activity (Pap and Cooper, 1998). An additional mechanism could involve an IGF-I-mediated increase of neuronal excitability (Núñez et al., 2003), as formation and secretion of Aβ in neurons that over-express APP is enhanced by neuronal activity. In turn, Aβ peptides depress excitatory synaptic transmission in the same neurons (Kamenetz et al., 2003).

At the organ level, IGF-I appears to be a major regulator of brain Aβ peptide concentrations through a mechanism involving a network of Aβ transport proteins interacting at the blood-brain barrier (Carro et al., 2002). IGF-I enhances brain levels of carrier proteins such as transthyretin, albumin and apolipoprotein J and promotes clearance of Aβ from the brain parenchyma (Fig. 1). This IGF-I induced reduction in Aβ levels appears to be due to an increase in the Aβ clearance rate, as IGF-I administration increases Aβ levels in cerebrospinal fluid, enhances the formation of Aβ/carrier protein complexes and does not modify APP concentrations (Carro and Torres-Alemán, 2004a). Treatment with IGF-I reduces amyloidosis in advanced stages of the process, not only because its administration inhibits Aβ accumulation, but it also eliminates previously developed deposits (Carro et al., 2005).

4. POTENTIAL FUNCTION OF IGF-I IN THE TREATMENT OF ALZHEIMER'S DISEASE

Currently cholinesterase inhibitors are the only drugs widely recommended for clinical use in AD, although the average benefit appears to be small. These drugs provide moderate cognitive protection and induce unwanted sleep disturbances (Potyk, 2005) and anti-amyloidal vaccine trials had undesirable effects. It is known that chronic neuroinflammation contributes to AD and vaccination with Aβ in mouse models of AD stimulated phagocytosis of the toxic peptides and gave optimal results, but similar vaccination in humans led to meningoencephalitis in some cases (McGeer and McGeer, 2004).

The amyloid overload in AD patients may be regarded as the result of increased amyloid deposition as compared to age-matched normal subjects. In this context, decreased IGF-I signaling and lower levels of Aβ carrier proteins in AD patients (Serot et al., 1997) could lead to increased Aβ levels in these patients. Stimulating IGF-I signaling could provide neuroprotection by increasing Aβ clearance, counteracting TNFα effects and decreasing inflammatory reactions (Gasparini and Xu, 2003). In addition, IGF-I could be safely administered systemically and continuous subcutaneous infusion would avoid hypoglycemia (Woodall et al., 1991). However, one of the major counter-indication is that IGF-I, due the its typical growth promoting actions, could advance the progression of some forms of cancer (Yakar et al., 2005). Because endogenous IGF-I could be an approach for Alzheimer's treatment the use of GH, the classical stimulator of IGF-I production, should be further investigated. In contrast to exogenous IGF-I treatment this hormone has been used therapeutically for over 20 years and the therapeutic profile is well verified. Indeed, GH treatment has been used in patients with amyotrophic lateral sclerosis, showing a benefit with scarce adverse effects (Smith et al., 1993). Other possible approaches are the use of IGF-I analogs that bind to IGFBPs displacing IGF-I and increasing its bioavailability or alternative secretagogues, such as imidazolines (Hoy et al., 2003), that appear to increase circulating IGF-I levels (Loddick et al., 1998; Carro and Torres-Alemán, 2004b).

Studies of the mechanisms implicated in Aβ-induced neuronal degeneration have led to the discovery of several peptides reported to prevent neuron death. One of these peptides is GPE, which is cleaved from the IGF-I molecule and protects against hypoxic-ischemic injury (Sizonenko et al., 2001). We have previously demonstrated that GPE protects the somatostatinergic system from Aβ insult (Aguado-Llera et al., 2004) and interestingly, a GPE analog reduces neuronal degeneration by preventing Aβ-mediated apoptosis (Ioudina and Uemura, 2003). These neuroprotective effects, independent of interaction with IGF receptors (Sara et al., 1993), suggest a novel and promising strategy toward treatment of patients with AD.

Recently, a clinical trial with donepezil, a cerebral selective cholinesterase inhibitor that improves cognitive symptoms in AD patients (Harry and Zakzanis, 2005), was reported to increase circulating IGF-I levels (Obermayr et al., 2005). This result suggests that the age-related down-regulation of the IGF-I axis, reverted by this drug, could be of importance in the treatment of AD patients. Clinical trials have also suggested beneficial effects of IGF-I in the treatment of peripheral neurodegenerative disease, with little side effects seen (Vlachopapadopoulou et al., 1995; Lai et al., 1997). In addition, treatment with IGF-I is effective in animal models of cerebellar ataxia (Fernández et al., 2005) and ischemia (Loddick et al., 1998). These promising results suggest that the IGF-I signaling pathway could represent a therapeutic target for the treatment of neurodegenerative diseases.

CONCLUSIONS

Disruption of normal IGF-I signaling is involved in the deposition of Aβ (Wei et al.,, 2002), which appears to initiate the pathological cascade leading to AD (Selkoe, 2001). Indeed, IGF-I has shown its efficacy in the treatment of some peripheral neurodegenerative disorders (Lai et al., 1997). As this growth factor modulates neurotoxicity and levels of Aβ (Carro et al., 2002), it could be considered as a possible therapeutic agent for the treatment of AD.

Acknowledgments

The authors wish to thank Dr. Julie A. Chowen for the critical review of the manuscript. Supported by Grant PI021241 from the Fondo de Investigación Sanitaria and Fundación Endocrinología y Nutrición.

References

Aguado-Llera, D., Arilla-Ferreiro, E., Campos-Barros, A., Puebla-Jiménez, L. and Barrios, V. Protective effects of insulin-like growth factor-I on the somatostatinergic system in the temporal cortex of β-amyloid-treated rats. J. *Neurochem*. 92: 607-615, 2005.

Aguado-Llera, D., Martín-Martínez, M., García-López, M.T., Arilla-Ferreiro, E. and Barrios V. Gly-Pro-Glu protects β-amyloid-induced somatostatin depletion in the rat cortex. *Neuroreport* 15: 1979-1982, 2004.

Blass, J.P., Gibson, G.E. and Hoyer S. The role of the metabolic lesion in Alzheimer's disease. *J. Alzheimers Dis*. 4: 225-232, 2002.

Bondy, C.A. and Cheng C.M. Insulin-like growth factor-1 promotes neuronal glucose utilization during brain development and repair processes. *Int. Rev. Neurobiol*. 51: 189-217, 2002.

Bondy, C.A. and Cheng C.M. Signaling by insulin-like growth factor 1 in brain. *Eur. J. Pharmacol*. 490: 25-31, 2004.

Busiguina, S., Fernández, A.M., Barrios, V., Clark, R., Tolbert, D.L., Berciano, J. and Torres-Alemán, I. Neurodegeneration is associated to changes in serum insulin-like growth factors. *Neurobiol. Dis*. 7: 657-665, 2000.

Carro, E. and Torres-Alemán, I. The role of insulin and insulin-like growth factor I in the molecular and cellular mechanisms underlying the pathology of Alzheimer's disease. *Eur. J. Pharmacol*. 490: 127-133, 2004a.

Carro, E. and Torres-Alemán, I. Insulin-like growth factor I and Alzheimer's disease: therapeutic prospects?. *Expert Rev. Neurother*. 4: 79-86, 2004b.

Carro, E., Trejo, J.L., Gerber, A., Loetscher, H., Torrado, J., Metzger, F. and Torres-Alemán, I. Therapeutic actions of insulin-like growth factor I on APP/PS2 mice with severe brain amyloidosis. *Neurobiol. Aging*. 27: 1250-1257, 2006.

Carro, E., Trejo, J.L., Gómez-Isla, T., LeRoith, D. and Torres-Alemán, I. Serum insulin-like growth factor I regulates brain amyloid-β levels. *Nat. Med*. 8: 1390-1397, 2002.

Condorelli, F., Salomoni, P., Cotteret, S., Cesi, V., Srinivasula, S.M., Alnemri, E.S. and Calabretta, B. Caspase cleavage enhances the apoptosis-inducing effects of BAD. *Mol. Cell. Biol*. 21: 3025-3036, 2001.

De la Monte, S.M., Ganju, N., Banerjee, K., Brown, N.V., Luong, T. and Wands, J.R. Partial rescue of ethanol-induced neuronal apoptosis by growth factor activation of phosphoinositol-3-kinase. *Alcohol Clin. Exp*. Res. 24: 716-726, 2000.

De la Monte, S.M. and Wands, J.R. Review of insulin and insulin-like growth factor expression, signaling, and malfunction in the central nervous system: relevance to Alzheimer's disease. *J. Alzheimers Dis*. 7: 45-61, 2005.

Doré, S., Kar, S. and Quirion, R. Insulin-like growth factor I protects and rescues hippocampal neurons against β-amyloid- and human amylin-induced toxicity. *Proc. Natl. Acad. Sci. USA* 94: 4772-4777, 1997.

Dudek, H., Datta, S.R., Franke, T.F., Birnbaum, M.J., Yao, R., Cooper, G.M., Segal, R.A., Kaplan, D.R. and Greenberg M.E. Regulation of neuronal survival by the serine-threonine protein kinase *Akt. Science* 275: 661-665, 1997.

Fernández, A.M., Carro, E.M., López-López, C. and Torres-Alemán, I. Insulin-like growth factor I treatment for cerebellar ataxia: Addressing a common pathway in the pathological cascade? *Brain Res. Brain Res.* Rev. 50: 134-141, 2005.

Frolich, L., Blum-Degen, D., Riederer, P. and Hoyer, S. A disturbance in the neuronal insulin receptor signal transduction in sporadic Alzheimer's disease. *Ann. N.Y. Acad. Sci.* 893: 290-293, 1999.

Gasparini, L., Gouras, G.K., Wang, R., Gross, R.S., Beal, M.F., Greengard, P. and Xu, H. Stimulation of β-amyloid precursor protein trafficking by insulin reduces intraneuronal β-amyloid and requires mitogen-activated protein kinase signaling. *J. Neurosci.* 21: 2561-2570, 2001.

Gasparini, L. and Xu, H. Potential roles of insulin and IGF-1 in Alzheimer's disease. *Trends Neurosci.* 26: 404-406, 2003.

Goate, A., Chartier-Harlin, M.C., Mullan, M., Brown, J., Crawford, F., Fidani, L., Giuffra, L., Haynes A., Irving, N., James, L., Mant, R., Newton, P., Rooke, K., Roques, P., Talbot C., Pericak-Vance, M., Roses, A., Williamson, R., Rossor, M., Owen, M. and Hardy, J. Segregation of a missense mutation in the amyloid precursor protein gene with familial Alzheimer's disease. *Nature* 349: 704-706, 1991.

Hardy, J. and Selkoe, D.J. The amyloid hypothesis of Alzheimer's disease: progress and problems on the road of therapeutics. *Science* 297: 353-356, 2002.

Harry, R.D. and Zakzanis K.K. A comparison of donepezil and galantamine in the treatment of cognitive symptoms of Alzheimer's disease: a meta-analysis. *Hum. Psychopharmacol.* 20: 183-187, 2005.

Hoy, M., Olsen, H.L., Andersen, H.S., Bokvist, K., Buschard, K., Hansen, J., Jacobsen, P., Petersen, J.S., Rorsman, P. and Gromada J. Imidazoline NNC77-0074 stimulates insulin secretion and inhibits glucagon release by control of Ca(2+)-dependent exocytosis in pancreatic α- and β-cells. *Eur. J. Pharmaco*l. 466: 213-221, 2003.

Ioudina, M. and Uemura, E. A three amino acid peptide, Gly-Pro-Arg, protects and rescues cell death induced by amyloid β-peptide. *Exp. Neurol.* 184: 923-929, 2003.

Jafferali, S., Dumont, Y., Sotty, F., Robitaille, Y., Quirion, R. and Kar S. Insulin-like growth factor-I and its receptor in the frontal cortex, hippocampus, and cerebellum of normal human and alzheimer disease brains. *Synapse* 38: 450-459, 2000.

Jones, J.I. and Clemmons, D.R. Insulin-like growth factors and their binding proteins: biological actions. *Endocr. Rev.* 16: 3-34, 1995.

Kamenetz, F., Tomita, T., Hsieh, H., Seabrook, G., Borchelt, D., Iwatsubo, T., Sisodia, S. and Malinow, R. APP processing and synaptic function. *Neuron* 37: 925-937, 2003.

Kerr, M.L. and Small, D.H. Cytoplasmic domain of the beta-amyloid protein precursor of Alzheimer's disease: function, regulation of proteolysis, and implications for drug development. *J. Neurosci. Res.* 80: 151-159, 2005.

Kubo, T., Nishimura, S., Kumagae, Y. and Kaneko I. In vivo conversion of racemized β-amyloid ([D-Ser 26]A β 1-40) to truncated and toxic fragments ([D-Ser 26]A β 25-35/40) and fragment presence in the brains of Alzheimer's patients. *J. Neurosci. Res.* 70: 474-483, 2002.

Lai, E.C., Felice, K.J., Festoff, B.W., Gawel, M.J., Gelinas, D.F., Kratz, R., Murphy, M.F., Natter H.M., Norris, F.H. and Rudnicki, S.A. Effect of recombinant human insulin-like growth factor-I on progression of ALS. A placebo-controlled study. The North America ALS/IGF-I Study Group. *Neurology* 49: 1621-1630, 1997.

LeRoith, D., Werner, H., Beitner-Johnson, D. and Roberts C.T. Jr. Molecular and cellular aspects of the insulin-like growth factor I receptor. *Endocr. Rev.* 16: 143-163, 1995.

Lobie, P,E, Zhu, T., Graichen, R. and Goh, E.L. Growth hormone, insulin-like growth factor I and the CNS: localization, function and mechanism of action. *Growth Horm. IGF Res.* 10 (Suppl B): S51-56, 2000.

Loddick, S.A., Liu, X.J., Lu, Z.X., Liu, C., Behan, D.P., Chalmers, D.C., Foster, A.C., Vale, W.W., Ling, N. and De Souza, E.B. Displacement of insulin-like growth factors from their binding proteins as a potential treatment for stroke. *Proc. Natl. Acad. Sci. USA 95*: 1894-1898, 1998.

Mattson, M.P. Brain evolution and lifespan regulation: conservation of signal transduction pathways that regulate energy metabolism. *Mech. Ageing Dev.* 123: 947-953, 2002.

Mattson, M.P. Pathways towards and away from Alzheimer's disease. *Nature* 430: 631-639, 2004.

Mattson, M.P., Gary, D.S., Chan, S.L. and Duan, W. Perturbed endoplasmic reticulum function, synaptic apoptosis and the pathogenesis of Alzheimer's disease. *Biochem. Soc. Symp.* 67: 151-162, 2001.

McGeer, P.L. and McGeer, E. Immunotherapy for Alzheimer's disease. Sci. *Aging Knowledge Environ.* 27: 29, 2004.

Morishima, Y., Gotoh, Y., Zieg, J., Barrett, T., Takano, H., Flavell, R., Davis, R.J., Shirasaki, Y. and Greenberg M.E. β-amyloid induces neuronal apoptosis via a mechanism that involves the c-Jun N-terminal kinase pathway and the induction of Fas ligand. *J. Neurosci.* 21: 7551-7560, 2001.

Mrak, R.E., Griffin, S.T. and Graham, D.I. Aging-associated changes in human brain. *J. Neuropathol. Exp. Neurol.* 56: 1269-1275, 1997.

Mustafa, A., Lannfelt, L., Lilius, L., Islam, A., Winblad, B. and Adem, A. Decreased plasma insulin-like growth factor-I level in familial Alzheimer's disease patients carrying the Swedish APP 670/671 mutation. *Dement. Geriatr. Cogn. Disord.* 10: 446-451, 1999.

Núñez, A., Carro, E. and Torres-Alemán, I. Insulin-like growth factor I modifies electrophysiological properties of rat brain stem neurons. *J Neurophysiol.* 89: 3008-3017, 2003.

Obermayr, R.P., Mayerhofer, L., Knechtelsdorfer, M., Mersich, N., Huber, E.R., Geyer, G. and Tragl, K.H. The age-related down-regulation of the growth hormone/insulin-like growth factor-1 axis in the elderly male is reversed considerably by donepezil, a drug for Alzheimer's disease. *Exp. Gerontol.* 40: 157-163, 2005.

Pap, M. and Cooper G.M. Role of glycogen synthase kinase-3 in the phosphatidylinositol 3-Kinase/Akt cell survival pathway. *J. Biol. Chem.* 273: 19929-19932, 1998.

Potyk, D. Treatments for Alzheimer disease. *South Med. J.* 98: 628-635, 2005.

Ryder, J., Su, Y. and Ni, B. Akt/GSK3beta serine/threonine kinases: evidence for a signalling pathway mediated by familial Alzheimer's disease mutations. *Cell Signal.* 16: 187-200, 2004.

Saltiel, A.R. and Pessin, J.E. Insulin signaling pathways in time and space. *Trends Cell. Biol.* 12: 65-71, 2002.

Sara, V.R., Carlsson-Skwirut, C., Bergman, T., Jornvall, H., Roberts, P.J., Crawford, M., Hakansson, L.N., Civalero, I. and Nordberg, A. Identification of Gly-Pro-Glu (GPE), the aminoterminal tripeptide of insulin-like growth factor 1 which is truncated in brain, as a novel neuroactive peptide. *Biochem. Biophys. Res. Commun.* 165: 766-771, 1989.

Sara, V.R., Carlsson-Skwirut, C., Drakenberg, K., Giacobini, M.B., Hakansson, L., Mirmiran, M., Nordberg, A., Olson, L., Reinecke, M., Stahlbom, P.A. and Nordqvist, A.C.S. The biological role of truncated insulin-like growth factor-1 and the tripeptide GPE in the central nervous system. *Ann. NY Acad. Sci.* 692: 183-191, 1993.

Saura, C.A., Chen, G., Malkani, S., Choi, S.Y., Takahashi, R.H., Zhang, D., Gouras, G.K., Kirkwood, A., Morris R.G. and Shen J. Conditional inactivation of presenilin 1 prevents amyloid accumulation and temporarily rescues contextual and spatial working memory impairments in amyloid precursor protein transgenic mice. *J. Neurosci.* 25: 6755-6764, 2005.

Selkoe, D.J. Normal and abnormal biology of the β-amyloid precursor protein. *Annu. Rev. Neurosci.* 17: 489-517, 1994.

Selkoe, D.J. Clearing the brain's amyloid cobwebs. Neuron 32: 177-180, 2001.

Serot, J.M., Christmann, D., Dubost, T. and Couturier, M. Cerebrospinal fluid transthyretin: aging and late onset al., zheimer's disease. *J. Neurol. Neurosurg. Psychiatry* 63: 506-508, 1997.

Sizonenko, S.V., Sirimanne, E.S., Williams, C.E. and Gluckman P.D. Neuroprotective effects of the N-terminal tripeptide of IGF-1, glycine-proline-glutamate, in the immature rat brain after hypoxic-ischemic injury. *Brain Res.* 922: 42-50, 2001.

Smith, R.A., Melmed, S., Sherman, B., Frane, J., Munsat, T.L. and Festoff B.W. Recombinant growth hormone treatment of amyotrophic lateral sclerosis. *Muscle Nerve* 16: 624-633, 1993.

Steen, E., Terry, B.M., Rivera, E.J., Cannon, J.L., Neely, T.R., Tavares, R., Xu, X.J., Wands, J.R. and de la Monte, S.M. Impaired insulin and insulin-like growth factor expression and signaling mechanisms in Alzheimer's disease-is this type 3 diabetes? *J. Alzheimers Dis.* 7: 63-80, 2005

Sun, X.J., Crimmins, D.L., Myers, M.G. Jr, Miralpeix, M. and White M.F. Pleiotropic insulin signals are engaged by multisite phosphorylation of IRS-1. *Mol. Cell. Biol.* 13: 7418-7428, 1993.

Tham, A., Nordberg, A., Grissom, F.E., sulin signals are engaged by multisite phosphorylation of IRS-1. MolCarlsson-Skwirut, C., Viitanen, M. and Sara, V.R. Insulin-like growth factors and insulin-like growth factor binding proteins in cerebrospinal fluid and serum of patients with dementia of the Alzheimer type. *J. Neural Transm. Park. Dis. Dement. Sect.* 5: 165-176, 1993.

Torres-Alemán, I. Insulin-like growth factors as mediators of functional plasticity in the adult brain. *Horm. Metab. Res.* 31: 114-119, 1999.

Torres-Alemán, I. Role of insulin-like growth factors in neuronal plasticity and neuroprotection. Valera-Nieto I., Chowen J.A. (Eds.), The growth hormone/insulin-like growth factor axis during development. *Advances in Experimental Medicine and Biology*, Vol. 567, pp. 243-258, Springer, New York, 2005.

Torres-Alemán, I., Barrios, V. and Berciano, J. The peripheral insulin-like growth factor system in amyotrophic lateral sclerosis and in multiple sclerosis. *Neurology* 50: 772-776, 1998.

Torres-Alemán, I., Barrios, V., Lledó, A. and Berciano, J. The insulin-like growth factor I system in cerebellar degeneration. *Ann. Neurol.* 39: 335-342, 1996.

Venters, H.D., Tang, Q., Liu, Q., VanHoy, R.W., Dantzer, R. and Kelley, K.W. A new mechanism of neurodegeneration: a proinflammatory cytokine inhibits receptor signaling by a survival peptide. *Proc. Natl. Acad. Sci. USA* 96: 9879-9884, 1999.

Vlachopapadopoulou, E., Zachwieja, J.J., Gertner, J.M., Manzione, D., Bier, D.M., Matthews, D.E. and Slonim A.E. Metabolic and clinical response to recombinant human insulin-like growth factor I in myotonic dystrophy: a clinical research center study. *J. Clin. Endocrinol. Metab.* 80: 3715-3723, 1995.

Walsh, D.M. and Selkoe, D.J. Deciphering the molecular basis of memory failure in Alzheimer's disease. *Neuron* 44: 181-193, 2004.

Wei, W., Wang, X. and Kusiak, J.W. Signaling events in amyloid β-peptide-induced neuronal death and insulin-like growth factor I protection. *J. Biol. Chem.* 277: 17649-17656, 2002.

Willis, P.E., Chadan, S., Baracos, V. and Parkhouse, W.S. Acute exercise attenuates age-associated resistance to insulin-like growth factor I. *Am. J. Physiol.* 272: E397-404, 1997.

Woodall, S.M., Breier, B.H., O'Sullivan, U. and Gluckman, P.D. The effect of the frequency of subcutaneous insulin-like growth factor-1 administration on weight gain in growth hormone deficient mice. Horm. Metab. Res. 23: 581-584, *1991.*

Yakar, S., Leroith, D. and Brodt, P. The role of the growth hormone/insulin-like growth factor axis in tumor growth and progression: Lessons from animal models. *Cytokine Growth Factor Rev.* 16: 407-420, 2005.

In: Cognitive Science Compendium, Volume 1
Editor: Miao-Kun Sun
ISBN: 978-1-60456-597-3

Chapter 11

HOW AND WHERE DOES Aβ EXERT ITS TOXIC EFFECTS IN ALZHEIMER'S DISEASE?

Damian C. Crowther*, Richard M. Page*†, Leila Luheshi and David A. Lomas

Department of Medicine, University of Cambridge,
Cambridge Institute for Medical
Research, Wellcome Trust/MRC Building, Hills Road,
Cambridge, CB2 2XY, U.K. and
†University Chemical Laboratory, Lensfield Road,
Cambridge CB2 1EW, U.K.

INTRODUCTION

Protein aggregation is the basis for many of the common human neurodegenerative diseases such as Alzheimer's disease (AD), Parkinson's disease and a family of disorders that includes Huntington's disease. In AD the aggregatory species is termed amyloid β (Aβ), a peptide derived from the proteolytic cleavage of amyloid precursor protein (APP), a ubiquitous transmembrane protein. The aggregatory properties of Aβ are determined by variations in the position of the proteolytic cleavage that generates the C-terminus. In healthy elderly individuals the ratio of the 40 amino acid peptide ($A\beta_{1-40}$) to the 42 amino acid species ($A\beta_{1-42}$) favours the less aggregatory $A\beta_{1-40}$ resulting in effective clearance of the peptide from the brain. In contrast, individuals who go on to develop the common sporadic form of AD have elevated $A\beta_{1-42}$ concentrations, or have a molar ratio of $A\beta_{1-40}$ to $A\beta_{1-42}$ that favours aggregation. In the five percent of AD cases that are inherited as an autosomal dominant trait all the causal mutations have been shown to favour Aβ aggregation, mostly by altering APP processing, either increasing $A\beta_{1-42}$ in absolute terms or in comparison to $A\beta_{1-40}$. In rare examples, where $A\beta_{1-42}$ levels are not elevated, mutations are found within the Aβ sequence

* These authors contributed equally. Correspondence should be addressed to D.C.C.:Email: dcc26@cam.ac.uk. Telephone: 44 1223 336825. Fax: 44 1223 336827

that accelerate the intrinsic rate of peptide aggregation and stabilise particularly toxic subpopulations of aggregates, a clear example of this is the Arctic APP mutation [1, 2].

In the context of cognitive decline, the demonstration of Aβ deposition in the brain in combination with intraneuronal aggregates of a microtubule-associated protein, tau, comprise the diagnostic criteria for AD. Mature deposits of Aβ are composed of ordered amyloid fibrils and it is their distinctive microscopic appearance and their affinity for dyes such as Congo red that favoured their early characterisation. However there is a poor correlation between the burden of amyloid plaques and the degree of cognitive impairment, indeed elderly individuals may have many plaques without showing signs of cognitive impairment [3-5]. In contrast, it is the intracellular tau pathology that has been shown to correlate more closely with clinical deficits. The location and progression of the tau lesions correlates well with the brain areas, such as the hippocampus, that are particularly impaired in AD [6].

The poor correlation between extracellular amyloid plaques and dementia has been used to detract from the significance of Aβ in the pathogenesis of AD. However recent evidence has clarified the situation, emphasising the toxic role of small Aβ aggregates rather than the amyloid fibrils. The finding that soluble Aβ correlates better with synaptic changes and cognitive deficits than plaque count [7-9] has prompted the investigation of soluble aggregates of Aβ. These small aggregates can be purified by column chromatography and are composed of as few as 4 [10] or as many as 180 [2] Aβ molecules. When applied to cell cultures the oligomers are toxic whereas in most cases amyloid fibrils and Aβ monomers are not [11, 12]. When oligomers are visualised under electron or atomic force microscopes they are heterogeneous, including spheres, beads-on-a-string and doughnuts [2], but it seems that the spherical species are most toxic [13]. Toxic oligomers may also be specifically detected, *in vitro* and *in vivo*, using rabbit antisera raised against Aβ immobilised on gold beads. The antiserum, described by Kayed and colleagues [14], binds specifically to small toxic aggregates of Aβ and neutralises their toxicity, in contrast the serum fails to detect monomeric or fibrillar forms of Aβ. Subsequent work has shown that the antiserum recognises an epitope on Aβ oligomers that is common to the oligomeric aggregates of a range of pathological proteins. The interesting corollary of this observation is that a common structural motif predicts a common mechanism of toxicity. This prediction is supported by work by Bucciantini *et al.*, . showing that oligomeric aggregates of a non disease related protein can elicit toxicity similar to that of Aβ oligomers in cell culture [15]. Further work done in cell culture by Demuro and colleagues [16] has shown that a shared ability to disturb membrane conductivity may underlie at least part of the toxicity of soluble protein aggregates.

However the hypothesis that soluble aggregates of Aβ represent a stable neurotoxic species has had to be reconsidered in the light of recent work showing that it is the ongoing *process* of aggregation that is toxic. It seems now that the soluble aggregates may simply be an efficient seed that can promote further addition of Aβ monomers. In their recent study, Wogulis and colleagues showed that, as expected, neither monomeric nor fibrillar Aβ were toxic to human or rat neuronal cell cultures. Their novel observation was that pre-treatment of cells with fibrillar Aβ, followed by a wash to remove unbound fibrils, primed the cells to die when they were subsequently treated with monomeric Aβ. The stability of the interaction of the fibrils with the cells was a surprise; following exposure to fibrils for only one hour the cells were still sensitized to the toxic effects of monomeric Aβ one week later [17].

With emphasis being placed on the oligomeric aggregates and the initial stages of the aggregation process, the mature plaques and tangles are increasingly being viewed as tombstones of pathological protein aggregation. Indeed there is evidence from cell-based models of Parkinson's disease that inclusions may be protective, reducing the rate of apoptosis [18] possibly by providing a sink for the disposal of toxic oligomers.

INTRANEURONAL Aβ $_{1-42}$ ACCUMULATION AND AGGREGATION

The classical view of APP processing is that Aβ is generated and released at the cell surface, resulting in extracellular amyloid plaque deposition and neurotoxicity. However it well documented that the machinery for generating Aβ exists intracellularly [19-22]. Some investigators have emphasised the importance of the secretory pathway in generating Aβ by showing that the treatment of cells with inhibitors of vesicular transport that effectively block APP export from the endoplasmic reticulum [20] or trans-Golgi network [21] do not abolish Aβ generation. Moreover APP processing in the endoplasmic reticulum preferentially yields $A\beta_{1-42}$ [22-24] that remains intracellular, whereas $A\beta_{1-40}$ is preferentially generated in the trans-Golgi network and packaged into secretory vesicles. There is also evidence that APP is processed after it has reached the plasma membrane and that endocytosis is important for the generation of Aβ [25]. This is of particular note because the low pH of the endosomes/lysosomes compartment will predictably favour oligomer formation [26, 27]. However there is the possibility that intracellular Aβ exerts at least part of its toxicity, not from the aqueous environment of vesicle lumen, but from within the membrane itself. There is evidence that Aβ peptides are sequestered in membranes predominantly as dimers [28] and some workers have proposed that specific intramembraneous protein-protein interactions may mediate some of the toxic effects of Aβ [29].

Aβ oligomerisation has been shown to start intracellularly in cell culture [30] and oligomers are present in the brains of patients with AD [10]. Clinical specimens have also shown that Aβ is intracellular during the early stages of AD but becomes predominantly extracellular as the patient develops advanced disease [31]. It may be that intracellular Aβ disappears with time because heavily-burdened neurones die, releasing their aggregates; indeed studies looking at the distribution and morphology of amyloid plaques suggest that each amyloid plaque is the result of a single neuronal lysis event [32].

The history of AD research has shown that good animal models have helped enormously to accelerate our understanding. The most recent animal models of Alzheimer's disease are providing strong support for the role of intracellular Aβ in generating the earliest symptoms of Alzheimer's disease [33, 34]. Triple transgenic mice that express disease-causing mutants of human APP, presenilin-1 and tau demonstrate clearly that intraneuronal accumulation of Aβ is sufficient to cause the earliest cognitive deficits [35]. At an age of four months the mice exhibit impaired long-term memory retention at a stage when plaques, tau pathology and neuronal death are entirely absent but intracellular Aβ is present. The presence of intracellular accumulation of Aβ may also explain why these triple transgenic mice are the first to show convincing neuronal loss as well as dysfunction [34]. This work in mouse models is supported by recent *Drosophila* models of AD that demonstrate non-amyloid intracellular Aβ aggregates are sufficient to cause locomotor deficits before extracellular Aβ deposits or cell death are seen [36]. Treating model organisms with anti-aggregatory compounds [36] or antibodies to Aβ [35] can ameliorate or even reverse the neuronal dysfunction that results from intraneuronal Aβ.

What Is the Role of Extracellular Aβ?

Extracellular Aβ has a wide range of effects that can be divided into two main categories. Firstly, Aβ has sequence-specific interactions with other proteins; notable is the binding of Aβ to receptors that are normally involved in the clearance of the peptide from the brain. Aβ also interacts specifically with receptors involved in neurotransmission and may cause some of the early, potentially reversible, symptoms of AD. Secondly, Aβ has biophysical effects on the electrical properties of membranes and also promotes oxidative stress, both of which contribute to the neuronal death seen in established cases of AD.

The Interaction of Extracellular Aβ with the Clearance Pathways

The concentration of Aβ in the brain depends both on the rate of production of the peptide and on the efficiency of the clearance mechanisms. Although the bulk of human genetic evidence points to the primary importance of Aβ synthesis in causing familial disease (presenilin 1 and 2 and APP mutations), there is some evidence that polymorphisms in genes related to clearance, such as the degrading enzyme neprilysin [37, 38], may influence an individual's risk of AD. It is thought that plaques may be cleared by the phagocytic activity of microglia and it is known that Aβ binds specifically to the plasma membranes of both microglia and neurones. On microglia a receptor complex has been identified that mediates the binding to Aβ fibrils [39]. Components of this receptor complex includes the B-class scavenger receptor CD36, the integrin-associated protein/CD47, and the alpha(6)beta(1)-integrin. It has also been reported that the receptor for advanced glycosylation end-products (RAGE) and FPRL1 (formyl peptide receptorlike 1) are able to bind both the monomeric and fibrillar forms of Aβ [40]. Microglia are found around neuritic plaques in the brains of patients with AD and the binding of Aβ to the receptors may stimulate an inflammatory response and mediate peptide clearance. Although these receptors may have a purely beneficial role in delivering peptide to the endosomes for degradation, however in the light of the discussion above, the internalisation of Aβ may in fact result in enhanced aggregation and toxicity.

Aβ and Long Term Potentiation

LTP is a phenomenon whereby stimulus-dependent enhancement of synaptic efficacy may encode memories. The role of LTP in the memory deficits of AD has been studied in transgenic mice that express the AD-causing Swedish mutant of human APP. Experiments in brain slices showed loss of LTP in the absence of neuronal death; moreover the loss of LTP correlated with deteriorating performance in behavioural tests of learning [41]. Similar loss of LTP was seen in rat brain slices treated with soluble Aβ aggregates of laboratory-synthesised peptides [42]. Walsh and colleagues have gone on to show that the oligomeric Aβ, secreted from CHO cells expressing a disease-causing APP mutant, can interfere with LTP in intact rat hippocampus. Intracerebroventricular injection of the conditioned medium containing Aβ oligomers was shown to completely abolish LTP, an effect that was not seen when control conditioned medium was injected [43]. Fractionation of Aβ species in the conditioned media into monomers and monomer-free oligomers demonstrated that it was the oligomers that were responsible for the learning deficits in rats following intracerebroventricular injection [44]. The straightforward explanation for the effects on LTP are that extracellular application of Aβ results in an effect mediated at the plasma membrane. However the demonstration of Aβ aggregates in endosomes implies that an intracellular mechanism remains a possibility [43].

Aβ and Nicotinic Neurotransmission

The disruption of cholinergic neurotransmission occurs early in AD and the elevation of synaptic acetylcholine (ACh) concentrations remains the main therapeutic approach in the clinic. Wang and colleagues have shown that Aβ binds with high affinity to alpha7n Ach receptors [45] resulting in the inhibition of receptor-dependent calcium signalling and acetylcholine release, two processes critically involved in neurotransmission and synaptic plasticity. The binding interaction between exogenous Aβ and the alpha 7 receptor may well facilitate the internalization and intracellular accumulation of Aβ in Alzheimer's disease brains. Indeed intracellular accumulation of Aβ in neurons has been shown in a cell culture model to correlate with the level of this receptor [46] and Aβ internalization can be blocked by alpha-bungarotoxin, an alpha 7 receptor antagonist. Moreover the high levels of the alpha 7 receptor found in the hippocampus and cortex [47] may account for the early involvement of these brain regions in AD. Although nicotinic stimulation has traditionally been seen as beneficial in AD (reviewed by Buccafusco *et al.*, . [48]) because of improved LTP and reduced amyloid deposition [49, 50] there is concern that up regulation of ACh receptors in smokers [51] may increase the proportion of the total Aβ that is soluble [50] and available for internalisation. This may account for the exacerbation of alpha 7 receptor-mediated tau phosphorylation [52] in transgenic mice treated with nicotine [53].

Aβ and Electrical Properties of Membranes

The electrical integrity of biological membranes is particularly important for the correct functioning and survival of neurones. Disruption of the resting potential across the plasma membrane may contribute to the toxic effects of Aβ as described in the preceding sections, namely LTP deficits and aberrant neurotransmission. In a similar way, disruption of the mitochondrial membrane may result in the oxidative stress [54] component of AD pathogenesis. There is however conflicting data about how oligomers cause membrane disruption; some groups have shown pore-like Aβ aggregates that insert into membranes;

others have evidence that membrane conductivity is increased but in the absence of discrete ion channel activity [55].

Pore-like aggregates of Aβ have a doughnut-shaped appearance, [56] being composed of 30-60 peptides monomers [2, 57] and resembling pore-forming bacterial toxins. These pore-like aggregates are proposed to insert into membranes forming a pathological ion channel, causing depolarisation of membranes and possibly calcium influx into the cell. In support of this hypothesis Lin and colleagues can demonstrate that pore-like aggregates have discrete ion channel activity that can be inhibited by zinc ions [56].

However, Aβ may share a channel-independent mechanism of membrane disruption with other aggregation-prone proteins. A recent study using fluorescently-loaded SHSY5Y cells showed that application of $A\beta_{1-42}$ and other oligomeric aggregates elevated intracellular Ca^{2+}, an effect that persisted even after depletion of intracellular Ca^{2+} stores [16]. The fact that the potent Ca^{2+} channel blocker cobalt failed to affect this response combined with the rapid leakage of anionic fluorescent dyes, point to a generalized increase in membrane permeability. This study provided evidence that the unregulated flux of ions across "leaky" membranes may provide a common mechanism for oligomer-mediated toxicity in many amyloid diseases. The dysregulation of calcium metabolism is likely to play an important role [58] because of the strong transmembrane concentration gradient and the involvement of calcium in intracellular signalling.

The Role of Oxidative Stress

It is well established that oxidative stress, as measured by oxidation products of proteins [59-61], lipids [62-64] and nucleic acids [65-68], has an important role in the pathogenesis of AD. Many of these studies have been performed in clinical specimens from late disease, however recent work has shown that isoprostane levels, a biochemical marker of lipid oxidation, are elevated even at the earliest stages of clinical AD [69].

How the reactive oxygen species are generated is hotly debated, however it is known that Aβ peptides are able to bind copper ions (Cu(II)) and reduce them (to Cu(I)), releasing hydrogen peroxide. Others metals may also be involved in similar reactions with aluminium and iron being possible candidates. Other workers have suggested that mitochondrial membrane disruption may release oxidative species. Interestingly, in concordance with the idea that it is the process of Aβ aggregation that is toxic, Tabner and colleagues have shown that, in the absence of redox-active metal ions, the earliest stages of Aβ aggregation can generate a brief burst of hydrogen peroxide [70].

Conclusion

There is an ongoing debate as to how Aβ causes neuronal dysfunction and death. The focus has recently been on small, soluble aggregates of Aβ, as the pathogenic agent however recent work suggest that the process of aggregation may be toxic, perhaps by generating oxidative species. A better understanding of the significance of intracellular Aβ may provide us with new therapeutic strategies for Alzheimer's disease.

References

[1] Nilsberth, C.; Westlind-Danielsson, A.; Eckman, C. B.; Condron, M. M.; Axelman, K.; Forsell, C.; Stenh, C.; Luthman, J.; Teplow, D. B.; Younkin, S. G.; Naslund, J.; Lannfelt, L. The 'Arctic' APP mutation (E693G) causes Alzheimer's disease by enhanced Abeta protofibril formation. *Nat Neurosci* 4: 887-893., 2001.

[2] Lashuel, H. A.; Hartley, D. M.; Petre, B. M.; Wall, J. S.; Simon, M. N.; Walz, T.; Lansbury, P. T., Jr. Mixtures of wild-type and a pathogenic (E22G) form of Abeta40 in vitro accumulate protofibrils, including amyloid pores. *J Mol Biol* 332: 795-808, 2003.

[3] Katzman, R.; Terry, R.; DeTeresa, R.; Brown, T.; Davies, P.; Fuld, P.; Renbing, X.; Peck, A. Clinical, pathological, and neurochemical changes in dementia: a subgroup with preserved mental status and numerous neocortical plaques. *Ann Neurol* 23: 138-144, 1988.

[4] Delaere, P.; Duyckaerts, C.; Masters, C.; Beyreuther, K.; Piette, F.; Hauw, J. J. Large amounts of neocortical beta A4 deposits without neuritic plaques nor tangles in a psychometrically assessed, non-demented person. *Neurosci Lett* 116: 87-93, 1990.

[5] Dickson, D. W.; Crystal, H. A.; Bevona, C.; Honer, W.; Vincent, I.; Davies, P. Correlations of synaptic and pathological markers with cognition of the elderly. *Neurobiol Aging* 16: 285-298; discussion 298-304, 1995.

[6] Braak, H.; Braak, E. Morphological criteria for the recognition of Alzheimer's disease and the distribution pattern of cortical changes related to this disorder. *Neurobiol Aging* 15: 355-356; discussion 379-380, 1994.

[7] Lue, L. F.; Kuo, Y. M.; Roher, A. E.; Brachova, L.; Shen, Y.; Sue, L.; Beach, T.; Kurth, J. H.; Rydel, R. E.; Rogers, J. Soluble amyloid beta peptide concentration as a predictor of synaptic change in Alzheimer's disease. *Am J Pathol* 155: 853-862, 1999.

[8] McLean, C. A.; Cherny, R. A.; Fraser, F. W.; Fuller, S. J.; Smith, M. J.; Beyreuther, K.; Bush, A. I.; Masters, C. L. Soluble pool of Abeta amyloid as a determinant of severity of neurodegeneration in Alzheimer's disease. *Ann Neurol* 46: 860-866, 1999.

[9] Naslund, J.; Haroutunian, V.; Mohs, R.; Davis, K. L.; Davies, P.; Greengard, P.; Buxbaum, J. D. Correlation between elevated levels of amyloid beta-peptide in the brain and cognitive decline. *JAMA* 283: 1571-1577, 2000.

[10] Gong, Y.; Chang, L.; Viola, K. L.; Lacor, P. N.; Lambert, M. P.; Finch, C. E.; Krafft, G. A.; Klein, W. L. Alzheimer's disease-affected brain: presence of oligomeric A beta ligands (ADDLs) suggests a molecular basis for reversible memory loss. *Proc Natl Acad Sci U S A* 100: 10417-10422, 2003.

[11] Lambert, M. P.; Barlow, A. K.; Chromy, B. A.; Edwards, C.; Freed, R.; Liosatos, M.; Morgan, T. E.; Rozovsky, I.; Trommer, B.; Viola, K. L.; Wals, P.; Zhang, C.; Finch, C. E.; Krafft, G. A.; Klein, W. L. Diffusible, nonfibrillar ligands derived from Abeta1-42 are potent central nervous system neurotoxins. *Proc Natl Acad Sci U S A* 95: 6448-6453., 1998.

[12] Hartley, D. M.; Walsh, D. M.; Ye, C. P.; Diehl, T.; Vasquez, S.; Vassilev, P. M.; Teplow, D. B.; Selkoe, D. J. Protofibrillar intermediates of amyloid beta-protein induce acute electrophysiological changes and progressive neurotoxicity in cortical neurons. *J Neurosci* 19: 8876-8884, 1999.

[13] Hoshi, M.; Sato, M.; Matsumoto, S.; Noguchi, A.; Yasutake, K.; Yoshida, N.; Sato, K. Spherical aggregates of beta-amyloid (amylospheroid) show high neurotoxicity and activate tau protein kinase I/glycogen synthase kinase-3beta. *Proc Natl Acad Sci U S A* 100: 6370-6375, 2003.

[14] Kayed, R.; Head, E.; Thompson, J. L.; McIntire, T. M.; Milton, S. C.; Cotman, C. W.; Glabe, C. G. Common structure of soluble amyloid oligomers implies common mechanism of pathogenesis. *Science* 300: 486-489, 2003.

[15] Bucciantini, M.; Giannoni, E.; Chiti, F.; Baroni, F.; Formigli, L.; Zurdo, J.; Taddei, N.; Ramponi, G.; Dobson, C. M.; Stefani, M. Inherent toxicity of aggregates implies a common mechanism for protein misfolding diseases. *Nature* 416: 507-511, 2002.

[16] Demuro, A.; Mina, E.; Kayed, R.; Milton, S. C.; Parker, I.; Glabe, C. G. Calcium dysregulation and membrane disruption as a ubiquitous neurotoxic mechanism of soluble amyloid oligomers. *J Biol Chem* 280: 17294-17300, 2005.

[17] Wogulis, M.; Wright, S.; Cunningham, D.; Chilcote, T.; Powell, K.; Rydel, R. E. Nucleation-dependent polymerization is an essential component of amyloid-mediated neuronal cell death. *J Neurosci* 25: 1071-1080, 2005.

[18] Tanaka, M.; Kim, Y. M.; Lee, G.; Junn, E.; Iwatsubo, T.; Mouradian, M. M. Aggresomes formed by alpha-synuclein and synphilin-1 are cytoprotective. *J Biol Chem* 279: 4625-4631, 2004.

[19] Hartmann, T.; Bieger, S. C.; Bruhl, B.; Tienari, P. J.; Ida, N.; Allsop, D.; Roberts, G. W.; Masters, C. L.; Dotti, C. G.; Unsicker, K.; Beyreuther, K. Distinct sites of intracellular production for Alzheimer's disease A beta40/42 amyloid peptides. *Nat Med* 3: 1016-1020, 1997.

[20] Wild-Bode, C.; Yamazaki, T.; Capell, A.; Leimer, U.; Steiner, H.; Ihara, Y.; Haass, C. Intracellular generation and accumulation of amyloid beta-peptide terminating at amino acid 42. *J Biol Chem* 272: 16085-16088, 1997.

[21] Xu, H.; Sweeney, D.; Wang, R.; Thinakaran, G.; Lo, A. C.; Sisodia, S. S.; Greengard, P.; Gandy, S. Generation of Alzheimer beta-amyloid protein in the trans-Golgi network in the apparent absence of vesicle formation. *Proc Natl Acad Sci U S A* 94: 3748-3752, 1997.

[22] Cook, D. G.; Forman, M. S.; Sung, J. C.; Leight, S.; Kolson, D. L.; Iwatsubo, T.; Lee, V. M.; Doms, R. W. Alzheimer's A beta(1-42) is generated in the endoplasmic reticulum/intermediate compartment of NT2N cells. *Nat Med* 3: 1021-1023, 1997.

[23] Skovronsky, D. M.; Doms, R. W.; Lee, V. M. Detection of a novel intraneuronal pool of insoluble amyloid beta protein that accumulates with time in culture. *J Cell Biol* 141: 1031-1039, 1998.

[24] Greenfield, J. P.; Tsai, J.; Gouras, G. K.; Hai, B.; Thinakaran, G.; Checler, F.; Sisodia, S. S.; Greengard, P.; Xu, H. Endoplasmic reticulum and trans-Golgi network generate distinct populations of Alzheimer beta-amyloid peptides. *Proc Natl Acad Sci U S A* 96: 742-747, 1999.

[25] Perez, R. G.; Soriano, S.; Hayes, J. D.; Ostaszewski, B.; Xia, W.; Selkoe, D. J.; Chen, X.; Stokin, G. B.; Koo, E. H. Mutagenesis identifies new signals for beta-amyloid precursor protein endocytosis, turnover, and the generation of secreted fragments, including Abeta42. *J Biol Chem* 274: 18851-18856, 1999.

[26] Huang, T. H.; Yang, D. S.; Plaskos, N. P.; Go, S.; Yip, C. M.; Fraser, P. E.; Chakrabartty, A. Structural studies of soluble oligomers of the Alzheimer beta-amyloid peptide. *J Mol Biol* 297: 73-87, 2000.

[27] Carrotta, R.; Manno, M.; Bulone, D.; Martorana, V.; San Biagio, P. L. Protofibril formation of amyloid beta-protein at low pH via a non-cooperative elongation mechanism. *J Biol Chem* 280: 30001-30008, 2005.

[28] Kawarabayashi, T.; Shoji, M.; Younkin, L. H.; Wen-Lang, L.; Dickson, D. W.; Murakami, T.; Matsubara, E.; Abe, K.; Ashe, K. H.; Younkin, S. G. Dimeric amyloid beta protein rapidly accumulates in lipid rafts followed by apolipoprotein E and phosphorylated tau accumulation in the Tg2576 mouse model of Alzheimer's disease. *J Neurosci* 24: 3801-3809, 2004.

[29] Marchesi, V. T. An alternative interpretation of the amyloid Abeta hypothesis with regard to the pathogenesis of Alzheimer's disease. *Proc Natl Acad Sci U S A* 102: 9093-9098, 2005.

[30] Walsh, D. M.; Tseng, B. P.; Rydel, R. E.; Podlisny, M. B.; Selkoe, D. J. The oligomerization of amyloid beta-protein begins intracellularly in cells derived from human brain. *Biochemistry* 39: 10831-10839, 2000.

[31] Gouras, G. K.; Tsai, J.; Naslund, J.; Vincent, B.; Edgar, M.; Checler, F.; Greenfield, J. P.; Haroutunian, V.; Buxbaum, J. D.; Xu, H.; Greengard, P.; Relkin, N. R. Intraneuronal Abeta42 accumulation in human brain. *Am J Pathol* 156: 15-20, 2000.

[32] D'Andrea, M. R.; Nagele, R. G.; Wang, H. Y.; Peterson, P. A.; Lee, D. H. in *Histopathology* 120-134 (2001).

[33] Oddo, S.; Caccamo, A.; Shepherd, J. D.; Murphy, M. P.; Golde, T. E.; Kayed, R.; Metherate, R.; Mattson, M. P.; Akbari, Y.; LaFerla, F. M. Triple-transgenic model of Alzheimer's disease with plaques and tangles: intracellular Abeta and synaptic dysfunction. *Neuron* 39: 409-421, 2003.

[34] Casas, C.; Sergeant, N.; Itier, J. M.; Blanchard, V.; Wirths, O.; van der Kolk, N.; Vingtdeux, V.; van de Steeg, E.; Ret, G.; Canton, T.; Drobecq, H.; Clark, A.; Bonici, B.; Delacourte, A.; Benavides, J.; Schmitz, C.; Tremp, G.; Bayer, T. A.; Benoit, P.; Pradier, L. Massive CA1/2 neuronal loss with intraneuronal and N-terminal truncated Abeta42 accumulation in a novel Alzheimer transgenic model. *Am J Pathol* 165: 1289-1300, 2004.

[35] Billings, L. M.; Oddo, S.; Green, K. N.; McGaugh, J. L.; Laferla, F. M. Intraneuronal Abeta causes the onset of early Alzheimer's disease-related cognitive deficits in transgenic mice. *Neuron* 45: 675-688, 2005.

[36] Crowther, D. C.; Kinghorn, K. J.; Miranda, E.; Page, R.; Curry, J. A.; Duthie, F. A.; Gubb, D. C.; Lomas, D. A. Intraneuronal Aβ, non-amyloid aggregates and neurodegeneration in a Drosophila model of Alzheimer's disease. *Neuroscience* 132: 123-135, 2005.

[37] Clarimon, J.; Munoz, F. J.; Boada, M.; Tarraga, L.; Sunyer, J.; Bertranpetit, J.; Comas, D. Possible increased risk for Alzheimer's disease associated with neprilysin gene. *J Neural Transm* 110: 651-657, 2003.

[38] Sakai, A.; Ujike, H.; Nakata, K.; Takehisa, Y.; Imamura, T.; Uchida, N.; Kanzaki, A.; Yamamoto, M.; Fujisawa, Y.; Okumura, K.; Kuroda, S. Association of the Neprilysin gene with susceptibility to late-onset al., zheimer's disease. *Dement Geriatr Cogn Disord* 17: 164-169, 2004.

[39] Bamberger, M. E.; Harris, M. E.; McDonald, D. R.; Husemann, J.; Landreth, G. E. A cell surface receptor complex for fibrillar beta-amyloid mediates microglial activation. *J Neurosci* 23: 2665-2674, 2003.

[40] Verdier, Y.; Zarandi, M.; Penke, B. Amyloid beta-peptide interactions with neuronal and glial cell plasma membrane: binding sites and implications for Alzheimer's disease. *J Pept Sci* 10: 229-248, 2004.

[41] Chapman, P. F.; White, G. L.; Jones, M. W.; Cooper-Blacketer, D.; Marshall, V. J.; Irizarry, M.; Younkin, L.; Good, M. A.; Bliss, T. V.; Hyman, B. T.; Younkin, S. G.; Hsiao, K. K. Impaired synaptic plasticity and learning in aged amyloid precursor protein transgenic mice. *Nat Neurosci* 2: 271-276, 1999.

[42] Wang, H. W.; Pasternak, J. F.; Kuo, H.; Ristic, H.; Lambert, M. P.; Chromy, B.; Viola, K. L.; Klein, W. L.; Stine, W. B.; Krafft, G. A.; Trommer, B. L. Soluble oligomers of beta amyloid (1-42) inhibit long-term potentiation but not long-term depression in rat dentate gyrus. *Brain Res* 924: 133-140, 2002.

[43] Walsh, D. M.; Klyubin, I.; Fadeeva, J. V.; Cullen, W. K.; Anwyl, R.; Wolfe, M. S.; Rowan, M. J.; Selkoe, D. J. Naturally secreted oligomers of amyloid beta protein potently inhibit hippocampal long-term potentiation *in vivo*. *Nature* 416: 535-539, 2002.

[44] Cleary, J. P.; Walsh, D. M.; Hofmeister, J. J.; Shankar, G. M.; Kuskowski, M. A.; Selkoe, D. J.; Ashe, K. H. Natural oligomers of the amyloid-beta protein specifically disrupt cognitive function. *Nat Neurosci* 8: 79-84, 2005.

[45] Wang, H. Y.; Lee, D. H.; D'Andrea, M. R.; Peterson, P. A.; Shank, R. P.; Reitz, A. B. beta-Amyloid(1-42) binds to alpha7 nicotinic acetylcholine receptor with high affinity. Implications for Alzheimer's disease pathology. *J Biol Chem* 275: 5626-5632, 2000.

[46] Nagele, R. G.; D'Andrea, M. R.; Anderson, W. J.; Wang, H. Y. Intracellular accumulation of beta-amyloid(1-42) in neurons is facilitated by the alpha 7 nicotinic acetylcholine receptor in Alzheimer's disease. *Neuroscience* 110: 199-211, 2002.

[47] Breese, C. R.; Adams, C.; Logel, J.; Drebing, C.; Rollins, Y.; Barnhart, M.; Sullivan, B.; Demasters, B. K.; Freedman, R.; Leonard, S. Comparison of the regional expression of nicotinic acetylcholine receptor alpha7 mRNA and [125I]-alpha-bungarotoxin binding in human postmortem brain. *J Comp Neurol* 387: 385-398, 1997.

[48] Buccafusco, J. J.; Letchworth, S. R.; Bencherif, M.; Lippiello, P. M. Long-lasting cognitive improvement with nicotinic receptor agonists: mechanisms of pharmacokinetic-pharmacodynamic discordance. *Trends Pharmacol Sci* 26: 352-360, 2005.

[49] Hellstrom-Lindahl, E.; Mousavi, M.; Ravid, R.; Nordberg, A. Reduced levels of Abeta 40 and Abeta 42 in brains of smoking controls and Alzheimer's patients. *Neurobiol Dis* 15: 351-360, 2004.

[50] Nordberg, A.; Hellstrom-Lindahl, E.; Lee, M.; Johnson, M.; Mousavi, M.; Hall, R.; Perry, E.; Bednar, I.; Court, J. Chronic nicotine treatment reduces beta-amyloidosis in the brain of a mouse model of Alzheimer's disease (APPsw). *J Neurochem* 81: 655-658, 2002.

[51] Mousavi, M.; Hellstrom-Lindahl, E.; Guan, Z. Z.; Shan, K. R.; Ravid, R.; Nordberg, A. Protein and mRNA levels of nicotinic receptors in brain of tobacco using controls and patients with Alzheimer's disease. *Neuroscience* 122: 515-520, 2003.

[52] Wang, H. Y.; Li, W.; Benedetti, N. J.; Lee, D. H. Alpha 7 nicotinic acetylcholine receptors mediate beta-amyloid peptide-induced tau protein phosphorylation. *J Biol Chem* 278: 31547-31553, 2003.
[53] Oddo, S.; Caccamo, A.; Green, K. N.; Liang, K.; Tran, L.; Chen, Y.; Leslie, F. M.; LaFerla, F. M. Chronic nicotine administration exacerbates tau pathology in a transgenic model of Alzheimer's disease. *Proc Natl Acad Sci U S A* 102: 3046-3051, 2005.
[54] Abramov, A. Y.; Canevari, L.; Duchen, M. R. Beta-amyloid peptides induce mitochondrial dysfunction and oxidative stress in astrocytes and death of neurons through activation of NADPH oxidase. *J Neurosci* 24: 565-575, 2004.
[55] Kayed, R.; Sokolov, Y.; Edmonds, B.; McIntire, T. M.; Milton, S. C.; Hall, J. E.; Glabe, C. G. Permeabilization of lipid bilayers is a common conformation-dependent activity of soluble amyloid oligomers in protein misfolding diseases. *J Biol Chem* 279: 46363-46366, 2004.
[56] Lin, H.; Bhatia, R.; Lal, R. Amyloid beta protein forms ion channels: implications for Alzheimer's disease pathophysiology. *Faseb J* 15: 2433-2444, 2001.
[57] Lashuel, H. A.; Hartley, D.; Petre, B. M.; Walz, T.; Lansbury, P. T., Jr. Neurodegenerative disease: amyloid pores from pathogenic mutations. *Nature* 418: 291., 2002.
[58] Mattson, M. P.; Cheng, B.; Davis, D.; Bryant, K.; Lieberburg, I.; Rydel, R. E. beta-Amyloid peptides destabilize calcium homeostasis and render human cortical neurons vulnerable to excitotoxicity. *J Neurosci* 12: 376-389, 1992.
[59] Castegna, A.; Aksenov, M.; Thongboonkerd, V.; Klein, J. B.; Pierce, W. M.; Booze, R.; Markesbery, W. R.; Butterfield, D. A. Proteomic identification of oxidatively modified proteins in Alzheimer's disease brain. Part II: dihydropyrimidinase-related protein 2, alpha-enolase and heat shock cognate 71. *J Neurochem* 82: 1524-1532, 2002.
[60] Castegna, A.; Aksenov, M.; Aksenova, M.; Thongboonkerd, V.; Klein, J. B.; Pierce, W. M.; Booze, R.; Markesbery, W. R.; Butterfield, D. A. Proteomic identification of oxidatively modified proteins in Alzheimer's disease brain. Part I: creatine kinase BB, glutamine synthase, and ubiquitin carboxy-terminal hydrolase L-1. *Free Radic Biol Med* 33: 562-571, 2002.
[61] Castegna, A.; Thongboonkerd, V.; Klein, J. B.; Lynn, B.; Markesbery, W. R.; Butterfield, D. A. Proteomic identification of nitrated proteins in Alzheimer's disease brain. *J Neurochem* 85: 1394-1401, 2003.
[62] Markesbery, W. R.; Carney, J. M. Oxidative alterations in Alzheimer's disease. *Brain Pathol* 9: 133-146, 1999.
[63] Keller, J. N.; Mattson, M. P. Roles of lipid peroxidation in modulation of cellular signaling pathways, cell dysfunction, and death in the nervous system. *Rev Neurosci* 9: 105-116, 1998.
[64] Mattson, M. P. Modification of ion homeostasis by lipid peroxidation: roles in neuronal degeneration and adaptive plasticity. *Trends Neurosci* 21: 53-57, 1998.
[65] Gabbita, S. P.; Lovell, M. A.; Markesbery, W. R. Increased nuclear DNA oxidation in the brain in Alzheimer's disease. *J Neurochem* 71: 2034-2040, 1998.
[66] Lovell, M. A.; Gabbita, S. P.; Markesbery, W. R. Increased DNA oxidation and decreased levels of repair products in Alzheimer's disease ventricular CSF. *J Neurochem* 72: 771-776, 1999.

[67] Nunomura, A.; Perry, G.; Pappolla, M. A.; Wade, R.; Hirai, K.; Chiba, S.; Smith, M. A. RNA oxidation is a prominent feature of vulnerable neurons in Alzheimer's disease. *J Neurosci* 19: 1959-1964, 1999.

[68] Wang, J.; Xiong, S.; Xie, C.; Markesbery, W. R.; Lovell, M. A. Increased oxidative damage in nuclear and mitochondrial DNA in Alzheimer's disease. *J Neurochem* 93: 953-962, 2005.

[69] Markesbery, W. R.; Kryscio, R. J.; Lovell, M. A.; Morrow, J. D. Lipid peroxidation is an early event in the brain in amnestic mild cognitive impairment. *Ann Neurol* 58: 730-735, 2005.

[70] Tabner, B. J.; El-Agnaf, O. M.; Turnbull, S.; German, M. J.; Paleologou, K. E.; Hayashi, Y.; Cooper, L. J.; Fullwood, N. J.; Allsop, D. Hydrogen Peroxide Is Generated during the Very Early Stages of Aggregation of the Amyloid Peptides Implicated in Alzheimer Disease and Familial British Dementia. *J Biol Chem* 280: 35789-35792, 2005.

In: Cognitive Science Compendium, Volume 1
Editor: Miao-Kun Sun
ISBN: 978-1-60456-597-3

Chapter 12

BRAIN FUNCTION IN ALTERED STATES OF CONSCIOUSNESS: COMPARISON BETWEEN ALZHEIMER DEMENTIA AND VEGETATIVE STATE

Mélanie Boly, Eric Salmon and Steven Laureys*
Neurology Department and Cyclotron Research Centre,
University of Liège, Belgium

ABSTRACT

Disorder of consciousness is not an all-or-none phenomenon but it rather represents a continuum. Alzheimer's disease (AD) is the most common cause of dementia among people aged 65 and older, and patients are frequently unaware of the importance of their cognitive deficits (Derouesne *et al.*, 1999). Vegetative state (VS) is a clinical entity with a complete lack of behavioural signs of awareness, but preserved arousal (ANA Committee on Ethical Affairs, 1993; The Multi-Society Task Force on PVS, 1994). Both clinical entities share a certain level of consciousness alteration, and a certain similarity in brain metabolic impairment. Here, we review differences and similarities in brain function between these two types of disorders of consciousness, as revealed by functional neuroimaging studies.

Keywords: vegetative state, Alzheimer dementia, functional brain imaging, consciousness.

* Corresponding author: Steven Laureys, MD, PhD. Cyclotron Research Centre, University of Liège, Sart Tilman B30. 4000 Liège, Belgium. Email: steven.laureys@ulg.ac.be; Phone number : 00324/3662316; Fax number : 00324/3662946

INTRODUCTION

Consciousness is a multifaceted concept that can be divided into two main components: arousal (i.e., wakefulness, or vigilance) and awareness (e.g., awareness of the self and the environment) (Plum and Posner, 1983; Zeman *et al.,* 1997). Arousal is supported by several brainstem neuron populations that directly project to both thalamic and cortical neurons (Steriade *et al.,* 1997). Awareness is thought to be dependent upon the functional integrity of the cerebral cortex and its sub-cortical connections; each of its many parts are located, to some extent, in anatomically defined regions in the brain (Dehaene and Naccache, 2001; Zeman, 2001). Usually, consciousness is assessed in neurologically disabled patients by the presence of voluntary or meaningful behaviours (in contrast to automatic, reflexive reactions) in response to various external stimulations.

"CONSCIOUSNESS" IN AD

AD is clinically characterized by a dementia syndrome. Current criteria for dementia include a deterioration in cognitive functions, sufficient to impair daily living activities, with mood and behaviour disturbances (APA, 1987). Dementia stages may be defined using a Mini-Mental State Examination (MMSE) score range (Folstein *et al.,* 1975). Usually, a mild stage corresponds to a MMSE score greater than 20, dementia is moderate between scores 20 and 12, and the deficit is severe when MMSE value is below 12.

There are different aspects of impaired consciousness in AD. An episodic memory trouble is the most frequent early clinical symptom (Perry and Hodges, 1996; Fleischman and Gabrieli, 1999). Episodic memory refers to personal episodes that can be associated to precise contextual information, concerning time and place for example. AD patients cannot vividly recollect and re-experience a number of episodes of their (recent) life, in which they were however deeply involved. Such an impairment in processing contextual characteristics of an episode that make it unique for a subject corresponds to a decrease of autonoetic consciousness (Wheeler *et al.,* 1997; Levine *et al.,* 1998; Tulving, 2002).

More generally, AD is characterised by a deficit of controlled cognitive processes that require conscious processing of information. Neuropsychological studies have demonstrated that AD patients show impairment in controlled processes during memory and executive tasks, while automatic activities may be more preserved (Fabrigoule *et al.,* 1998). It is frequently observed that AD patients fail to consciously recollect information whereas they provide target memories in implicit conditions. Several neuropsychological data show that AD patients, because of diminished control capacities, base their daily functioning upon automatic processes (Adam *et al.,* 2005). They remain efficient in routine situations, but are not able to face unexpected situations.

Beside cognitive impairment, dementia symptoms include personality change and altered judgment (APA, 1987). Behavioural and psychological impairments are well described in AD (Cummings *et al.,* 1994; Neary *et al.,* 1998). Importantly, there is an early lack of awareness for self-cognitive or self-behavioural difficulties (an anosognosia for deterioration) in the disease (Kalbe *et al.,* 2005b). AD patients may show different degrees of awareness in different domains (e.g., reasoning, memory, affect; Figure 1), and there happens to be a continuum in levels of anosognosia concerning the cognitive or behavioural deficits in the disease. (Damasio, 1994; Salmon *et al.,* 2005a). The level of anosognosia has been related to many clinical variables, and not always to dementia severity (Salmon *et al.,* 2005b).

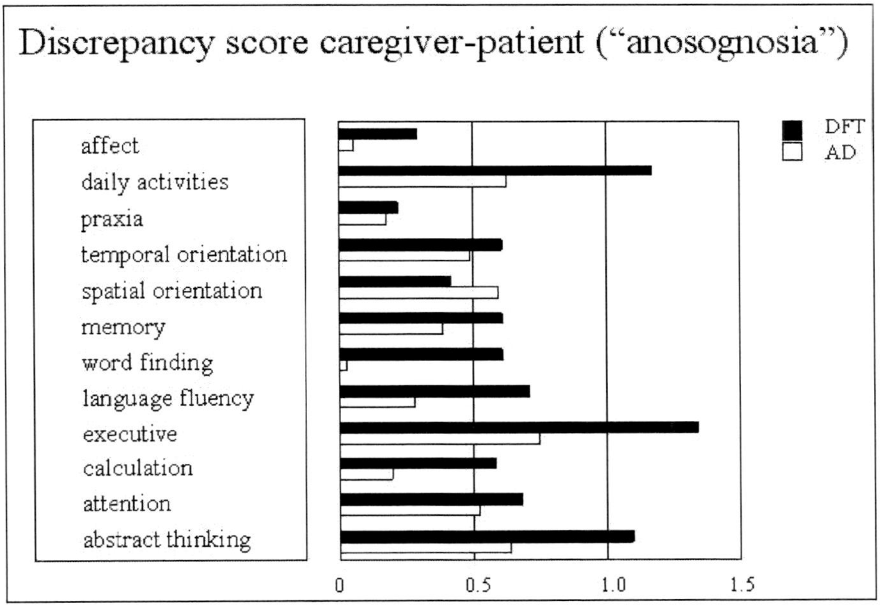

Figure 1. Anosognosia levels in several aspects of cognitive impairment in demented patients, as measured by the discrepancy score between caregivers and patients judgment in Alzheimer disease (AD) and fronto-temporal dementia (DFT)

As dementia develops, patients lose the awareness of time, place and other people (Fishback, 1977). Progression of AD gradually decreases the ability of patients to interact with their environment. Patients may develop aphasia, become unable to recognize friends and family members, and eventually lose the ability to maintain eye contact with their caregivers (Volicer *et al.,* 1987, 1997). As mental oblivion intervenes, the last thing the patient forgets is his/her own name (Fishback, 1977). In later stages, awareness completely disappears, marking the start of a sort of vegetative state (The Multi-Society Task Force on PVS, 1994).

Vegetative State

In 1972, Jennet and Plum defined the vegetative state as a clinical condition of 'wakefulness without awareness' (Jennett and Plum, 1972). These patients have preserved sleep-wake cycles but they do not show clinical signs of awareness of self or environment. They usually present reflex or spontaneous eye opening and breathing, they occasionally move their limbs or trunk in a meaningless way. They may be aroused by painful or prominent stimuli, opening their eyes if they are closed, increasing their respiratory rate, heart rate and blood pressure and occasionally grimacing or moving. Vegetative state patients can make a range of spontaneous movements including chewing, teeth-grinding and swallowing.

More distressingly, they can even show rage, cry, grunt, moan, scream or smile reactions spontaneously or to non-specific stimulation. Their head and eyes sometimes, inconsistently, turn fleeting towards new sounds or sights (Laureys *et al.,* 2000, 2002, 2004). Emergence of vegetative state is defined by the minimally conscious state (MCS). MCS patients show minimal but definite evidence of self or environment awareness but are unable to communicate (Giacino *et al.,* 2002).

Degenerative disorders such as AD are considered as a classical aetiology of vegetative state (The Multi-Society Task Force on PVS, 1994). In patients with degenerative diseases, a persistent vegetative state usually evolves over a period of several months or years (Walshe and Leonard, 1985). Those who remain in a vegetative state may die of superimposed infection or illness. Those who survive such an illness remain in a vegetative state or go into a coma (The Multi-Society Task Force on PVS, 1994). End-of-life questions and ethical debates are similar for VS and bed-ridden, latest stage AD. The question is what action is consistent with the ethical principles of proportionality in balancing the benefits and burdens of medical intervention, and how best to respect the autonomy and self-determination of the patient (Jennett, 1972). It is always difficult to know demented patients' awareness of the end of life. It is also really difficult to accompany these patients, with whom communication is essentially nonverbal (Michel *et al.,* 2002). Many patients with dementia lose the ability to feed themselves in the advanced stages of the disease (McNamara and Kennedy, 2001). Once a patient is considered permanently vegetative, the ethical debate revolves largely around the decision about continuing or withdrawing artificial nutrition and hydration (Jennett, 2005).

Global Cerebral Metabolism in VS and in AD

Since the beginnings of positron emission tomography (PET) imaging, quantified studies of regional consumption of oxygen and glucose have been performed in AD. They showed a global diminution of cerebral activity (Demetriades, 2002). This diminution of metabolism was proportional to dementia severity (Figure 2; Cutler et al., 1985), being 20% decreased in patients with mild to moderate dementia, and 40% decreased in patients with severe dementia, compared to normal age-matched controls (Frackowiak et al., 1981).

PET has also shown a substantial reduction in global brain metabolism in vegetative patients. Studies of our group and others have shown a 50 to 60% decrease in brain metabolism in vegetative state of different aetiology and duration (Figure 3; Levy et al., 1987; Momose et al., 1989; De Volder et al., 1990; Beuret et al., 1993; Tommasino et al., 1995; Laureys et al., 1999a,b, Rudolf et al., 1999, 2002; Schiff et al., 2002; Edgren et al., 2003). In "permanent" vegetative state (i.e., 12 months after a trauma or 3 months after non-traumatic brain damage), brain metabolism values drop to 30–40% of the normal mean activity (Rudolf et al., 1999, 2002).

A global decrease in cerebral metabolism is not specific to AD or VS. When different anaesthetics are titrated to the point of unresponsiveness, the reduction in brain metabolism is similar to that in comatose patients (Alkire et al., 1995, 1997, 1999). The lowest values of brain metabolism have been reported during propofol anaesthesia (to 28% of the normal range). (Alkire et al., 1995) A transient decrease in brain metabolism also takes place during deep sleep (stage III and IV), (Maquet et al., 1996, 1997, 1999) where cortical cerebral metabolism can drop to nearly 40% of the normal range of values.

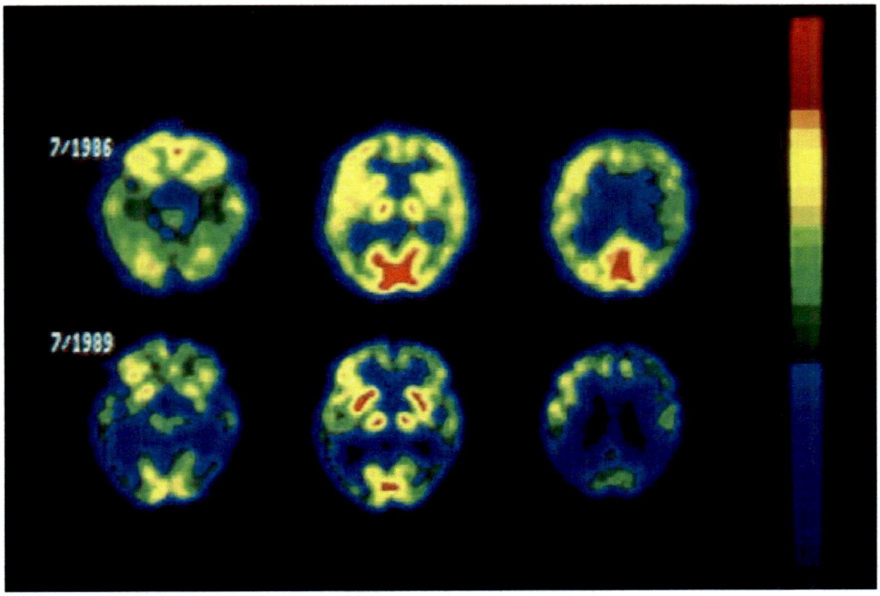

Figure 2. Quantified brain glucose metabolism in a single patient examined twice while in mild dementia stage (1986, upper panel) then in moderate dementia (1989, lower panel). This figure illustrate progressive brain metabolism reduction correlated with the clinical evolution of the Alzheimer disease

Regional Metabolic Distribution

Neuroimaging studies in AD revealed a characteristic regional distribution of decreased activity, showing consistent impairment of metabolism in temporo-parietal cortices, postero-medial regions (posterior cingulate cortex, precuneus), and lateral frontal associative cortices (Salmon, 2002). There is, however, in these patients a relative preservation of primary neo-cortical structures, such as the sensori-motor and primary visual cortex, and also of subcortical structures, like the basal ganglia, brainstem, and thalamus (Herholz *et al.,* 2002). Beside hypometabolism in cortical associative regions, a significant functional impairment of medial temporal regions was also reported in AD (Cutler *et al.,* 1985). However, when voxel-based analysis were used, medial temporal regions were found less diminished than associative cortical regions in these patients (Minoshima *et al.,* 1997). This is even more striking given that brain morphometric studies (voxel-based morphometry) show a major structural temporal medial atrophy (Busatto *et al.,* 2003), and that correlation studies showed a relationship between diminution of amnesic performances and alteration of medial temporal cortex activity (Lekeu *et al.,* 2003). Noteworthy, the involvement of medial temporal structures in autonoetic consciousness remains a matter of debate. It has been suggested that after a stage of diminished hippocampal and entorhinal activity, in the first stages of AD, complexity of medial temporal circuitry would explain local maintenance of activity despite atrophy severity and neurofibrillary histological lesions. One can recognize in AD different 'pathological poles' involving (1) medial posterior regions, often in relationship with frontal and parietal lateral associative regions, or (2) medial temporal regions, but also (3) medial frontal regions, like the anterior cingulate cortex. These poles would be involved variably depending on individuals, and this variability could explain individual clinical profiles (Salmon, 2002).

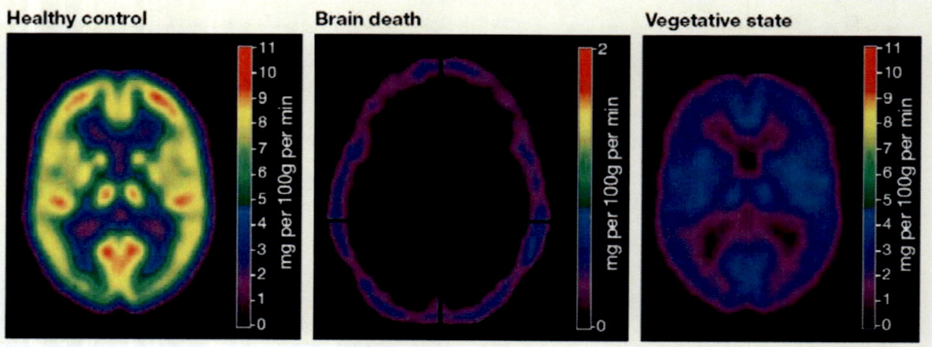

Figure 3. Illustration of the differences in resting brain metabolism measured in brain death and in the vegetative state, compared with controls. The image in patients with brain death shows a clear-cut 'hollow-skull sign', which is tantamount to a 'functional decapitation'. This is markedly different from the situation seen in patients in a vegetative state, in whom cerebral metabolism is massively and globally decreased (to 50% of normal value) but not absent. The colour scale shows the amount of glucose metabolized per 100 g of brain tissue per minute (reproduced with permission from Nature Reviews Neuroscience (Laureys, 2005) copyright 2005 Macmillan Magazines Ltd.)

In AD patients, reduction of metabolism in associative cortical regions is correlated to dementia severity (Salmon *et al.,* 2000, 2005b). This severity is evaluated by means of cognitive capacities scales, like the Mini Mental State Exam or of daily functionality, like the Instrumental Activity of Daily Living scale (Lawton and Brody, 1969). Lower scores mainly reflect lower (consciously) controlled capacities, related to lower metabolism in the first pathological pole just described. Longitudinal PET studies in AD patients showed an expansion as well as an increased severity of hypometabolism in association cortical areas and subcortical structures (Mielke *et al.,* 1996), and a close correlation between progressive metabolic reduction and impaired cognitive performance has been shown (Minoshima *et al.,* 1997; Demetriades, 2002). Some of the fronto-posterior regions included in the first AD pathological pole have been linked, in healthy volunteers, to autonoetic information retrieval, i.e. a remembering with full awareness of re-lived events, which is particularly impaired in AD patients (Whalen and Liberman, 1987).

On the other hand, the metabolic activity of the temporo-parietal junction is in relationship with a measure of anosognosia, i.e. the differential score between a relative and a patient evaluation of cognitive capacities of this patient (Salmon, *et al.,* 2005b). Thus, the anosognosia, or 'decreased of awareness of cognitive impairment' that happens very early in AD reflects a lack of conscious access to daily reality in AD patients, and is related to the activity in a portion of the fronto-posterior associative pathological pole (Kalbe *et al.,* 2005b).

The hallmark of the vegetative state is a complete loss of awareness and a systematic impairment of metabolism in the polymodal associative cortices (bilateral prefrontal regions, Broca's area, parieto-temporal, posterior parietal areas and precuneus and posterior cingulate)

(Laureys *et al.,* 2002). These regions are important in various functions that are necessary for consciousness, such as attention, memory, and language (Baars *et al.,* 2003). On the other hand, VS patients show a relative preservation of metabolism in brainstem, thalamus and posterior hypothalamic regions. In rare cases where patients in a vegetative state recover awareness of self and environment, PET shows a functional recovery of metabolism in these same cortical regions (Laureys *et al.,* 1999b).

The pattern of metabolic impairment in fronto-parietal associative areas found in VS is quite similar to that found in advanced AD (Figure 4). In AD as in VS, the medial posterior cortex has received great attention. The posterior cingulate, retrosplenial cortex and precuneus have all been involved in early stages of AD (Neunzig and Kunze, 1987; Minoshima *et al.,* 1997). In AD, posterior cingulate cortex metabolism is inversely correlated to score reduction in MMSE examination (Figure 5; Salmon *et al.,* 2000). Similarly, the medial parietal cortex (precuneus) and adjacent posterior cingulate cortex seem to be the brain regions that differentiate patients in minimally conscious state from those in vegetative state (Laureys *et al.,* 2004). Interestingly, these areas are among the most active brain regions in conscious waking (Maquet *et al.,* 1996; Gusnard and Raichle, 2001) and are among the least active regions in altered states of consciousness such as halothane (Alkire *et al.,* 1999) or propofol (Fiset *et al.,* 1999) -induced general anaesthesia, sleep (Maquet *et al.,* 1996, 1999), hypnotic state (Maquet *et al.,* 1997; Rainville *et al.,* 1999), and also in Wernicke–Korsakoff's or postanoxic amnesia (Aupee *et al.,* 2001). In a recent study, Vogt et al., . suggested that the ventral portion of the posterior cingulate cortex would be related to processing of events for their self/emotional significance (Vogt *et al.,* 2005).

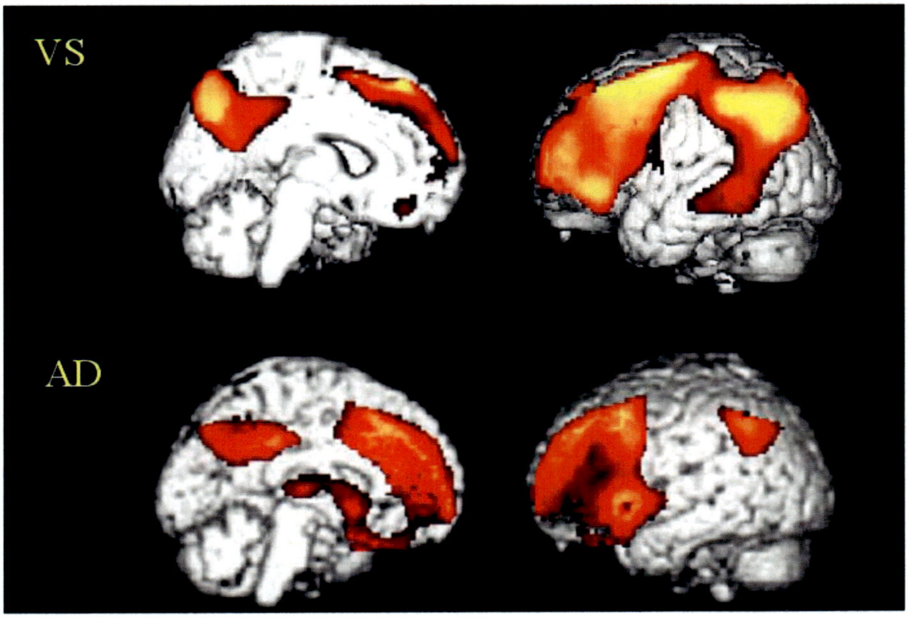

Figure 4. Regional metabolic alteration as compared to healthy subjects in patients in vegetative state (VS, upper panel) and Alzheimer demented patients (AD, lower panel) reveals a striking similarity in medial and lateral fronto-parietal associative areas impairment between both populations

Other PET and fMRI studies showed involvement of precuneus in reflective self-awareness (Kjaer *et al.,* 2002) and processing of one's own name compared to other names (Vogeley *et al.,* 2004; Perrin *et al.,* 2005). Precuneus and posterior cingulate cortex may thus be part of the neural network subserving human consciousness (Baars *et al.,* 2003), especially of a midline parieto-frontal core involved in self-awareness (Lou *et al.,* 2005).

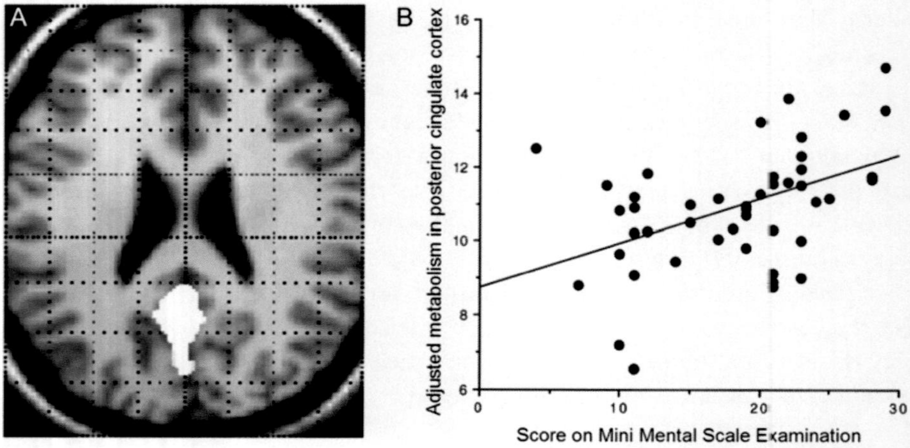

Figure 5. A. Early decrease of activity in posteromedial areas (encompassing posterior cingulate cortex and precuneus) in patients with Alzheimer's disease. B. Linear correlation between dementia severity (score on mini mental state examination) and metabolism in posterior cingulate cortex (area 31, coordinates: x = 0, y =-56, z = 28 mm) in Alzheimer's disease. (From Salmon et al., 2000, reprinted with permission of Wiley-Liss, Inc., a subsidiary of John Wiley and Sons, Inc)

FUNCTIONAL CONNECTIVITY

In AD, some studies showed a functional disconnection between certain pathological poles, i.e. a diminished functional correlation between parahippocampal and frontal regions (Lekeu *et al.,* 2003), or between prefrontal and other associative brain areas involved in short term memory tasks (Grady *et al.,* 2001). EEG studies also showed decreased interhemispheric EEG coherence in AD, reflecting both a lower mean level of functional connectivity as well as diminished fluctuations in the level of synchronization (Stam *et al.,* 2005). Other studies show a decreased complexity of EEG patterns and reduced information transmission among cortical areas in AD (Jeong, 2004). Diffusion tensor magnetic resonance imaging was also used to compare the integrity of several white matter fibre tracts in patients with probable AD (Rose *et al.,* 2004; Bozzali *et al.,* 2002). Relative to normal controls, patients with probable AD showed a highly significant reduction in the integrity of the association white matter fibre tracts, such as the splenium of the corpus callosum, superior longitudinal fasciculus, and cingulum. By contrast, pyramidal tract integrity seemed unchanged (Rose *et al.,* 2000). In another study, strong correlations were found between the mini mental state examination score and the average overall white matter integrity (Bozzali *et al.,* 2002). All these data are in line with the hypothesis that AD is a disconnection pathology, linked to the distribution of histological lesions in the different cortical layers (Knowles *et al.,* 1999). Activation studies in AD have broadly showed two types of patterns. In some memory studies, there is a decrease of activation in pathological poles such as the medial temporal area, probably related to impaired performances in memory tasks. In very early cases, it is however possible to observe a "normal" hippocampal activation, that might be related to compensatory functioning. In most reports, AD patients show activation in more cortical areas than their controls. The hypotheses are (1) that supplementary activation correspond to recruitment of further resources to reach the current performance or (2) that a non-efficient and useless activation occurs because selection and inhibition processes are deficient.

In vivo brain imaging data show multifocal brain atrophy (Juengling *et al.,* 2005) and impaired functional connectivity in patients in VS (Laureys *et al.,* 1999a, 2000). The resumption of long range functional connectivity between different associative cortices (Laureys *et al.,* 2004) and between some of these and the intralaminar thalamic nuclei parallels the restoration of the functional integrity of these patients (Laureys *et al.,* 2000b). An EEG study in a single VS patient showed markedly asymmetrically reduced EEG coherence in relationship with subcortical structural damage (Davey *et al.,* 2000). In cohort studies of patients unambiguously meeting the clinical diagnosis of vegetative state, simple noxious somatosensory (Laureys *et al.,* 2002) and auditory (Laureys *et al.,* 2000a; Boly *et al.,* 2004) stimuli have shown systematic activation of primary sensory cortices and lack of activation in higher order associative cortices from which they were functionally disconnected. High intensity noxious electrical stimulation activated midbrain, contralateral thalamus, and primary somatosensory cortex (Laureys *et al.,* 2002). However, secondary somatosensory, insular, posterior parietal, and anterior cingulate cortices, which were activated in all control individuals, did not show significant activation in any patient. Moreover, in patients in a vegetative state, the activated primary somatosensory cortex was functionally disconnected from higher-order associative cortices of the pain matrix (Laureys *et al.,* 2002). Similarly, although simple auditory stimuli activated bilateral primary auditory cortices, higher-order multimodal association cortices were not activated. A cascade of functional disconnections were also observed along the auditory cortical pathways, from primary auditory areas to multimodal and limbic areas (Laureys *et al.,* 2000a), suggesting that the observed residual cortical processing in the vegetative state does not lead to integrative processes, which are thought to be necessary for awareness. In contrary, functional connectivity analysis for similar simple auditory stimuli, performed in patients in a minimally conscious state, showed preserved functional connections between secondary auditory cortex and a large set of cortical areas (encompassing frontal and temporal association cortices) compared to VS patients (Boly *et al.,* 2004).

CONCLUSIONS

Disorder of consciousness should be considered as a continuum, not as an all-or-none phenomenon. Our review of correlation and longitudinal studies suggests that this is true in dementia as well as in severely brain injured patients like in vegetative and minimally conscious states. However, progression to VS is rarely present in AD patients (Volicer *et al.,* 1997). Even in the end stage dementia, some behavioural signs of consciousness remain in most cases. Experienced caregivers can detect a discomfort even in patients with very advanced dementia who are unable to maintain their posture in a chair and are mute (Hurley *et al.,* 1992). Late-stage patients are in fact quite different from one another and in most cases continue to interact with their environment (Boller *et al.,* 2002).

Looking at functional imaging data, both patients' populations show diminished global metabolism, diminished regional metabolism in associative areas with a relative preservation of brainstem, thalamus and primary sensory cortices, and impaired functional connectivity. These similarities in brain activity impairment patterns could reflect the general mechanisms of alteration of consciousness present in both clinical conditions, with impairment of the parieto-frontal associative network thought to be necessary to reach conscious perception (global workspace theory (Baars *et al.,* 2003)).

The level of metabolic alteration in the hippocampus and its role in autonoetic consciousness impairment in AD patients remain a matter of debate, even if its structural atrophy was related to episodic memory impairment early in the evolution of AD. Medial temporal regions seem not to be part of the more metabolically impaired regions in VS and were not related to impaired consciousness in this population of patients. A direct comparisons of metabolic disturbances in both conditions might add to the discussion on altered consciousness in neurological diseases and its physiopathological correlations.

Acknowledgments

The work in the Cyclotron Research Centre in Liege was made possible by grants from the National Fund for Scientific Research (FNRS), Fondation Medical Reine Elisabeth, a 5th European Framework project (NEST-DD) and InterUniversity Attraction Pole 5/04 – Belgian Science Policy. SL is also supported by FNRS grants 3.4517.04 and by grants from the Centre Hospitalier Universitaire Sart Tilman, the University of Liège, and the Mind Science Foundation. MB and SL are respectively Research Fellow and Research Associate at the Fonds National de la Recherche Scientifique (FNRS).

References

Adam, S., Van der Linden, M., Collette, F., Lemauvais, L. and Salmon, E. Further exploration of controlled and automatic memory processes in early Alzheimer's disease. *Neuropsychology* 19: 420-427, 2005.

Alkire, M.T., Haier, R.J., Barker, S.J., Shah, N.K., Wu, J.C. and Kao, Y.J. Cerebral metabolism during propofol anesthesia in humans studied with positron emission tomography. *Anesthesiology* 82: 393-403, 1995.

Alkire, M.T., Haier, R.J., Shah, N.K. and Anderson, C.T. Positron emission tomography study of regional cerebral metabolism in humans during isoflurane anesthesia. *Anesthesiology* 86: 549-557, 1997.

Alkie, M.T., Pomfrett, C.J., Haier, R.J., Gianzero, M.V., Chan, C.M., Jacobsen, B.P. and Fallon, J.H. Functional brain imaging during anesthesia in humans: effects of halothane on global and regional cerebral glucose metabolism. *Anesthesiology* 90: 701-709, 1999.

ANA Committee on Ethical Affairs. Persistent vegetative state: report of the American Neurological Association Committee on Ethical Affairs. *Ann. Neurol.* 33: 386-390, 1993.

APA. *Diagnostic and statistical manual of mental disorders (3d edition revised).* American Psychiatric Association, Washington DC, 1987.

APA (1994) *Diagnostic and statistical manual of mental disorders (4th edition).* American Psychiatric Association, Washington DC, 1987.

Aupee, A.M., Desgranges, B., Eustache, F., Lalevee, C., de la Sayette, V., Viader, F. and Baron, J.C. Voxel-based mapping of brain hypometabolism in permanent amnesia with PET. *Neuroimage* 13: 1164-1173, 2001.

Baars, B.J., Ramsoy, T.Z. and Laureys, S. Brain, conscious experience and the observing self. *Trends Neurosci.* 26: 671-675, 2003.

Beuret, P., Feihl, F., Vogt, P., Perret, A., Romand, J.A. and Perret, C. Cardiac arrest: prognostic factors and outcome at one year. *Resuscitation* 25: 171-179, 1993.

Boller, F., Verny, M., Hugonot-Diener, L. and Saxton, J. Clinical features and assessment of severe dementia. A review. *Eur. J. Neurol.* 9: 125-136, 2002.

Boly, M., Faymonville, M.E., Peigneux, P., Lambermont, B., Damas, P., Del Fiore, G., Degueldre, C., Franck, G., Luxen, A., Lamy, M., Moonen, G., Maquet, P. and Laureys, S. Auditory processing in severely brain injured patients: differences between the minimally conscious state and the persistent vegetative state. *Arch. Neurol.* 61: 233-238, 2004.

Bozzali, M., Falini, A., Franceschi, M., Cercignani, M., Zuffi, M., Scotti, G., Comi, G. and Filippi, M. White matter damage in Alzheimer's disease assessed in vivo using diffusion tensor magnetic resonance imaging. *J. Neurol. Neurosurg. Psychiatry* 72: 742-746, 2002.

Busatto, G.F., Garrido, G.E., Almeida, O.P., Castro, C.C., Camargo, C.H., Cid, C.G., Buchpiguel, C.A., Furuie, S. and Bottino, C.M. A voxel-based morphometry study of temporal lobe gray matter reductions in Alzheimer's disease. *Neurobiol. Aging.* 24: 221-231, 2003.

Cummings, J.L., Mega, M., Gray, K., Rosenberg-Thompson, S., Carusi, D.A. and Gornbein, J. The Neuropsychiatric Inventory: comprehensive assessment of psychopathology in dementia. *Neurology* 44: 2308-2314, 1994.

Cutler, N.R., Haxby, J.V., Duara, R., Grady, C.L., Kay, A.D., Kessler, R.M., Sundaram, M. and Rapoport, S.I. Clinical history, brain metabolism, and neuropsychological function in Alzheimer's disease. *Ann. Neurol.* 18: 298-309, 1985.

Damasio, A.R. *Descartes' error : emotion, reason, and the human brain.* G.P. Putnam, New York, 1994.

Davey, M.P., Victor, J.D. and Schiff, N.D. Power spectra and coherence in the EEG of a vegetative patient with severe asymmetric brain damage. *Clin. Neurophysiol.* 111: 1949-1954, 2000.

De Volder, A.G., Goffinet, A.M., Bol, A., Michel, C., de, B.T. and Laterre, C. Brain glucose metabolism in postanoxic syndrome. Positron emission tomographic study. *Arch. Neurol.* 47: 197-204, 1990.

Dehaene, S. and Naccache, L. Towards a cognitive neuroscience of consciousness: basic evidence and a workspace framework. *Cognition* 79: 1-37, 2001.

Demetriades, A.K. Functional neuroimaging in Alzheimer's type dementia. *J. Neurol. Sci.* 203-204: 247-251, 2002.

Derouesne, C., Lacomblez, L., Thibault, S. and LePoncin, M. Memory complaints in young and elderly subjects. *Int. J. Geriatr. Psychiatry* 14: 291-301, 1999.

Edgren, E., Enblad, P., Grenvik, A., Lilja, A., Valind, S., Wiklund, L., Hedstrand, U., Stjernstrom, H., Persson, L., Ponten, U. and Langstrom, B. Cerebral blood flow and metabolism after cardiopulmonary resuscitation. A pathophysiologic and prognostic positron emission tomography pilot study. *Resuscitation* 57: 161-170, 2003.

Fabrigoule, C., Rouch, I., Taberly, A., Letenneur, L., Commenges, D., Mazaux, J.M., Orgogozo, J.M. and Dartigues, J.F. Cognitive process in preclinical phase of dementia. *Brain* 121 (Pt 1): 135-141, 1998.

Fiset, P., Paus, T., Daloze, T., Plourde, G., Meuret, P., Bonhomme, V., Hajj-Ali, N., Backman, S.B. and Evans, A.C. Brain mechanisms of propofol-induced loss of consciousness in humans: a positron emission tomographic study. *J. Neurosci.* 19: 5506-5513, 1999.

Fishback, D.B. Mental status questionnaire for organic brain syndrome, with a new visual counting test. *J. Am. Geriatr. Soc.* 25: 167-170, 1977.

Fleischman, D.A. and Gabrieli, J. Long-term memory in Alzheimer's disease. *Curr. Opin. Neurobiol.* 9: 240-244, 1999.

Folstein, M.F., Folstein, S.E. and McHugh, P.R. "Mini-mental state". A practical method for grading the cognitive state of patients for the clinician. *J. Psychiatr.* 12: 189-198, 1975.

Frackowiak, R.S., Pozzilli, C., Legg, N.J., Du Boulay, G.H., Marshall, J., Lenzi, G.L. and Jones, T. Regional cerebral oxygen supply and utilization in dementia. A clinical and physiological study with oxygen-15 and positron tomography. *Brain* 104: 753-778, 1981.

Giacino, J.T., Ashwal, S., Childs, N., Cranford, R., Jennett, B., Katz, D.I., Kelly, J.P., Rosenberg, J.H., Whyte, J., Zafonte, R.D. and Zasler, N.D. The minimally conscious state: definition and diagnostic criteria. *Neurology* 58: 349-353, 2002.

Grady, C.L., Furey, M.L., Pietrini, P., Horwitz, B. and Rapoport, S.I. Altered brain functional connectivity and impaired short-term memory in Alzheimer's disease. *Brain* 124: 739-756, 2001.

Gusnard, D.A. and Raichle, M.E. Searching for a baseline: functional imaging and the resting human brain. *Nat. Rev. Neurosci.* 2: 685-694, 2001.

Herholz, K., Salmon, E., Perani, D., Baron, J.C., Holthoff, V., Frolich, L., Schonknecht, P., Ito, K., Mielke, R., Kalbe, E., Zundorf, G., Delbeuck, X., Pelati, O., Anchisi, D., Fazio, F., Kerrouche, N., Desgranges, B., Eustache, F., Beuthien-Baumann, B., Menzel, C., Schroder, J., Kato, T., Arahata, Y., Henze, M. and Heiss, W.D. Discrimination between Alzheimer dementia and controls by automated analysis of multicenter FDG PET. *Neuroimage* 17: 302-316, 2002.

Hurley, A.C., Volicer, B.J., Hanrahan, P.A., Houde, S. and Volicer, L. Assessment of discomfort in advanced Alzheimer patients. *Res. Nurs. Health.* 15: 369-377, 1992.

Jennett, B. Thirty years of the vegetative state: clinical, ethical and legal problems. *Prog. Brain Res.* 150: 537-543, 2005.

Jennett, B. and Plum, F. Persistent vegetative state after brain damage. A syndrome in search of a name. *Lancet* 1: 734-737, 1972.

Jeong, J. EEG dynamics in patients with Alzheimer's disease. *Clin. Neurophysiol.* 115: 1490-1505, 2004.

Juengling, F.D., Kassubek, J., Huppertz, H.J., Krause, T. and Els, T. Separating functional and structural damage in persistent vegetative state using combined voxel-based analysis of 3-D MRI and FDG-PET. *J. Neurol. Sci.* 228: 179-184, 2005.

Kalbe, E., Reinhold, N., Brand, M., Markowitsch, H.J. and Kessler, J. A new test battery to assess aphasic disturbances and associated cognitive dysfunctions -- German normative data on the aphasia check list. *J. Clin. Exp. Neuropsychol.* 27: 779-794, 2005a.

Kalbe, E., Salmon, E., Perani, D., Holthoff, V., Sorbi, S., Elsner, A., Weisenbach, S., Brand, M., Lenz, O., Kessler, J., Luedecke, S., Ortelli, P. and Herholz, K. Anosognosia in very mild Alzheimer's disease but not in mild cognitive impairment. *Dement. Geriatr. Cogn. Disord.* 19: 349-356, 2005b.

Kjaer, T.W., Nowak, M. and Lou, H.C. Reflective self-awareness and conscious states: PET evidence for a common midline parietofrontal core. *Neuroimage* 17: 1080-1086, 2002.

Knowles, R.B., Wyart, C., Buldyrev, S.V., Cruz, L., Urbanc, B., Hasselmo, M.E., Stanley, H.E. and Hyman, B.T. Plaque-induced neurite abnormalities: implications for disruption of neural networks in Alzheimer's disease. *Proc. Natl. Acad. Sci. U S A* 96: 5274-5279, 1999.

Laureys, S. Science and society: death, unconsciousness and the brain. *Nat. Rev. Neurosci.* 6: 899-909, 2005.

Laureys, S., Faymonville, M.E., Degueldre, C., Fiore, G.D., Damas, P., Lambermont, B., Janssens, N., Aerts, J., Franck, G., Luxen, A., Moonen, G., Lamy, M. and Maquet, P. Auditory processing in the vegetative state. *Brain* 123: 1589-1601, 2000a.

Laureys, S., Faymonville, M.E., Luxen, A., Lamy, M., Franck, G. and Maquet, P. Restoration of thalamocortical connectivity after recovery from persistent vegetative state. *Lancet* 355: 1790-1791, 2000b.

Laureys, S., Faymonville, M.E., Peigneux, P., Damas, P., Lambermont, B., Del Fiore, G., Degueldre, C., Aerts, J., Luxen, A., Franck, G., Lamy, M., Moonen, G. and Maquet, P. Cortical processing of noxious somatosensory stimuli in the persistent vegetative state. *Neuroimage* 17: 732-741, 2002.

Laureys, S., Goldman, S., Phillips, C., Van Bogaert, P., Aerts, J., Luxen, A., Franck, G. and Maquet, P. Impaired effective cortical connectivity in vegetative state: preliminary investigation using PET. *Neuroimage* 9: 377-382, 1999a.

Laureys, S., Lemaire, C., Maquet, P., Phillips, C. and Franck, G. Cerebral metabolism during vegetative state and after recovery to consciousness. *J. Neurol. Neurosurg. Psychiatry* 67: 121, 1999b.

Laureys, S., Owen, A.M. and Schiff, N.D. Brain function in coma, vegetative state, and related disorders. *Lancet Neurol.* 3: 537-546, 2004.

Lawton, M.P. and Brody, E.M. Assessment of older people: self-maintaining and instrumental activities of daily living. *Gerontologist* 9: 179-186, 1969.

Lekeu, F., Van der Linden, M., Chicherio, C., Collette, F., Degueldre, C., Franck, G., Moonen, G. and Salmon, E. Brain correlates of performance in a free/cued recall task with semantic encoding in Alzheimer disease. *Alzheimer Dis. Assoc. Disord.* 17: 35-45, 2003.

Levine, B., Black, S.E., Cabeza, R., Sinden, M., McIntosh, A.R., Toth, J.P., Tulving, E. and Stuss, D.T. Episodic memory and the self in a case of isolated retrograde amnesia. *Brain* 121 (Pt 10): 1951-1973, 1998.

Levy, D.E., Sidtis, J.J., Rottenberg, D.A., Jarden, J.O., Strother, S.C., Dhawan, V., Ginos, J.Z., Tramo, M.J., Evans, A.C. and Plum, F. Differences in cerebral blood flow and glucose utilization in vegetative versus locked-in patients. *Ann. Neurol.* 22: 673-682, 1987.

Lou, H.C., Nowak, M. and Kjaer, T.W. The mental self. *Prog. Brain Res.* 150: 197-204, 2005.

Maquet, P., Degueldre, C., Delfiore, G., Aerts, J., Peters, J.M., Luxen, A. and Franck, G. Functional neuroanatomy of human slow wave sleep. *J. Neurosci.* 17: 2807-2812, 1997.

Maquet, P., Faymonville, M.E., Degueldre, C., Delfiore, G., Franck, G., Luxen, A. and Lamy, M. Functional neuroanatomy of hypnotic state. *Biol. Psychiatry* 45: 327-333, 1999.

Maquet, P., Peters, J., Aerts, J., Delfiore, G., Degueldre, C., Luxen, A. and Franck, G. Functional neuroanatomy of human rapid-eye-movement sleep and dreaming. *Nature* 383: 163-166, 1996.

McNamara, E.P. and Kennedy, N.P. Tube feeding patients with advanced dementia: an ethical dilemma. *Proc. Nutr. Soc.* 60: 179-185, 2001.

Michel, J.P., Pautex, S., Zekry, D., Zulian, G. and Gold, G. End-of-life care of persons with dementia. *J. Gerontol. A. Biol. Sci. Med. Sci.* 57: M640-644, 2002.

Mielke, R., Schroder, R., Fink, G.R., Kessler, J., Herholz, K. and Heiss, W.D. Regional cerebral glucose metabolism and postmortem pathology in Alzheimer's disease. *Acta Neuropathol. (Berl.)* 91: 174-179, 1996.

Minoshima, S., Giordani, B., Berent, S., Frey, K.A., Foster, N.L. and Kuhl, D.E. Metabolic reduction in the posterior cingulate cortex in very early Alzheimer's disease. *Ann. Neurol.* 42: 85-94, 1997.

Momose, T., Matsui, T. and Kosaka, N. Effect of cervical spinal cord stimulation (cSCS) on cerebral glucose metabolism and blood flow in a vegetative patient assessed by positron emission tomography (PET) and single photon emission computed tomography (SPECT). *Radiat. Med.* 7: 243-246, 1989.

Neary, D., Snowden, J.S., Gustafson, L., Passant, U., Stuss, D., Black, S., Freedman, M., Kertesz, A., Robert, P.H., Albert, M., Boone, K., Miller, B.L., Cummings, J. and Benson, D.F. Frontotemporal lobar degeneration: a consensus on clinical diagnostic criteria. *Neurology* 51: 1546-1554, 1998.

Neunzig, H.P. and Kunze, K. [Clinical aspects and prognosis following severe craniocerebral trauma]. *Fortschr. Neurol. Psychiatr.* 55: 223-230, 1987.

Perrin, F., Maquet, P., Peigneux, P., Ruby, P., Degueldre, C., Balteau, E., Del Fiore, G., Moonen, G., Luxen, A. and Laureys, S. Neural mechanisms involved in the detection of our first name: a combined ERPs and PET study. *Neuropsychologia* 43: 12-19, 2005.

Perry, R.J. and Hodges, J.R. Spectrum of memory dysfunction in degenerative disease. *Curr. Opin. Neurol.* 9: 281-285, 1996.

Plum, F. and Posner, J.B. *The diagnosis of stupor and coma.* Davis,F.A., Philadelphia, 1983.

Rainville, P., Hofbauer, R.K., Paus, T., Duncan, G.H., Bushnell, M.C. and Price, D.D. Cerebral mechanisms of hypnotic induction and suggestion. *J. Cogn. Neurosci.* 11: 110-125, 1999.

Rose, S.E., Chen, F., Chalk, J.B., Zelaya, F.O., Strugnell, W.E., Benson, M., Semple, J. and Doddrell, D.M. Loss of connectivity in Alzheimer's disease: an evaluation of white matter tract integrity with colour coded MR diffusion tensor imaging. *J. Neurol. Neurosurg. Psychiatry* 69: 528-530, 2000.

Rudolf, J., Ghaemi, M., Haupt, W.F., Szelies, B. and Heiss, W.D. Cerebral glucose metabolism in acute and persistent vegetative state. *J. Neurosurg. Anesthesiol.* 11: 17-24, 1999.

Rudolf, J., Sobesky, J., Ghaemi, M. and Heiss, W.D. The correlation between cerebral glucose metabolism and benzodiazepine receptor density in the acute vegetative state. *Eur. J. Neurol.* 9: 671-677, 2002.

Salmon, E. Functional brain imaging applications to differential diagnosis in the dementias. *Curr. Opin. Neurol.* 15: 439-444, 2002.

Salmon, E., Collette, F., Degueldre, C., Lemaire, C. and Franck, G. Voxel-based analysis of confounding effects of age and dementia severity on cerebral metabolism in Alzheimer's disease. *Hum. Brain Mapp.* 10: 39-48, 2000.

Salmon, E., Ruby, P., Perani, D., Kalbe, E., Laureys, S., Adam, S. and Collette, F. Two aspects of impaired consciousness in Alzheimer's disease. *Prog. Brain Res.* 150: 287-298, 2005a.

Salmon, E., Perani, D., Herholz, K., Marique, P., Kalbe, E., Holthoff, V., Delbeuck, X., Beuthien-Baumann, B., Pelati, O., Lespagnard, S., Collette, F. and Garraux, G. Neural correlates of anosognosia for cognitive impairment in Alzheimer's disease. *Hum Brain Mapp.* in press, 2005b.

Schiff, N.D., Ribary, U., Moreno, D.R., Beattie, B., Kronberg, E., Blasberg, R., Giacino, J., McCagg, C., Fins, J.J., Llinas, R. and Plum, F. Residual cerebral activity and behavioural fragments can remain in the persistently vegetative brain. *Brain* 125: 1210-1234, 2002.

Stam, C.J., Montez, T., Jones, B.F., Rombouts, S.A., van der Made, Y., Pijnenburg, Y.A. and Scheltens, P. Disturbed fluctuations of resting state EEG synchronization in Alzheimer's disease. *Clin. Neurophysiol.* 116: 708-715, 2005.

Steriade, M., Jones, E.G. and McCormick, D. *Thalamus*. Elsevier, Amsterdam ; New York, 1997.

The Multi-Society Task Force on PVS. Medical aspects of the persistent vegetative state (1). *N. Engl. J. Med.* 330: 1499-1508, 1994.

Tommasino, C., Grana, C., Lucignani, G., Torri, G. and Fazio, F. Regional cerebral metabolism of glucose in comatose and vegetative state patients. *J. Neurosurg. Anesthesiol.* 7: 109-116, 1995.

Tulving, E. Episodic memory: from mind to brain. *Annu. Rev. Psychol.* 53: 1-25, 2002.

Vogeley, K., May, M., Ritzl, A., Falkai, P., Zilles, K. and Fink, G.R. Neural correlates of first-person perspective as one constituent of human self-consciousness. *J. Cogn. Neurosci.* 16, 817-827, 2004.

Vogt, B.A., Vogt, L. and Laureys, S. Cytology and functionally correlated circuits of human posterior cingulate areas. *Neuroimage* 29: 452-466, 2005.

Volicer, L., Berman, S.A., Cipolloni, P.B. and Mandell, A. Persistent vegetative state in Alzheimer disease. Does it exist? *Arch. Neurol.* 54: 1382-1384, 1997.

Volicer L, Rheaume Y, et al., . Progression of Alzheimer-type dementia in institutionalized patients: a cross-sectional study. *J. Appl. Gerontol.* 6: 83-94, 1987.

Walshe, T.M. and Leonard, C. Persistent vegetative state. Extension of the syndrome to include chronic disorders. *Arch. Neurol.* 42: 1045-1047, 1985.

Whalen, D.H. and Liberman, A.M. Speech perception takes precedence over nonspeech perception. *Science* 237: 169-171, 1987.

Wheeler, M.A., Stuss, D.T. and Tulving, E. Toward a theory of episodic memory: the frontal lobes and autonoetic consciousness. *Psychol. Bull.* 121: 331-354, 1997.

Zeman, A. Consciousness. *Brain* 124: 1263-1289, 2001.

Zeman, A.Z., Grayling, A.C. and Cowey, A. Contemporary theories of consciousness. *J. Neurol. Neurosurg. Psychiatry* 62: 549-552, 1997.

In: Cognitive Science Compendium, Volume 1
Editor: Miao-Kun Sun
ISBN: 978-1-60456-597-3

Chapter 13

Categorization of Species-specific Vocalizations in the Non-human Primate: Features Guiding Behavioral Discrimination and Neural Processing in the Ventrolateral Prefrontal Cortex

Gordon W. Gifford III[1,2*] and Yale E. Cohen[3,4]
[1]Department of Psychology
[2]Institute for Systems Research
University of Maryland at College Park
College Park, MD 20742
[3]Department of Psychological and Brain Sciences
[4]Center for Cognitive Neuroscience
Dartmouth College
Hanover, NH 03755

Abstract

Communication is one of the fundamental components of both human and non-human animal behavior. Whereas the benefits of language in human evolution are obvious, other communication systems have also evolved to transmit information that is critical for survival. This review focuses on auditory communication signals, specifically species-specific vocalizations, and the underlying neural processes that may support their use in guiding goal-directed behavior. We first highlight the fundamental role that species-specific vocalizations play in the socioecology of several species of non-human primates, with a focus on rhesus monkeys (Macaca mulatta). Next, we review the structure and function of auditory cortical processing streams involved in spatial and non-

* Corresponding author: Gordon W. Gifford III, Ph.D., Department of Psychology, University of Maryland at College Park, College Park, MD 20742, Phone: 301 405 2877, FAX: 301 314 9566, E-mail: ggifford@umd.edu

spatial processing. Finally, we discuss the role that the ventrolateral prefrontal cortex may play in the categorization of species-specific vocalizations.

Keywords: Auditory system, categorization, non-human primate, and vocalization

The Behavioral Relevance of Species-specific Vocalizations

The species-specific vocalizations of many primate species, as well as other animals, contain information that can be used by a listener as a source of information about objects, events and the status of peers in their environment (See Hauser, 1997, Chapter 1 for a comprehensive review). The information transmitted by a vocalization can be classified into three classes – "cues", "signs", and "signals" – based on their social use by a caller and how the information transmitted by the vocalization can be used by a listener (Cheney and Seyfarth, 1992; Hauser, 1997).

Cues are directly linked to the phenotype of the call producer, such as body weight and body size (Fitch, 1997). These cues provide a listener with information about the identity (Rendall et al., 1998) and sex (Rendall et al., 2004) of a caller. For example, female macaques use vocal cues to identify different matrilineal relatives (Rendall et al., 1996). In another example, baboons use cues to identify female and male conspecifics (Rendall et al., 2004). This capacity is important since it provides a listener the means to identify individuals with whom they can engage in beneficial social behaviors such as food foraging and social grooming (Rendall et al., 1996) as well as information on individuals they wish to avoid. Finally, since vocal cues are shaped by natural selection (Hauser, 1997) (i.e., a product a caller's physical characteristics), they cannot be readily manipulated and provide a listener with a veridical source of information about a specific caller. A listener's ability to use cues is based on their ability to analyze and determine the harmonic structure of a vocalization (Rendall et al., 1998). Consequently, certain vocalizations provide better information about caller identity than others (Owren and Rendall, 2003). Coos, since they have a clear harmonic structure (Fig. 1A) are useful in caller identification. Grunts (Fig. 1B) or screams, however, are not that useful since these vocalizations are characterized by a more noisy acoustic structure.

The second type of information contained in vocalizations is a "sign". The information contained in signs, unlike cues, is not shaped by natural selection (Hauser, 1997). Instead, the information transmitted by a sign is due to a listener's capacity to associate the acoustic features of a particular call with the status of a caller or the caller's location. One example of a this capacity is the copulation call of the male rhesus macaque (Hauser, 1993). The copulation call provides potential mates with information about the caller's fitness. However, the production of this call can also be used by rival males as a sign of the presence and the location of a mating competitor. As a result, as males increase their production of copulation calls, they not only increase the likelihood of coming into contact with receptive females but also increase the likelihood of encountering aggressive mating competitors (Hauser, 1993). In this way, a listener can use the information contained in vocalizations in behavioral contexts outside the purpose for which it was produced.

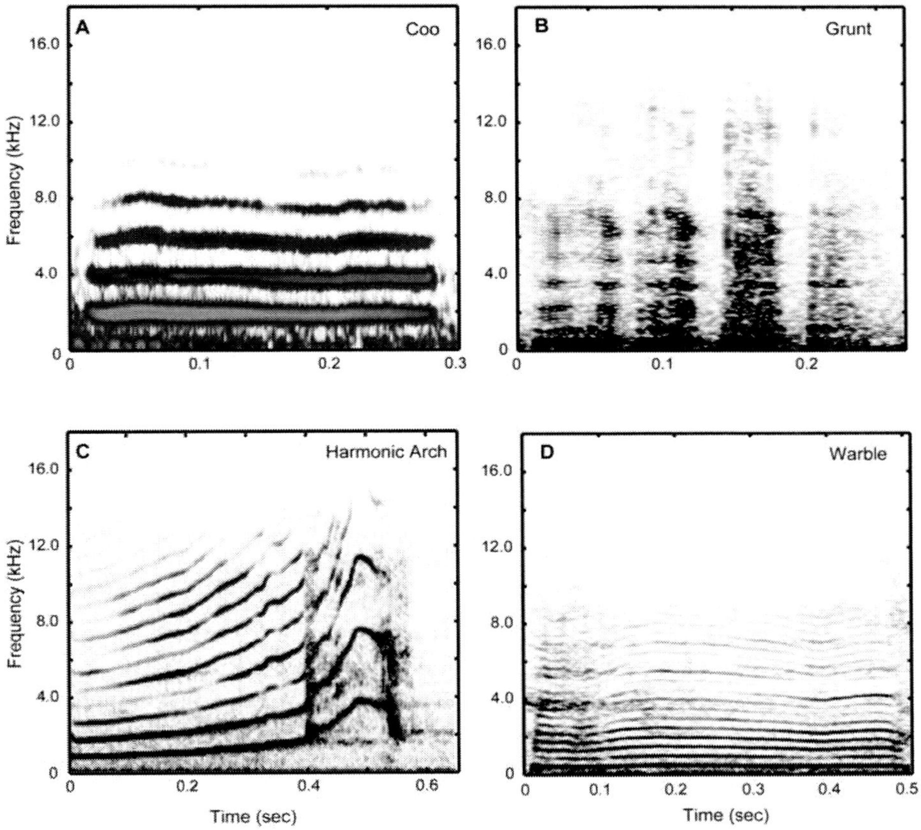

Figure 1: Sound spectrograms of the four call types: (A) coos, (B) grunts, and (C) harmonic arches and (D) warbles. Note that the time scale on the abscissa is different in each panel.

Third, vocalizations can transmit functional (i.e., "referential") information about an event or the presence of an object in the environment. That is, based on signal morphology alone, and regardless of the mechanisms that might underlie vocalization production, listeners are able to extract information about a vocalization's putative meaning. A classic example of referential communication signaling is the predator alarm calls of vervet monkeys (Seyfarth et al., 1980). Vervets produce unique alarm calls for three different predators, a snake, a leopard, and an eagle. The information that is transmitted to the listeners initiates specific patterns of species-specific behaviors. Specifically, when they hear an eagle-alarm call, vervets scan the sky for visual cues of the airborne predator, and in some cases, run to locations that provide overhead coverage. In contrast, when they hear a snake-alarm call, vervets stand up and scan the ground. Finally, a leopard-alarm call initiates a third distinct behavior: vervets run up the nearest tree while scanning the horizon for the presence of a leopard.

The capacity to successfully use referents also allows an animal to use the referent information that is transmitted by another species' vocalizations. For example, female Diana monkeys produce a predator alarm call in response to hearing a leopard-alarm call from conspecific males or the leopard-alarm call of the crested guinea fowl (Zuberbuhler, 2000; Zuberbuhler, 2000; Zuberbuhler and Seyfarth, 1997). Diana monkeys also form mixed species associations with putty-nose monkeys based on the ability of putty-nose monkeys to provide vocal warnings of eagle predation (Eckardt and Zuberbuhler, 2004). These observations suggests that Diana monkeys can form abstract categorical representations of a vocalization's functional meaning that is independent of acoustics and the species generating the signal.

Another example of categorization based on the referent meaning of vocalizations is the behavioral responses of rhesus macaques to food-associated calls (Hauser and Marler, 1993; Hauser and Marler, 1993). In the context of food, wild rhesus produce four acoustically distinct, vocalizations to signal the presence of food items: coos (Fig. 1A), grunts (Fig. 1B), harmonic arches (Fig. 1C), and warbles (Fig. 1D). Warbles and harmonic arches are produced in the presence of rare, high-quality food items (e.g., a piece of coconut). In contrast, in the presence of common, low-quality food items, rhesus produce coos or grunts. These two vocalizations are also produced in the context of positive social interactions, but the acoustic structure of the grunt vocalizations differs from those produced in food contexts (Hauser, 1998; Hauser and Marler, 1993). Analogous to Diana monkeys, wild and laboratory-housed rhesus categorize these food-associated calls based on their referential information and not their acoustic features (Hauser, 1998; Gifford III et al., 2003). That is, rhesus monkeys do not discriminate between call types that transmit the same information about high-quality food (i.e., harmonic arches and warbles), even though these vocalizations have distinct acoustic features. In contrast, rhesus monkeys do discriminate between call types that transmit different referential information (i.e., harmonic arches/warbles and grunts). Taken together, these findings from Diana and rhesus monkeys illustrate how a primate's response to a vocalization can depend more on the functional category that a vocalization represents than its acoustic morphology.

Finally, referential signaling may be formed by combining different vocalizations from multiple species. For example, Diana monkeys, while responding appropriately to the leopard-alarm and eagle-alarm calls of Campbell's monkeys, also respond appropriately to these alarm calls when they are preceded by the Campbell's monkeys' "boom" vocalization that signifies a lower-level threat (Zuberbuhler, 2000). More recent studies suggests that putty-nosed monkeys can also combine referent alarm calls - for leopards and eagles - together to form a new referent that transmits information about group movement, independent of the occurrence of a predator (Arnold and Zuberbuhler, 2006).

Neural Pathways Involved in Vocalization Processing

Many, if not most, recent investigations into the neural basis of vocalization processing in awake and behaving non-human primates have focused on processing in the temporal lobe (Barbour and Wang, 2003; Bendor and Wang, 2005; Kadia and Wang, 2003; Lu et al., 2001; Lu and Wang, 2004; Wang et al., 2003). These studies have highlighted many important features that relate to vocalization processing. For example, neurons in the primary auditory cortex and regions of the belt and parabelt of marmosets code the temporal features found in vocalizations (Lu et al., 2001; Lu and Wang, 2004; Wang et al., 2003). Also, many neurons in the primate auditory cortex and the surrounding belt and parabelt areas of the marmoset are sensitive to spectral contrast (Barbour and Wang, 2003) and the pitch (Bendor and Wang,

2005) of an auditory stimulus, which may be used to infer the affective content of a vocalization (Bendor and Wang, 2005) .

While insights into how the auditory system represents the acoustic and perceptual features of stimuli are critical for understanding vocalization processing, this level of investigation can not fully explain how the auditory system represents the of more abstract qualities of a vocalization that are used to guide behavior (Wang, 2000; Wang and Kadia, 2001; see Wang, 2000 for a comprehensive review). How is the affective quality of a vocalization represented beyond the neural coding of percepts like pitch? Does the auditory system represent the referent information that is transmitted by a vocalization separate from its acoustic properties? To address these questions we have begun a series of experiments that test how the more abstract qualities of vocalizations are represented in areas outside of the temporal lobe.

Prior to discussing these experiments and to put them into some context, we review an important model of auditory, as well as visual, function. This model posits that the spatial and non-spatial processing of auditory stimuli occurs in parallel (i.e., independent) pathways: a "dorsal" pathway that preferentially processes the spatial attributes of a stimulus (*where* is the sound?), and a "ventral" pathway that preferentially processes the non-spatial attributes (*what* is the sound?).

The two proposed pathways for auditory spatial and non-spatial processing originate in the auditory cortex (Divac et al., 1977; Hackett et al., 2001; Hackett et al., 1998; Hackett et al., 1998; Hackett et al., 1999; Hyvärinen, 1982; Kaas and Hackett, 1998; Kaas and Hackett, 1999; Pandya and Kuypers, 1969; Pandya and Yeterian, 1985; Poremba et al., 2003; Romanski et al., 1999; Romanski et al., 1999). In the auditory cortex, there are three anatomically defined areas - the core, belt, and parabelt. Subdivisions within the belt and parabelt form the bases of the spatial and non-spatial pathways. The auditory spatial pathway is defined by a series of projections that originate in the caudal belt and includes the posterior parietal cortex (particularly, the lateral intraparietal area [area LIP]) and the dorsolateral prefrontal cortex (i.e., the frontal eye fields; area 8a); auditory input to area LIP may also be due to input from multimodal cortical areas and the frontal eye fields (Baizer et al., 1991; Blatt et al., 1990; Schall et al., 1995; Seltzer and Pandya, 1991). In contrast, the auditory non-spatial pathway is defined by a series of projections that originate in the anterior belt and includes regions of the prefrontal cortex.

This parcelization of auditory function is not unique to non-human primates. In humans, there are several lines of anatomical (Galaburda and Sanides, 1980; Rivier and Clarke, 1997; Tardif and Clarke, 2001), neurophysiological (Alain et al., 2001; Anourova et al., 2001; Hart et al., 2004; Maeder et al., 2001; Rämä et al., 2004; Warren and Griffiths, 2003; Warren et al., 2002), and neuropsychological (Clarke et al., 2000) evidence suggesting that distinct pathways process different attributes of an auditory stimulus (Warren and Griffiths, 2003). A "posterior" pathway that includes regions of inferior parietal lobule is thought to process the spatial attributes of an auditory stimulus (Alain et al., 2001; Baumgart et al., 1999; Bremmer et al., 2001; Bremmer et al., 2001; Bushara et al., 2003; Bushara et al., 1999; Cusack et al., 2000; Deouell et al., 2000; Deouell and Soroker, 2000; Griffiths et al., 1998; Hart et al., 2004; Maeder et al., 2001; Warren et al., 2002). An "anterior" pathway is thought to process the non-spatial attributes of auditory stimuli (Griffiths et al., 1998; Hart et al., 2004; Maeder et al., 2001; Patterson et al., 2002; Scott et al., 2000; Vouloumanos et al., 2001; Zatorre et al., 1992). The areas of the auditory cortex, parietal cortex, and prefrontal cortex that constitute these two pathways may be analogous to those areas identified as important for spatial and non-spatial processing in rhesus monkeys (Rauschecker and Tian, 2000; Romanski et al., 1999).

Whereas this auditory parallel-processing scheme is conceptually elegant and forms an important hypothetical framework for neural function (Milner and Goodale, 1995), it has come under serious scrutiny. First, it is not clear whether the parcelization of auditory information into "what" and "where" streams is appropriate (Cohen and Wessinger, 1999; Griffiths et al., 2004; Middlebrooks 2002; Recanzone 2002; Warren et al., 2005; Zatorre et al., 2002). For instance, it has been suggested that these auditory pathways do not code the what/where attributes of an auditory stimulus but, instead, code sound identification (i.e., "what") and coding changes in frequency over time (i.e., "how") (Belin and Zatorre, 2000). It has also been observed that the anatomical connections terminating in multiple sites throughout the prefrontal cortex more likely represent multiple functional pathways that are not strictly parallel (Kaas and Hackett, 1999).

Second, data from our group suggest that even if the parcelization of auditory processing into "what" and "where" is appropriate, these pathways are not strictly parallel: LIP neurons and neurons in the ventrolateral prefrontal cortex (vPFC), an area that is at the apex of the non-spatial pathway, are modulated robustly, if not equally well, by both the spatial and non-spatial attributes of an auditory stimulus (Cohen et al., 2004; Gifford III, and Cohen, 2005). Figure 2 illustrates this degree of spatial and non-spatial modulation for a two vPFC neurons. Figure 2A shows a vPFC neuron that was modulated preferentially by a coo exemplar. However, the firing rate was also modulated by the location from which the coo exemplar is presented. Its firing rate was highest when the coo was presented at the more eccentric contralateral or ipsilateral locations. Figure 2B shows a neuron with a somewhat different response profile. This neuron was modulated primarily by auditory-stimulus location but also somewhat by auditory-stimulus type. For all seven vocalization exemplars, the neuron's firing rate increased as the exemplars were presented at more contralateral locations. The sensitivity to auditory-stimulus type can be seen when comparing the stimulus-period firing rates elicited by the copulation scream, the girney, and the grunt: the stimulus-period firing rate of the neuron was highest when the copulation scream was presented and lowest when the girney or grunt was presented. These two examples are representative of neural responses to both the identity and location of a sound throughout the vPFC. Importantly, similar evidence for such functional convergence can also be found in the visual system (Ferrera et al., 1992; Ferrera et al., 1994; Rizzuto et al., 2005; Sereno and Maunsell, 1998; Toth and Assad, 2002). In both the visual and auditory system, this convergence is most likely mediated by the anatomical connections that exist between the two spatial and non-spatial pathways (Kaas and Hackett, 1999; Kaas and Hackett, 2000; Sereno and Maunsell, 1998; Webster et al., 1994).

What purpose does spatial information serve in the non-spatial pathway? Similarly, what purpose does non-spatial information serve in the spatial pathway? The answers to these questions are not known, but one interesting hypothesis is that the mixture of spatial and non-spatial information may benefit those computations that create consistent perceptual representations that guide goal-directed behavior (Ferrera et al., 1994; Graziano et al., 1997; Kusunoki et al., 2000; Rao et al., 1997; Toth and Assad, 2002).

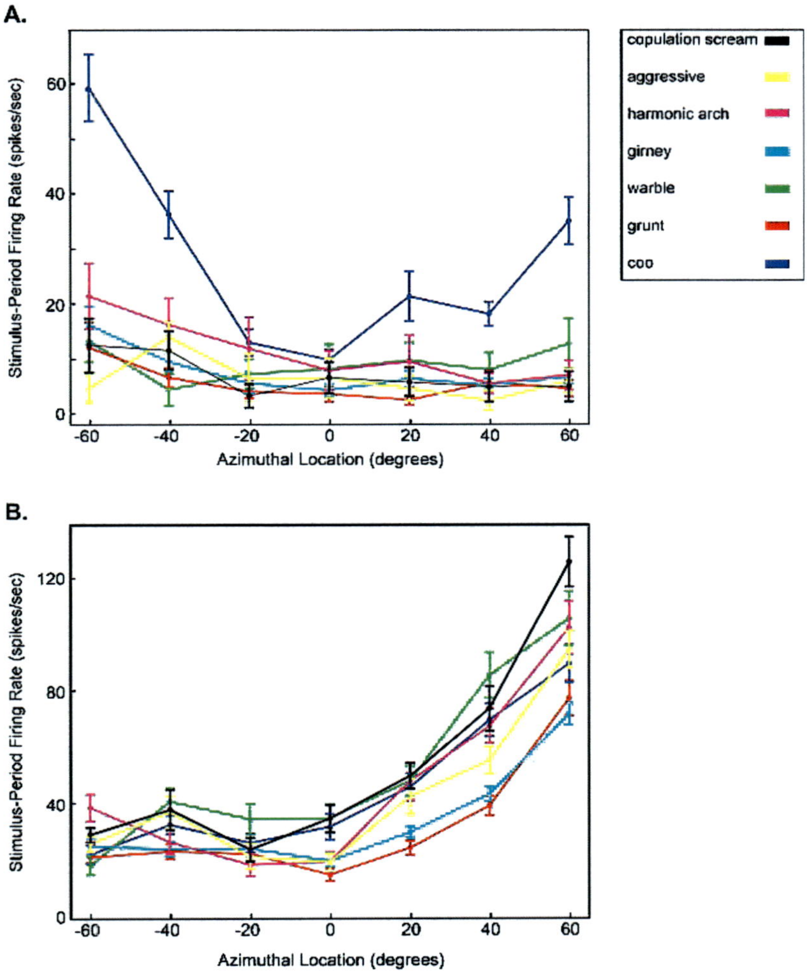

Figure 2: Responses of two vPFC neurons to species-specific vocalizations presented from different azimuthal locations. In both A and B, the mean stimulus-period firing rate of a vPFC neuron is plotted as a function of auditory-stimulus type (colored lines) and auditory-stimulus location (x-axis). The error bars in the panel represent SEM. Negative azimuthal locations are ipsilateral to the recording site, and positive values are contralateral.

Despite the functional interactions between these spatial and non-spatial pathways, we hypothesize for several reasons that vPFC neurons might be involved in preferentially processing the functional meaning (i.e., referent) of a vocalization, as opposed to acoustic structures. First, as noted above, the vPFC is thought to be at the apex of a pathway involved in processing auditory objects (Rauschecker and Tian, 2000; Romanski et al., 1999; Romanski et al., 2000). In addition, the vPFC is thought to play an important role in contextually-dependent vocalization production (Jurgens, 2002). Data from the human literature also suggests a role of the vPFC in processing the functional meaning of vocalizations. For instance, Bunge et al., . (2003) suggests that the vPFC is involved in the maintenance and retrieval of abstract rules and Gabrieli et al., . (1998) posits a role for this region in semantic processing and syntactic working memory.

The Categorization of Species-specific Vocalizations in vPFC

To test the hypothesis that vPFC neurons were preferentially modulated by the referential information that is transmitted by a vocalization (i.e., a vocalization's membership in a functional category), we utilized a paradigm that is relatively unique. vPFC activity was recorded while monkeys listened passively to repeated presentations of one vocalization (the "repeated" vocalization) that was interrupted randomly by the presentation of a second, different ("oddball") vocalization; this type of paradigm is analogous to human studies that test how event-related potentials are modulated by rare ("oddball") stimuli (Näätänen, 1992). We used this paradigm because we were interested in testing the neural mechanisms underlying the "spontaneous" categorization of natural (i.e., ethological) stimuli and not the neural mechanisms that may arise following training with artificial categories (Logothetis and Pauls, 1995; Miller et al., 2002).

Consistent with our hypothesis, we found that vPFC neurons were modulated preferentially by oddball stimuli that were preceded by repetitions of a stimulus that belonged to a different functional category. In contrast, vPFC activity was not modulated preferentially by oddball stimuli that belonged to the same category as the repeated stimulus, even if the oddball and repeated stimuli belonged to different acoustic classes; a vocalization's membership in an acoustic class is based on its acoustic features. These two observations are seen in more detail in Figure 3. In this figure, we plot the normalized population responses of vPFC neurons to the oddball stimuli when it was presented during the (oddball) paradigm described above and during a second paradigm, a "baseline" paradigm. In the baseline paradigm, monkeys listened passively while we presented a variety of auditory stimuli (including the repeated and oddball vocalizations) in a random order with equal probability. When the oddball and repeated vocalizations transmitted different types of referential information (e.g., one was a grunt [low-quality food] and the other was a harmonic arch [high-quality food]) and belonged to different acoustic classes, the average response of vPFC neurons to the oddball vocalization was significantly higher during the oddball paradigm than the response to the same vocalization presented during the baseline condition (Figs. 3Aa and 3Ab). In contrast, when the oddball and repeated vocalizations belonged to different acoustic classes but transmitted the same type of information (i.e., the harmonic arch and warble belong to different acoustic classes but both vocalizations transmit information about high-quality food), the average response of vPFC neurons to the oddball vocalization was not reliably different than its response in the baseline paradigm (Figs. 3Ba and 3Bb). Finally, when the oddball and repeated vocalizations belonged to the same acoustic class, and hence the same functional category (e.g., both are grunts), the average vPFC response to the oddball vocalization was not reliably different than its response to the same vocalization in the baseline condition (Figs. 3Ca-c).

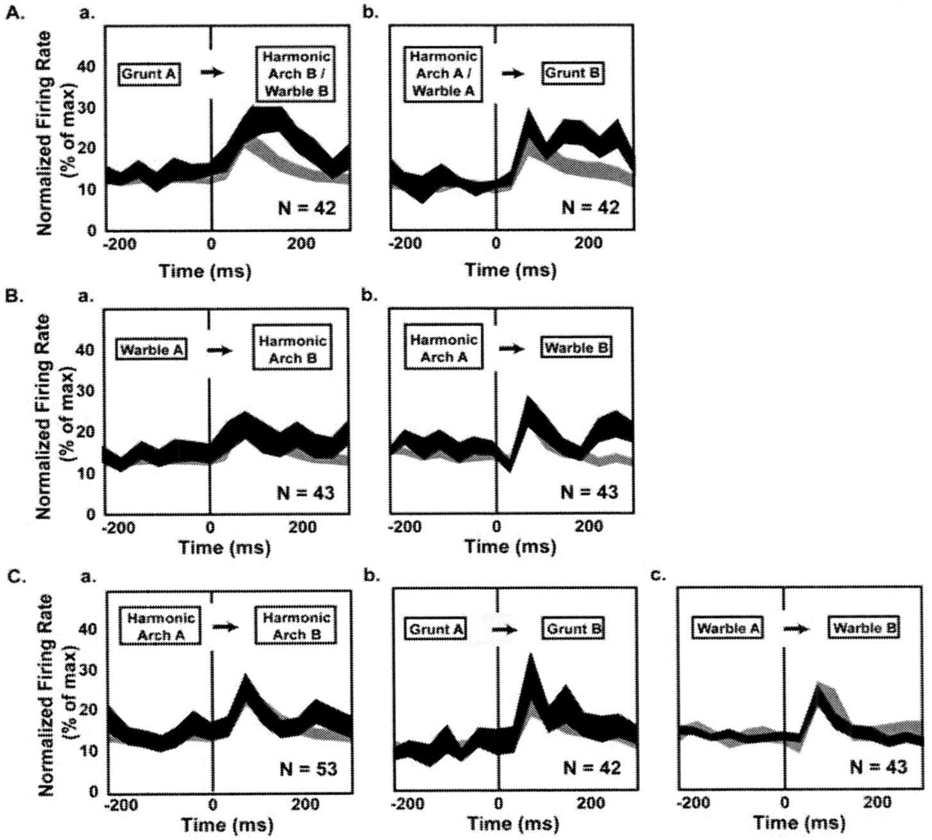

Figure 3: Population responses of vPFC activity in response to transitions between vocalization exemplars. A. Normalized vPFC population activity to a test vocalization that belongs to a different acoustic class and different functional category than the repeated vocalization. B. Normalized vPFC population activity to a test vocalization that belongs to a different acoustic class but the same functional category as the repeated vocalization. C. Normalized vPFC population activity to a test vocalization that belongs to the same acoustic class as the repeated vocalization. The thick grey line in each panel represents the normalized population firing rates in response to a test vocalization during the repetitive-presentation context. The thick black line in each panel represents normalized population firing rates in response to a test vocalization during the random-presentation context. The thickness of the lines indicates activity ranging two standard deviations from the mean response for the population. All data are aligned relative to onset of the test vocalization; the vertical black line indicates stimulus onset. In each panel's schematic, the repeated and test vocalizations for the data are respectively denoted to the left and right of the vertical black line indicating stimulus onset.

In a second, related study, we tested further the categorical nature of vPFC activity by directly examining spontaneous neural modulation as a function of a vocalizations category membership (Cohen et al., In Press). In this experiment vPFC neural responses to food calls (i.e., both high-quality and low-quality food) and calls associated with various non-food events were collected. Consistent with categorical processing, we found that the vPFC neurons responded differently to vocalizations that transmitted unrelated types of information and similarly to vocalizations that came from the same category,

However, unlike the results of the previous oddball study, the responses of vPFC neurons did not differentiate between the information transmitted by vocalizations regarding food quality. Specifically, vPFC neural modulation within the categories of low-quality food and high-quality food was not statistically different than vPFC neural modulation between the two categories. These observations are consistent with the hypothesis that vPFC neurons respond to food and non-food vocalizations based on category membership but do not differentiate between categories of food quality. These results contrast with those of our other study, which suggested that in the context of the oddball paradigm vPFC neural responses are modulated by the categorical membership of the non-oddball and oddball food vocalizations (i.e., food quality).

These somewhat different findings may highlight an important issue of experimental design that should be the focus of future study. Specifically, both studies confirm the hypothesis that vPFC responses are modulated by the categorical membership of vocalizations irrespective of their acoustics. However, the context in which vocalizations are presented (i.e., spontaneous responses versus oddball auditory responses) may alter how categorical boundaries are represented in vPFC activity.

Together, these two sets of experimental data are consistent with the proposition that the vPFC may be involved in the computations underlying the spontaneous categorization of natural categories that are based on the referent information transmitted by vocalizations, a key component of auditory-object processing, and not simple acoustic-feature extraction. What role do these categorical representations play in rhesus behavior? One possibility is that the vPFC is part of a circuit involved in the categorization of socially meaningful signals. Indeed, previous work has suggested that vPFC activity may reflect abstract rule-based processes associated with classifying vocal signals (Deacon, 1992; Jürgens, 2002). For instance, the vPFC might contribute to the production of "purposeful" vocalizations or perhaps to the contextually-dependent inhibition of vocalization production.

Conclusion

The results of our behavioral work in laboratory-bred macaques illustrate how rhesus macaques respond to "higher-order" features of vocalizations that are not directly related to their acoustic properties. In turn, we have also provided examples of how vPFC neurons at the apex of a hypothesized auditory-object processing stream responds to both the "higher-order" features of a sound and to a lesser extent the spatial attributes of a sound source. The core issue of these studies has been an attempt to understand how the auditory system spontaneously categorizes vocalizations and the potential role that the vPFC may play in that process. Overall, our findings are in line with previous work from Freedman and colleagues (Freedman et al., 2001; Freedman et al., 2002) who demonstrated that PFC neurons are involved in the categorical representation of visual objects.

It is important to note that not all vPFC studies are suggestive of a role of the vPFC in categorization. For instance, a recent study indicated that vPFC activity more closely correlates with the perceptual features of species-specific vocalizations rather than the transmitted information (Romanski et al., 2005). However, our studies and the Romanski et al., . study are consistent with the hypothesis that the vPFC is not involved in simple-feature extraction but instead plays an important role in the computations underlying the perception of auditory objects. Thus, it may be that the results of these studies differ only in the nature and extent of these computations and that these differences may be due to differences in recording locations and/or analysis techniques.

One limitation of our studies is the reliance on the "natural" categories associated with specific call types. While these experiments have provided an insight into vPFC's role in the categorization of vocalizations, further experiments are necessary to fully understand the operation of this process and a more general role that it may play in the categorization of conditioned stimuli. To this end, operant training will be necessary to specify the acoustic boundaries along which categories are defined. Operant training will also allow a broader set of acoustic features to be associated with a specific behavioral salience and the creation of novel stimuli that cross perceptual boundaries (Specht et al., 2005). In addition, the flexibility provided by operant training in assigning a categorical salience to a sound would also allow for the examination of how categories are formed and adjusted through experience or context. Finally, the use of this method will provide a greater understanding into whether the mechanisms of spontaneous categorization are fundamentally different from the mechanisms that mediate operantly defined categories (Freedman et al., 2002) and whether categorical representations are dependent on task demand (Freedman et al., 2005).

Beyond the relationship between how the vPFC codes natural and artificial categories, our studies raise several other intriguing and important questions. For example, since communication is often multimodal (Hauser, M. D 1993; Hinde and Rowell, 1962; Partan and Marler, 1999) and since vPFC neurons respond to visual stimuli (Romanski and Goldman-Rakic, 2002), an interesting question to consider is whether this circuit represents auditory as well as visual communication signals and represents this information in a modality-independent format (Ghazanfar and Logothetis, 2003). Also, we do not know whether these categorical representations are formed in the vPFC or just stored there during online processing (Bunge et al., 2003). To address this question, it will be important to test the categorical nature of auditory activity in areas that project to the vPFC (Hackett et al., 1999; Romanski et al., 1999; Romanski et al., 1997; Romanski et al., 1999). Finally, in order to understand how socially meaningful stimuli relate to and promote an animal's behavior, we need to understand how these stimuli are represented and coded across different cortical areas and the role that different areas play in representing different types of referential, affective, and conceptual information. This goal can be accomplished through a systems-level, circuit approach in which neural activity in multiple cortical areas (such as other regions of the PFC, limbic and paralimbic structures, and higher-order visual areas) is targeted to examine their role in the processing information that is critically related to animal's behavior (Barsalou, 1999; Gil-da-Costa et al., 2004; Martin and Chao, 2001; Miller, 2000; Warrington and Shallice, 1984).

ACKNOWLEDGMENTS

We thank M. LaFleur and B. Russ for their helpful discussions and comments in the preparation of this manuscript. YEC was supported by grants from the Whitehall Foundations, NIH and a Burke Award. GWG was supported by a Ruth L. Kirschstein National Research Service Award research training grant, provided through the Center for

Comparative and Evolutionary Biology of Hearing at the University of Maryland at College Park (T32-DC000046 NIDCD).

REFERENCES

Alain, C., Arnott, S. R., Hevenor, S., Graham, S. and Grady, C. L. "What" and "where" in the human auditory system. *Proc Natl Acad Sci U S A* 98: 12301-6, 2001.

Anourova, I., Nikouline, V. V., Ilmoniemi, R. J., Hotta, J., Aronen, H. J. and Carlson, S. Evidence for dissociation of spatial and nonspatial auditory information processing. *Neuroimage* 14: 1268-1277, 2001.

Arnold, K. and Zuberbuhler, K. Language evolution: semantic combinations in primate calls. *Nature* 441: 303, 2006.

Baizer, J. S., Ungerleider, L. G. and Desimone, R. Organization of visual inputs to the inferior temporal and posterior parietal cortex in macaques. *J Neurosci* 11: 168-90, 1991.

Barbour, D. L. and Wang, X. Contrast tuning in auditory cortex. *Science* 299: 1073-5, 2003.

Barsalou, L. W. Perceptual symbol systems. *Behav Brain Sci* 22: 577-609; discussion 610-60, 1999.

Baumgart, F., Gaschler-Markefski, B., Woldorff, M. G. and Scheich, H. A movement-sensitive area in auditory cortex. *Nature* 400: 724-725, 1999.

Belin, P. and Zatorre, R. J. 'What,' 'where' and 'how' in auditory cortex. *Nat Neurosci* 3: 965-966, 2000.

Bendor, D. and Wang, X. The neuronal representation of pitch in primate auditory cortex. *Nature* 436: 1161-5, 2005.

Blatt, G. J., Andersen, R. A. and Stoner, G. R. Visual receptive field organization and cortico-cortical connections of the lateral intraparietal area (area LIP) in the macaque. *J Comp Neurol* 299: 421-45, 1990.

Bremmer, F., Schlack, A., Duhamel, J. R., Graf, W. and Fink, G. R. Space coding in primate posterior parietal cortex. *Neuroimage* 14: S46-51, 2001.

Bremmer, F., Schlack, A., Jon Shah, N., Zafiris, O., Kubischik, M., Hoffman, K.-P., Zilles, K. and Fink, G. R. Polymodal motion processing in posterior parietal and premotor cortex: A human fMRI study strongly implies equivalencies between humans and monkeys. *Neuron* 29: 287-296, 2001.

Bunge, S. A., Kahn, I., Wallis, J. D., Miller, E. K. and Wagner, A. D. Neural circuits subserving the retrieval and maintenance of abstract rules. *J Neurophysiol* 90: 3419-28, 2003.

Bushara, K. O., Hanakawa, T., Immisch, I., Toma, K., Kansaku, K. and Hallett, M. Neural correlates of cross-modal binding. *Nat Neurosci* 6: 190-5, 2003.

Bushara, K. O., Weeks, R. A., Ishii, K., Catalan, M. J., Tian, B., Rauschecker, J. P. and Hallett, M. Modality-specific frontal and parietal areas for auditory and visual spatial localization in humans. *Nat Neurosci* 2: 759-66, 1999.

Cheney, D. L. and Seyfarth, R. M. in *Topics in Primartology, Vol 1: Human Origins* (eds. T. Nishida, W. C. McGrew, P. Marler, M. Pickford and F. de Waal) 315-330 (Tokyo University Press, Tokyo, 1992).

Clarke, S., Bellmann, A., Meuli, R. A., Assal, G. and Steck, A. J. Auditory agnosia and auditory spatial deficits following left hemispheric lesions: evidence for distinct processing pathways. *Neuropsychologia* 38: 797-807, 2000.

Cohen, Y., Hauser, M. and Russ, B. Spontaneous processing of abstract categorical information in the ventrolateral prefrontal cortex. *Biology Letters.*, (In Press).

Cohen, Y. E., Russ, B. E., Gifford III, G. W., Kiringoda, R. and MacLean, K. A. Selectivity for the spatial and nonspatial attributes of auditory stimuli in the ventrolateral prefrontal cortex. *J Neurosci* 24: 11307-16, 2004.

Cohen, Y. E. and Wessinger, C. M. Who goes there? *Neuron* 24: 769-71, 1999.

Cusack, R., Carlyon, R. P. and Robertson, I. H. Neglect between but not within auditory objects. *J Cogn Neurosci* 12: 1056-65, 2000.

Deacon, T. W. Cortical connections of the inferior arcuate sulcus cortex in the macaque brain. *Brain Res* 573: 8-26, 1992.

Deouell, L. Y., Hamalainen, H. and Bentin, S. Unilateral neglect after right-hemisphere damage: contributions from event-related potentials. *Audiol Neurootol* 5: 225-34., 2000.

Deouell, L. Y. and Soroker, N. What is extinguished in auditory extinction? *Neuroreport*: 3059-3062, 2000.

Divac, I., Lavail, J. H., Rakic, P. and Winston, K. R. Heterogenous afferents to the inferior parietal lobule of the rhesus monkey revealed by the retrograde transport method. *Brain Res*: 197-201, 1977.

Eckardt, W. and Zuberbuhler, K. Cooperation and competition in two forest monkeys. *Behavioral Ecology* 15: 400-411, 2004.

Ferrera, V. P., Nealey, T. A. and Maunsell, J. H. Mixed parvocellular and magnocellular geniculate signals in visual area V4. *Nature* 358: 756-761, 1992.

Ferrera, V. P., Rudolph, K. K. and Maunsell, J. H. Responses of neurons in the parietal and temporal visual pathways during a motion task. *Journal of Neuroscience* 14: 6171-86, 1994.

Fitch, W. T. Vocal tract length and formant frequency dispersion correlate with body size in rhesus macaques. *J Acoust Soc Am* 102: 1213-22, 1997.

Freedman, D. J., Riesenhuber, M., Poggio, T. and Miller, E. K. Categorical representation of visual stimuli in the primate prefrontal cortex. *Science* 291: 312-6, 2001.

Freedman, D. J., Riesenhuber, M., Poggio, T. and Miller, E. K. Visual categorization and the primate prefrontal cortex: Neurophysiology and behavior. *J Neurophysiol* 88: 929-941, 2002.

Freedman, D. J., Riesenhuber, M., Poggio, T. and Miller, E. K. Visual categorization and the primate prefrontal cortex: neurophysiology and behavior. *J Neurophysiol* 88: 929-41, 2002.

Freedman, D. J., Riesenhuber, M., Poggio, T. and Miller, E. K. Experience-Dependent Sharpening of Visual Shape Selectivity in Inferior Temporal Cortex. *Cereb Cortex*, 2005.

Galaburda, A. and Sanides, F. Cytoarchitectonic organization of the human auditory cortex. *J Comp Neurol* 190: 597-610, 1980.

Ghazanfar, A. A. and Logothetis, N. K. Neuroperception: facial expressions linked to monkey calls. *Nature* 424: 937-8, 2003.

Gifford III, G. W. and Cohen, Y. E. Spatial and non-spatial auditory processing in the lateral intraparietal area. *Exp Brain Res* 162: 509-512, 2005.

Gifford III, G. W., Hauser, M. D. and Cohen, Y. E. Discrimination of functionally referential calls by laboratory-housed rhesus macaques: Implications for neuroethological studies. *Brain Behav Evol* 61: 213-224, 2003.

Gil-da-Costa, R., Braun, A., Lopes, M., Hauser, M. D., Carson, R. E., Herscovitch, P. and Martin, A. Toward an evolutionary perspective on conceptual representation: species-specific calls activate visual and affective processing systems in the macaque. *Proc Natl Acad Sci U S A* 101: 17516-21, 2004.

Graziano, M. S., Hu, X. T. and Gross, C. G. Coding the locations of objects in the dark. *Science* 277: 239-41, 1997.

Griffiths, T. D., Buchel, C., Frackowiak, R. S. and Patterson, R. D. Analysis of temporal structure in sound by the human brain. *Nat Neurosci* 1: 422-7, 1998.

Griffiths, T. D., Rees, G., Rees, A., Green, G. G. R., Witton, C., Rowe, D., Buchel, C., Turner, R. and Frackwiak, R. S. J. Right parietal cortex is involved in the perception of sound movement in humans. *Nat Neurosci* 1: 74-79, 1998.

Griffiths, T. D., Warren, J. D., Scott, S. K., Nelken, I. and King, A. J. Cortical processing of complex sound: a way forward? *Trends Neurosci* 27: 181-185, 2004.

Hackett, T. A., Preuss, T. M. and Kaas, J. H. Architectonic identification of the core region in auditory cortex of macaques, chimpanzees, and humans. *J Comp Neurol* 441: 197-222, 2001.

Hackett, T. A., Stepniewska, I. and Kaas, J. H. Subdivisions of auditory cortex and ipsilateral cortical connections of the parabelt auditory cortex in macaque monkeys. *J Comp Neurol* 394: 475-95, 1998.

Hackett, T. A., Stepniewska, I. and Kaas, J. H. Thalamocortical connections of the parabelt auditory cortex in macaque monkeys. *J Comp Neurol* 400: 271-86, 1998.

Hackett, T. A., Stepniewska, I. and Kaas, J. H. Prefrontal connections of the parabelt auditory cortex in macaque monkeys. *Brain Res* 817: 45-58, 1999.

Hart, H. C., Palmer, A. R. and Hall, D. A. Different areas of human non-primary auditory cortex are activated by sounds with spatial and nonspatial properties. *Hum Brain Mapp* 21: 178-204, 2004.

Hauser, M. D. Rhesus monkey copulation calls: honest signals for female choice? *Proc R Soc Lond B Biol Sci* 254: 93-6., 1993.

Hauser, M. D. The Evolution of Communication. MIT Press, Cambridge, MA, 1997.

Hauser, M. D. Functional referents and acoustic similarity: field playback experiments with rhesus monkeys. *Anim Behav* 55: 1647-1658, 1998.

Hauser, M. D., Evans, C. S. and Marler, P. The role of articulation in the production of rhesus monkey, Macaca mulatta, vocalizations. *Anim Behav* 4: 423-433, 1993.

Hauser, M. D. and Marler, P. Food-associated calls in rhesus macaques (*Macaca mulatta*) 1. Socioecological factors influencing call production. *Behav Ecol* 4: 194-205, 1993.

Hauser, M. D. and Marler, P. Food-associated calls in rhesus macaques (*Macaca mulatta*) II. Costs and benefits of call production and suppression. *Behav Ecol* 4, 1993.

Hinde, R. A. and Rowell, T. E. Communication by postures and facial expressions in the rhesus monkey (macaca mulatta). *Proc Zool Soc Lond* 138: 1-21, 1962.

Hyvärinen, J. Posterior parietal lobule of the primate brain. *Physiol Rev* 62: 1060-1129, 982.

Jurgens, U. Neural pathways underlying vocal control. *Neurosci Biobehav Rev.* 26: 235-58, 2002.

Jürgens, U. Neural pathways underlying vocal control. *Neurosci Biobehav Rev* 26: 235-258, 2002.

Kaas, J. H. and Hackett, T. A. Subdivisions of auditory cortex and levels of processing in primates. *Audiol Neurootol* 3: 73-85., 1998.

Kaas, J. H. and Hackett, T. A. 'What' and 'where' processing in auditory cortex. *nature neuroscience* 2: 1045-1047, 1999.

Kaas, J. H. and Hackett, T. A. Subdivisions of auditory cortex and processing streams in primates. *Proc Natl Acad Sci U S A* 97: 11793-9., 2000.

Kadia, S. C. and Wang, X. Spectral integration in A1 of awake primates: neurons with single- and multipeaked tuning characteristics. *J Neurophysiol* 89: 1603-22, 2003.

Kusunoki, M., Gottlieb, J. and Goldberg, M. E. The lateral intraparietal area as a salience map: the representation of abrupt onset, stimulus motion, and task relevance. *Vision Res* 40: 1459-68, 2000.

Logothetis, N. K. and Pauls, J. Psychophysical and physiological evidence for viewer-centered object representations in the primate. *Cerebral Cortex* 5: 270-88, 1995.

Lu, T., Liang, L. and Wang, X. Temporal and rate representations of time-varying signals in the auditory cortex of awake primates. *Nat Neurosci* 4: 1131-8, 2001.

Lu, T. and Wang, X. Information content of auditory cortical responses to time-varying acoustic stimuli. *J Neurophysiol* 91: 301-13, 2004.

Maeder, P. P., Meuli, R. A., Adriani, M., Bellmann, A., Fornari, E., Thiran, J. P., Pittet, A. and Clarke, S. Distinct pathways involved in sound recognition and localization: a human fMRI study. *Neuroimage* 14: 802-16, 2001.

Martin, A. and Chao, L. L. Semantic memory and the brain: structure and processes. *Curr Opin Neurobiol* 11: 194-201, 2001.

Middlebrooks, J. C. Auditory space processing: here, there or everywhere? *Nat Neurosci* 5: 824-826, 2002.

Miller, E. K. The prefrontal cortex and cognitive control. *Nat Rev Neurosci* 1: 59-65., 2000.

Miller, E. K., Freedman, D. J. and Wallis, J. D. The prefrontal cortex: categories, concepts, and cognition. *Philos Trans R Soc Lond B Biol Sci* 29: 1123-1136, 2002.

Milner, A. D. and Goodale, M. A. The Visual Brain in Action. Oxford Univ. Pres, Oxford, 1995.

Näätänen, R. Attention and Brain Function. Lawrence Erlbaum Associates, Hillsdale, NJ, 1992.

Owren, M. J. and Rendall, D. Salience of caller identity in rhesus monkey (Macaca mulatta) coos and screams: perceptual experiments with human (Homo sapiens) listeners. *J Comp Psychol* 117: 380-90, 2003.

Pandya, D. N. and Kuypers, H. G. Cortico-cortical connections in the rhesus monkey. *Brain Res* 13: 13-36, 1969.

Pandya, D. N. and Yeterian, E. H. in *Cerebral Cortex, Volume IV: Association and auditory cortices* (eds. A. Peters and E. G. Jones) 3-61 (Plenum Press, New York, 1985).

Partan, S. and Marler, P. Communication goes multimodal. *Science* 283: 1272-3, 1999.

Patterson, R. D., Johnsrude, I. S., Uppenkamp, S. and Griffiths, T. D. The processing of temporal pitch and melody information in auditory cortex. *Neuron* 36: 767-776, 2002.

Poremba, A., Saunders, R. C., Crane, A. M., Cook, M., Sokoloff, L. and Mishkin, M. Functional mapping of the primate auditory system. *Science* 299: 568-72, 2003.

Rämä, P., Poremba, A., Sala, J. B., Yee, L., Malloy, M., Mishkin, M. and Courtney, S. M. Dissociable functional topographics for working memory maintenance of voice identity and location. *Cereb Cortex* 14: 768-780, 2004.

Rao, S. C., Rainer, G. and Miller, E. K. Integration of what and where in the primate prefrontal cortex. *Science* 276: 821-4., 1997.

Rauschecker, J. P. and Tian, B. Mechanisms and streams for processing of "what" and "where" in auditory cortex. *PNAS* 97: 11800-11806, 2000.

Recanzone, G. H. Where was that? - human auditory spatial processing. *Trends Cogn Sci* 6: 319-320, 2002.

Rendall, D., Owren, M. J. and Rodman, P. S. The role of vocal tract filtering in identity cueing in rhesus monkey (Macaca mulatta) vocalizations. *J Acoust Soc Am* 103: 602-14, 1998.

Rendall, D., Owren, M. J., Weerts, E. and Hienz, R. D. Sex differences in the acoustic structure of vowel-like grunt vocalizations in baboons and their perceptual discrimination by baboon listeners. *J Acoust Soc Am* 115: 411-21, 2004.

Rendall, D., Rodman, P. S. and Emond, R. E. Vocal recognition of individuals and kin in free-ranging rhesus monkeys. *Animal Behavior* 51: 1007-1015, 1996.

Rivier, F. and Clarke, S. Cytochrome oxidase, acetylcholinesterase and NADPH-diaphorase staining in human supratemporal and insular cortex: evidence for multiple auditory areas. *Neuroimage* 6: 288-304, 1997.

Rizzuto, D. S., Mamelak, A. N., Sutherling, W. W., Fineman, I. and Andersen, R. A. Spatial selectivity in human ventrolateral prefrontal cortex. *Nat Neurosci* 8: 415-7, 2005.

Romanski, L. M., Averbeck, B. B. and Diltz, M. Neural representation of vocalizations in the primate ventrolateral prefrontal cortex. *J Neurophysiol* 93: 734-47, 2005.

Romanski, L. M., Bates, J. F. and Goldman-Rakic, P. S. Auditory belt and parabelt projections to the prefrontal cortex in the rhesus monkey. *J Comp Neurol* 403: 141-57., 1999.

Romanski, L. M. and Goldman-Rakic, P. S. An auditory domain in primate prefrontal cortex. *Nat Neurosci* 5: 15-6., 2002.

Romanski, L. M., Ojima, H., Fritz, J., Tian, B., Mishkin, M., Goldman-Rakic, P. and Rauschecker, J. Physiologically defined regions of the auditory belt cortex of the superior temporal gyrus target specific prefrontal regions in the rhesus monkey. *Society for Neuroscience Abstracts* 23: 2073, 1997.

Romanski, L. M., Tian, B., Fritz, J., Mishkin, M., Goldman-Rakic, P. S. and Rauschecker, J. P. Dual streams of auditory afferents target multiple domains in the primate prefrontal cortex. *Nat Neurosci* 2: 1131-6., 1999.

Romanski, L. M., Tian, B., Fritz, J. B., Mishkin, M., Goldman-Rakic, P. S. and Rauschecker, J. P. Reply to "What', 'where' and 'how' in auditory cortex'. *Nat Neurosci* 3: 966, 2000.

Schall, J. D., Morel, A., King, D. J. and Bullier, J. Topography of visual cortex connections with frontal eye field in macaque: convergence and segregation of processing streams. *J Neurosci* 15: 4464-87, 1995.

Scott, S. K., Blank, C. C., Rosen, S. and Wise, R. J. S. Identification of a pathway for intelligible speech in the left temporal lobe. *Brain* 123: 2400-2406, 2000.

Seltzer, B. and Pandya, D. N. Post-rolandic cortical projections of the superior temporal sulcus in the rhesus monkey. *J Comp Neurol* 312: 625-40., 1991.

Sereno, A. B. and Maunsell, J. H. Shape selectivity in primate lateral intraparietal cortex. *Nature* 395: 500-3., 1998.

Seyfarth, R. M., Cheney, D. L. and Marler, P. Monkey responses to three different alarm calls: evidence of predator classification and semantic communication. *Science* 210: 801-3, 1980.

Specht, K., Rimol, L. M., Reul, J. and Hugdahl, K. "Soundmorphing": a new approach to studying speech perception in humans. *Neurosci Lett* 384: 60-5, 2005.

Tardif, E. and Clarke, S. Intrinsic connectivity of human auditory areas: a tracing study with DiI. *Eur J Neurosci* 13: 1045-1050, 2001.

Toth, L. J. and Assad, J. A. Dynamic coding of behaviourally relevant stimuli in parietal cortex. *Nature* 415: 165-8., 2002.

Vouloumanos, A., Kiehl, K. A., Werker, J. F. and Liddle, P. F. Detection of sounds in the auditory stream: event-related fMRI evidence for differential activation to speech and nonspeech. *J Cogn Neurosci* 13: 994-1005, 2001.

Wang, X. On cortical coding of vocal communication sounds in primates. *PNAS* 97: 11843-11849, 2000.

Wang, X. and Kadia, S. C. Differential representation of species-specific primate vocalizations in the auditory cortices of marmoset and cat. *J Neurophysiol* 86: 2616-20., 2001.

Wang, X., Lu, T. and Liang, L. Cortical processing of temporal modulations. *Speech Communication* 41: 107-121, 2003.

Warren, J. D. and Griffiths, T. D. Distinct mechanisms for processing spatial sequences and pitch sequences in the human auditory brain. *J Neurosci* 23: 5799-5804, 2003.

Warren, J. D., Zielinski, B. A., Green, G. G., Rauschecker, J. P. and Griffiths, T. D. Perception of sound-source motion by the human brain. *Neuron* 34: 139-48, 2002.

Warren, J. E., Wise, R. J. and Warren, J. D. Sounds do-able: auditory-motor transformations and the posterior temporal plane. *Trends Neurosci* 28: 636-43, 2005.

Warrington, E. K. and Shallice, T. Category specific semantic impairments. *Brain* 107 (Pt 3): 829-54, 1984.

Webster, M. J., Bachevalier, J. and Ungerleider, L. G. Connections of inferior temporal areas TEO and TE with parietal and frontal cortex in macaque monkeys. *Cereb Cortex* 4: 470-83., 1994.

Zatorre, R. J., Bouffard, M., Ahad, P. and Belin, P. Where is 'where' in the human auditory cortex? *Nat Neurosci* 5: 905-9, 2002.

Zatorre, R. J., Evans, A. C., Meyer, E. and Gjedde, A. Lateralization of phonetic and pitch discrimination in speech processing. *Science* 256: 846-9, 1992.

Zuberbuhler, K. Causal cognition in a non-human primate: field playback experiments with Diana monkeys. *Cognition* 76: 195-207., 2000.

Zuberbuhler, K. Interspecies semantic communication in two forest primates. *Proc R Soc Lond B Biol Sci* 267: 713-8., 2000.

Zuberbuhler, K. Interspecies semantic communication in two forest primates. *Proc Biol Sci* 267: 713-8, 2000.

Zuberbuhler, K. and Seyfarth, R. M. Diana monkey long-distance calls: messages for conspecifics and predators. *Anim Behav* 53: 589-604, 1997.

In: Cognitive Science Compendium, Volume 1 ISBN: 978-1-60456-597-3
Editor: Miao-Kun Sun

Chapter 14

ROLE OF ENDOPLASMIC RETICULUM-MEDIATED APOPTOTIC PATHWAY IN Aβ TOXICITY

Elisabete Ferreiro, Rui Costa, Sueli Marques, Catarina R. Oliveira and Cláudia M. F. Pereira*

Center for Neurosciences and Cell Biology,
Institute of Biochemistry, Faculty of Medicine,
University of Coimbra, 3004-504 Coimbra, Portugal

ABSTRACT

The endoplasmic reticulum (ER) plays an essential role in normal cell functioning and also in the pathogenesis of neuronal death. Irreparable perturbations in the homeostasis of the ER and alterations that lead to the accumulation of unfolded or misfolded protein in the ER lumen are thought to trigger several signalling pathways to restore the homeostasis of this organelle. These processes have been described to be involved in different important human diseases, including Alzheimer's disease (AD). Increasing evidences points to apoptosis as the major mechanism responsible for the neuronal death that occurs in this neurodegenerative disorder. In this context, mitochondrial dysfunction has been widely studied as a pathway integrated in the apoptosis that occurs in AD. It is an increasing assumption that the ER and the mitochondria cooperate in the mechanisms responsible for neuronal death and evidences suggest that mitochondria/ER crosstalk is also involved in AD. The recent and the novel research on ER stress and mitochondrial dysfunction and the cooperation between these two pathways can enhance the understanding of the molecular mechanisms that occur in AD and provide potential therapeutical targets.

* Correspondence and requests for reprints: Cláudia M.F. Pereira, Institute of Biochemistry, Faculty of Medicine, University of Coimbra, 3004-504 Coimbra, Portugal. Telephone: +351-239820190; Fax: +351-239822776. E-mail: cpereira@cnc.cj.uc.pt

Keywords: Alzheimer's disease, amyloid-β peptide, apoptosis, endoplasmic reticulum stress.

ABBREVIATIONS

Aβ	amyloid-beta peptide;
ABAD	Abeta-binding alcohol dehydrogenase;
AD	Alzheimer's disease;
AIF	apoptosis-inducing factor;
APP	amyloid precursor protein;
Ask1	apoptosis-signal-regulating-kinase-1;
ATF4	activating transcription factor 4;
ATF6	activating transcription factor 6;
CICR	Ca^{2+}-induced Ca^{2+} release;
eIF2α	eukaryotic initiation factor-2 alpha;
ER	endoplasmic reticulum;
ERAD	ER-associated protein degradation;
GRP78	glucose-regulated protein 78;
InsP3R	inositol 1,4,5-triphosphate receptor;
IRE1	inositol requiring transmembrane kinase/endonuclease;
JNK	c-Jun N-terminal kinase;
PERK	RNA-dependent protein kinase (PKR)-like ER membrane-localized kinase;
PS1	presenilin-1;
PS2	presenilin-2;
ROS	reactive oxygen species;
RyR	ryanodine receptor;
SERCA	sarco-endoplasmic reticulum ATPase;
TRAF2	TNF-associated factor 2;
UPR	Unfolded Protein Response.

1. ENDOPLASMIC RETICULUM AS AN ESSENTIAL ORGANELLE IN THE CELL

1.1. ER Function and Localization

The endoplasmic reticulum (ER) is an organelle found in all eukaryotic cells, and is an interconnected network of tubules, vesicles and cisternae, responsible for several specialized functions: protein translation, folding, and transport of proteins to be used in the cell membrane, or to be secreted from the cell; sequestration of calcium; and production and storage of glycogen, steroids, and other macromolecules (Baumann and Walz, 2001). The general structure of the ER is an extensive membrane network of cisternae held together by the cytoskeleton. The functions of the ER vary greatly depending on the exact type of ER and the type of cell in which it resides. Three types of ER can be found in cells: rough ER, smooth

ER, and sarcoplasmic reticulum. The last one is a subtype of smooth ER and can be found in striated muscle, the main function being the pump and store of Ca^{2+} ions. Both rough and smooth ER are continuous with the outer layer of the nuclear envelope. The rough ER owns is name to its association to ribossomes, giving a "rough" appearance and is responsible for protein synthesis. The rough ER works in concert with the Golgi complex to target new proteins to their proper destinations. The smooth ER has functions in several metabolic processes, including synthesis of lipids, metabolism of carbohydrates and Ca^{2+} concentration. The network of smooth ER allows increased surface area for the action or storage of several proteins (Baumann and Walz, 2001).

In neurons, ER is widely distributed, being present in dendrites and dendric spines, axons and presynaptic nerve terminals and in growth cones (Cheng and Reese, 1985; Deitch and banker, 1993; Kanaseki et al., 1998; Levesque et al., 1999; Dailey and Bridgman, 1989). Although smooth and rough ER coexist in neuronal cell bodies and proximal regions of axons and dendrites, the specialized endings of neuritis contain mainly smooth ER (Mattson et al., 2000).

1.2. When the ER Becomes Stressed

1.2.1. ER and the Unfolded Protein Response

As referred above, one of the main functions of ER is the synthesis, modification and delivery of proteins to their proper target sites within the secretory pathway and extracellular space. In the ER, proteins fold into their native conformation and undergo several post-translational modifications (Schröder and Kaufman, 2005). Only correctly folded proteins are exported to the Golgi complex. When unfolded proteins accumulate in the ER, the adaptive ER stress response is triggered and several signalling pathways are activated. This adaptive response is known as the Unfolded Protein Response (UPR) (Kaufman, 1999; Mori, 2000; Patil et al., 2001; Kaufman 2002). The UPR comprises a chain of reactions that attenuates protein translation, elevates expression of ER chaperones and upregulates proteins needed for ER-associated degradation (ERAD) (Rutkowski and Kaufman, 2004). Three distinct signalling pathways are activated in the UPR, mediated by the sensors: IRE1 (inositol requiring transmembrane kinase/endonuclease), composed of two identified homologues IRE1α (Tirasophon et al., 1998) and IRE1β (Wang et al., 1998), PERK (RNA-dependent protein Kinase (PKR)-like ER membrane-localized kinase) and ATF6 (activating transcription factor 6) (Harding et al., 1999; Liu et al., 2000; Nadanaka et al., 2007) (Fig.1). In normal conditions, the luminal domains of these sensors are occupied by BiP/GRP78, an ER HSP70 family protein, that as high affinity for protein substrates, and represses the three signalling pathways. When the unfolded proteins accumulate in the lumen of the ER, they competitively disrupt the interaction of BiP/GRP78 with IRE1, PERK and ATF6, activating the three signalling pathways (Shröder and Kaufman, 2005). IRE1 pathway regulates chaperone induction, expansion of the ER and ERAD in response to ER stress. After the dissociation of BiP/GRP78, IRE1 oligomerizes (Shamu and Walter, 1996; Liu et al., 2002) and initiates unconventional splicing of an mRNA encoding the basic leucine zipper (bZIP)-type transcription factor XBP1 that controls the up-regulation of a subset of ER-resident molecular chaperones (Iwakoshi et al., 2003; Mori 2000; Schröder and Kaufman, 2005; Lee et al., 2003). After being released by BiP/GRP78, PERK oligomerizes and phosphorylates the eIF2α (eukaryotic initiation factor-2 alpha) (Harding et al., 1999; Shi et al., 1998; Shi et al., 1999). Phosphorylartion of eIF2α results in the attenuation of global mRNA translation (Harding et al., 1999) and activates translation of ATF4 (activating transcription factor 4), a transcription factor that induces expression of genes, depending on the environmental stress, that function in amino acid metabolism, in antioxidant response and apoptosis (Harding et al., 2000; 2003; Lu et al., 2004; Vattem and Wek 2004). ATF6 is a Type 2 transmembrane protein, composed of a cytosolic NH2-terminal domain (a transcription factor of the basic-leucine zipper, bZip, family), followed by a transmembrane domain and an extracytoplasmic domain that projects into the ER lumen (Haze et al., 1999). Upon ER stress, cleaved ATF6 is transported from the ER to the Golgi complex to be processed by the sequential action of Site-1 protease (S1P) and Site-2 protease (S2P) (Ye et al., 2000; Okada et al., 2003). Cleaved ATF6 (p50 ATF6) translocates to the nucleus where it activates the transcription of genes encoding ER-localized molecular chaperones, like BiP/GRP78, GRP94, and calnexin and several folding enzymes (Yoshida et al., 2001; Yoshida et al., 2002; Nadanaka et al., 2007). If these signalling pathways fail to relieve the ER stress, the IRE1-mediated upregulation of ubiquitination enzymes belonging to the ERAD machinery will take place to target misfolded proteins for degradation (Yoshida et al., 2003; Hampton, 2003). Upregulation of ERAD components may be the last resort to deal with ER stress-induced accumulation of misfolded proteins in the ER (Shen et al., 2007). Finally, when cells are not able to adapt to the extensive ER stress, several mechanisms of cell death are activated.

1.2.2. ER and Calcium Homeostasis

ER is the major storage place of intracellular Ca^{2+} of the cell, which can be activated by both electrical and chemical cell stimulation (Bootman et al., 2002; Berridge, 2002; Verkhratsky and Petersen, 2002). The total concentration of Ca^{2+} in the ER lumen is estimated to be 1–3mM, similar to that in the extracellular space, and most of this is bound to chaperones, such as calreticulin, calnexin, BiP, GRP94 and calumin (Meldolesi and Pozzan, 1998; Michalak et al., 2002; Zhang et al., 2007). Nevertheless, a significant portion of Ca^{2+} in the lumen is free. The free concentration of Ca^{2+} in the ER ranges from 60 to 400 μM when stores are full, and from 1 to 50 μM when they are empty (Meldolesi and Pozzan, 1998; Corbett and Michalak, 2000). The release of Ca^{2+} from ER store regulates a variety of cellular events and processes, for example excitation-concentration coupling in muscle, regulation of genes expression, neuronal plasticity, neurotransmitter release and exocytosis (Niggli, 1999; Hardingham et al., 2001, Freguenlli et al., 1996; Rose and Konnerth, 2001). Release of Ca^{2+} from the ER is controlled by channels like the inositol-1,4,5-trisphosphate receptor ($InsP_3R$) and the ryanodine receptor (RyR) families (Berridge, 1993; Clapham, 1995) (Figure 1).

Ca^{2+} is the principal activator of these channels, in a process called Ca^{2+}-induced Ca^{2+} release (CICR) (Iino, 1990). Ca^{2+} is released from the $InsP_3R$ in response to agonists that activate receptors coupled, via a GTP-binding protein, to phospholipase C (PLC). Activation of PLC results in cleavage of PtdIns(4,5)P2, which in turn results in the liberation of diacylglycerol and Ins(1,4,5)P_3 that binds to $InsP_3R$. Opening of the $InsP_3$ receptor depends simultaneously of the presence of both $InsP_3$ and Ca^{2+} (Finch et al., 1991).

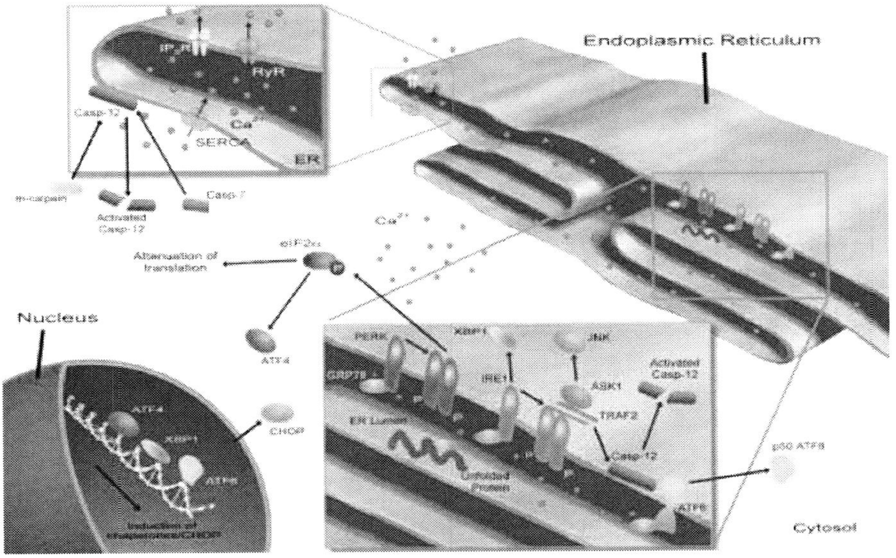

Figure 1. Mechanism of cell survival and cell death in response to ER stress. Accumulation of unfolded proteins in the ER activates the ER stress sensors, such as IRE1, ATF6 and PERK, which mediates ER stress response. Dissociation of BiP/GRP78 from stress transducers leads to oligomerization and autophosphorylation of IRE1 and PERK. Oligomerized IRE1 initiates unconventional splicing of an mRNA encoding the basic leucine zipper (bZIP)-type transcription factor XBP1. Phosphorylated PERK phosphorylates eIF2α, which attenuates global mRNA translation and activates translation of ATF4. Dissociation of BiP/GRP78 from ATF6 allows this protein to be cleaved by S1P and S2P. Cleaved ATF6 (p50 ATF6) translocates to the nucleus where, together with ATF4, XBP1 activate the transcription of genes encoding ER-localized molecular chaperones. Release of Ca^{2+} from the ER is controlled by channels like the RyR and InsP3R. SERCA is responsible for the return of Ca^{2+} to the internal stores. If ER Ca^{2+} homeostasis is altered, accumulation of unfolded proteins can lead to UPR. Prolonged ER stress triggers ER-mediated apoptotic pathway, leading to cell death. Activated IRE1α binds TRAF2, which in turn activates the Ask1 and subsequently JNK. ATF6 translocates to the nucleus and activates several genes, namely CHOP/GADD153, decreasing the expression of Bcl-2. On the other hand, PERK/eIF2α and IRE1/TRAF2/Ask1 pathways increase CHOP/GADD153 activity at a posttranscriptional level. TRAF-2 interacts with procaspase-12 and promotes its activation. Procaspase-12 can also be cleaved by Ca^{2+}-activated m-calpains and by caspase-7.

In the presence of low concentrations of Ca^{2+} (100–300 nM) the channel is more sensitive to $InsP_3$, whereas at concentrations above 300 nM, Ca^{2+} becomes inhibitory (Bootman and Lipp, 1999). RyR reside in close proximity to the cell membrane and respond to Ca^{2+} influx through voltage-gated Ca^{2+} channels and also to Ca^{2+} released from the $InsP_3R$. RyR channels contain two different Ca^{2+} binding sites, with high and low affinity for the cation. As for $InsP_3R$, at low Ca^{2+} concentrations RyR are activated and at high Ca^{2+} concentrations the activity is inhibited (Wrzosek, 2000).

The autocatalytic process of CICR enables the $InsP_3Rs$ and RyRs to communicate with each other to establish coordinated Ca^{2+} signals (Berridge et al., 2002). In neurons, RyR-mediated signals are more pronounced in dendritic spines and presynaptic terminals (Padua et al., 1996; Rose and Konnerth 2001), and $InsP_3R$ and $InsP_3$-evoked signals are predominantly localized in the soma and proximal dendrites of cortical and hippocampal neurons (Sharp et al., 1993; Nakamura et al., 1999; Stutzmann et al., 2003).

When the Ca^{2+} signalling has reached the purpose by which it was triggered, it is rapidly removed from the cytoplasm by various pumps (Pozzan et al., 1994) and exchangers (Blaustein and Lederer, 1999). The plasma membrane Ca^{2+}-ATPase (PMCA) pumps and Na^{+}/Ca^{2+} exchangers extrude Ca^{2+} to the outside. The sarco-endoplasmic reticulum ATPase (SERCA) pumps return Ca^{2+} to the internal stores (Fig. 1). The mitochondrion is another important component by sequestering Ca^{2+} rapidly during the development of the Ca^{2+} signal and releasing it back slowly during the recovery phase (Budd and Nicholls, 1996).

In the same way that ER Ca^{2+} is involved in Ca^{2+} signalling, the high Ca^{2+} concentration found in the ER is also important for the folding and processing of newly synthesized proteins, since these are strictly Ca^{2+}-dependent reactions requiring high Ca^{2+} activity for correct functioning (Kuznetsov et al., 1992; Lodish et al., 1992). If ER Ca^{2+} homeostasis is altered, accumulation of unfolded proteins can lead to UPR and consequently activate the ER-stress-induced apoptosis pathway (Kaufman, 1999; Pashen, 2001).

1.2.3. ER-Stress-Induced Apoptotic Pathway

Prolonged ER stress triggers ER-mediated apoptotic pathway, leading to cell death (Fig. 1). In the numerous studies performed to elucidate the ER-mediated cell death, pharmalogical agents have been used to induce ER stress, either by inducing UPR (like brefeldin A, an ER-Golgi transport inhibitor or tunicamycin, a N-linked glycosylation inhibitor) or by perturbating ER Ca^{2+} homeostasis (like thapsigargin, an inhibitor of SERCA/Ca^{2+}ATPase).

As a response of UPR, activated IRE1α can bind to the protein TRAF2 (TNF-associated factor 2), activating protein kinases, like Ask1 (apoptosis-signal-regulating-kinase-1), which can trigger the c-Jun N-terminal kinase (JNK) pathway (Urano et al., 2000; Nishitoh et al., 2002). Furthermore, ATF6 translocates to the nucleus and activates several genes, namely CHOP/GADD153, a transcription factor that decreases the expression of the anti-apoptotic protein Bcl-2 (McCullough et al., 2001). In addition, PERK/eIF2α pathway and IRE1/TRAF2/Ask1 pathway can increase CHOP/GADD153 activity at a posttranscriptional level (Wang and Ron, 1996). Activation of the JNK, triggered by IRE1, can also increase the levels of the proapoptotic protein Bim (Lei and Davis, 2003; Putcha et al., 2003) and decrease the Bcl-2 protein levels (Yamamoto, 1999). In rodents, Ire1-α, through the recruitment of TRAF-2, can lead to the activation of caspase-12. TRAF-2 interacts with procaspase-12 and promotes its clustering and its activation (Yoneda et al., 2001). Pro-caspase-12 is specifically localized on the cytoplasmic side of the ER and can also be proteolytically activated by m-calpain, a Ca^{2+}-responsive cytosolic cysteine protease (Nakagawa and Yuan, 2000). Caspase-12 may also by activated by caspase-7, which translocates this caspase from the ER to the cytosol (Rao et al., 2001). The involvement of caspase-12 in the apoptosis induced by ER stress has been widely contested, due to its absence in most humans (Fischer et al., 2002). The closest homolog of pro-caspase-12 in humans is pro-caspase-4 that has been suggested to be involved in ER stress-induced apoptosis (Hitomi et al., 2004). However, Obeng and Boise, in 2005, have shown that caspase-12 and caspase-4 are not required for the induction of ER stress-induced apoptosis and that caspase-4-like activity is not always associated with an initiating event. Recently, caspase-2 has been proposed to have an important role in the initiation of ER-induced apoptosis (Cheung et al., 2006). The authors propose that activation of caspase-2, which seems to be located partially in the ER, may integrate the apoptotic signalling between the ER and the Golgi complex and also acts as an upstream signalling to permeabilize the outer mitochondrial membrane. Furthermore, caspase-8, an initiator caspase that has been linked to the Death Receptors (TNF receptor and the Fas receptor) (Ashkenazi, 2002), was shown to be activated by thapsigargin, leading to cleavage of the pro-apoptotic BH3-only protein Bid, and thus to be involved in ER-stress-mediated cell death (He et al., 2002).

Several evidences suggest that mitochondria have an important role in the ER-mediated cell death and that these two organelles have a close cooperation in the process. First, apoptotic stimuli known to act through Ca^{2+} release from the ER induce a prolonged increase in the mitochondrial Ca^{2+} concentration, which can activate the permeability transition pore (PTP) (Szalai et al., 1999). This membrane channel can facilitate mitochondrial swelling and disruption of mitochondrial permeability transition (MPT) leading to the release of cytochrome c, AIF (apoptosis-inducing factor), several procaspases and Smac-α/DIABLO (Pinton et al., 2000; Bernardi et al., 2001), thereby facilitating the formation of the apoptosome that activates caspase-9, which ultimately activates caspase-3 and caspase-7 (Ferri and Kroemer, 2001). Beside having a role in the mitochondrial-mediated apoptosis, various proteins of the BCL-2 family modulate Ca^{2+} content regulation by both mitochondria and ER. Overexpression of the antiapoptotic protein Bcl-2 modulates Ca^{2+} signalling, by decreasing the ER Ca^{2+} load and consequently protecting the cells from death (Foyouzi-Youssefi et al., 2000; Pinton et al., 2000). Conversely, the pro-apoptotic Bax/Bak proteins overexpression favors the transfer of Ca^{2+} from ER to mitochondria and induces cell death (Nutt et al., 2002a,b; Scorrano et al., 2003). Cytochrome c, released from mitochondria, can also bind to and promote Ca^{2+} release through the $InsP_3R$ in the ER membrane. The released Ca^{2+} triggers the extrusion of a large amount of cytochrome c from all the mitochondria in the cell, amplifying the death signal (Boehning et al., 2003; 2004; Wang and El-Deiry, 2004).

In the cooperation between ER and mitochondria in inducing apoptosis, the role of reactive oxygen species (ROS) cannot be forgotten. ROS can be produced in mitochondria, which are thought to be the major source of ROS in vivo (Boveris and Chance, 1973) and by the ER (Chan, 1996). Several studies have demonstrated that the ER is sensitive to oxidative stress (Dreher et al., 1995; Racay et al., 1995) and oxidative stress may lead to perturbations of the ER Ca^{2+} homeostasis because many of the proteins involved in the regulation of Ca^{2+} in the ER, such as $InsP_3R$ and RyR, Ca^{2+} ATPases and ER-resident proteins, are sensitive to oxidants (Huang et al., 2004). Furthermore, CHOP/GADD153 can lead to the transcription of the ER oxiredutase ERO1α, which participates in protein disulfide bond during protein folding formation and protein refolding in the ER (to help relieve ER stress), and consequently lead to the transfer of electrons to molecular oxygen, forming ROS (Marciniak et al., 2004). In addition, Ca^{2+} that results from the depletion of ER Ca^{2+} stores can be taken up by juxtaposed mitochondria, inducing ROS formation (Tardif et al., 2005).

2. ER Stress in Alzheimer's Disease

ER stress as been associated to several diseases, including ischemia/reperfusion, inflammation, diabetes and neurodegerative diseases like Parkinson's, Prion's and Alzheimer's diseases (Hetz et al., 2003; Xu et al., 2005; Zhao and Ackerman, 2006).

Alzheimer´s disease (AD), the most common form of senile dementia, is a progressive neurodegenerative disease of the central nervous system, characterized by insidious, chronic and progressive memory impairment. A defining feature of AD is the accumulation of amyloid-β (Aβ; Mattson, 1997), a 39–43-aminoacid peptide derived from the amyloid precursor protein (APP; Wisniewski et al., 1997). The importance of Aβ in the pathogenesis of AD is suggested by several findings. Notably, mutations in APP or presenilin-1 and -2 (PS1 and PS2, respectively), two proteins that are implicated in familiar forms of AD, lead to an increase in the amyloidogenic forms of Aβ (Selkoe, 1999). APP is first cleaved by β-secretase at the N-terminus of Aβ, followed by an intramembranous cleavage at the C-terminus of Aβ by the γ-secretase, resulting two principal products, Aβ1–40 and Aβ1–42 that have been shown to be toxic to a wide variety of cells, including neurons (Yankner, 1996). Accumulating evidence suggest that PS1, a transmembrane domain aspartyl protease that is mainly located in the ER, together with three other membrane proteins (nicastrin, APH-1 and PEN-2) form the γ-secretase complex (reviewed in Steiner and Haass, 2000; Mattson, 2003). More recently, APP localization was reported to be altered under ER stress, being translocated from late compartments to early compartments of the secretory pathway, consequently leading to the decrease of the levels of Aβ1–40 and Aβ1–42 by β and γ-cutting, suggesting a possible early cellular defensive mechanism to reduce Aβ levels (Kudo et al., 2006). Furthermore, cells that express mutant mutant PS1 have been reported to be more sensitive to apoptotic stimuli, especially ER stress (Guo et al., 1997; 1999 a,b).

Several data have linked UPR to AD. In 2005, Hoozemans and co-workers have reported that the levels of Bip/GRP78 were increased in the temporal cortex and hippocampus of post-mortem brain tissues of AD patients. Furthemore, they found phosphorylated PERK (p-PERK, the active form of PERK) in AD brain tissues, suggesting UPR involvement in AD. Furthermore, Aβ1–42 was also shown to induce ER stress, by increasing the levels of the ER chaperones, Bip/GRP78 and GRP94 and also to increase the activity of the transcription factor CHOP/GADD153 (Ghribi et al., 2004). We have also recently showed that in primary cortical neurons the peptide Aβ1–40 was able to induce the increase the levels of Bip/GRP78, inducing in this manner ER stress (Ferreiro et al., 2006) (Figure 2). In addition, in cholinergic neurons, soluble oligomeric Aβ1–42 forms were shown to affect genes of proteins present in the ER and Golgi-complex involved in protein modification and degradation, like several chaperones and ATF6, being an additional proof for the involvement of the UPR and ER stress in Aβ toxicity (Heinitz et al., 2006).

In addition to the link of UPR to AD, several evidences show that perturbation of ER Ca^{2+} homeostasis is also an important step in the beginning or progress of the neuronal death that occurs in the disease. In 2000, Leissring find out that the presenilins, located in the ER membrane, not only function to process the APP but also modulate Ca^{2+} signalling. Mutations of presenilin result in overfilling of the ER leading to larger Ca^{2+} signals and a decrease in capacitative Ca^{2+} entry. Recently, overexpression of APP was shown to potently enhance cytosolic Ca^{2+} levels and cell death after ER Ca^{2+} store depletion and the transcription factor CHOP was significantly upregulated in these conditions (Copanaki et al., 2007). Furthermore, experiments in cortical neurons from mutant PS1 knock-in mice demonstrated the potentiation of ER Ca^{2+} liberation by photoreleased $InsP_3$ and altered membrane excitability (Stutzmann et al., 2003, 2004). More recently, Stutzmann and co-wokers have shown that neurons from young, adult, and aged AD mice have enhanced RyR recruitment, given by the higher levels of RyR expression, that contribute to the exaggerated Ca^{2+} signals (Stutzmann et al., 2006). In addition, in neurons from PS1 knock-in mice and 3xTg-AD mice that harbor mutant genes: presenilin-1 (PS1M146V), APP (bAPPswe) and tau P301L transgenes) enhanced Ca^{2+} response was associated with the increased levels of RyR (Smith et al., 2005). Several studies have shown that the Aβ peptide is also responsible for Ca^{2+} homeostasis perturbation (Figure 2). Aβ1-42 was observed to mediate changes in intracellular Ca^{2+} homeostasis, through a direct increase of the RyR3 isorfom expression and function (Supnet et al., 2006).

Several studies argue that the perturbation in Ca^{2+} homeostasis can be due to an early formation of membranar Aβ calcium channel, suggesting a possible initiating mechanism of Aβ toxicity as a death message that can be extended to the ER. In fact, morphological and intracellular Ca^{2+} changes in fibroblasts and endothelial cells are apparent 10-15 min after treatment with freshly prepared Aβ solutions. These early effects have been linked to the formation of nascent Aβ calcium channel (Bathia et al., 2000). In addition, Simakova and Arispe (2006) have shown that exposure of cells to Aβ results in the initial formation of oligomeric Aβ aggregates which assemble Ca^{2+} channels as they interact with the cell surface membrane (Figure 2). In cortical neurons treated with Aβ1-40, an increase in the intracellular Ca^{2+} concentration was observed, due to the early release of ER Ca^{2+} through RyR and InsP3R, activating ER stress (Ferreiro et al., 2004, 2006). As a response of Aβ-induced ER stress several mechanism are activated. Caspase-12 in rodents and human caspase-4 were shown to be activated by Aβ peptide (Nakagawa et al., 2000; Ferreiro et al., 2006; Hitomi et al., 2004). Furthermore, due to the Ca^{2+} release through RyR and $InsP_3R$ we have shown that Aβ1-40 induces the formation of ROS, cytochrome c release from mitochondria, caspase-3 activation and cell death (Ferreiro et al., 2006) (Figure 2).

Several studies have point to mitochondria dysfunction as fundamental phenomenon in AD (Cardoso et al., 2001; Cardoso et al., 2002; for review see Reddy and Beal, 2005 and Bubber et al., 2005) (Figure 2). In fact, accumulation of full-length APP in the mitochondria compartment can cause mitochondrial dysfunction and impaired energy metabolism (Anandatheerthavarada et al., 2003) In addition, Aβ was shown to accumulate within mitochondria and such accumulation to be associated to mitochondrial dysfunction (Caspersen et al., 2005). Furthermore, Aβ can enter mitochondria inducing free radicals generation and mitochondrial dysfunction (Reddy, 2006). Aβ can also interact with Aβ-binding alcohol dehydrogenase (ABAD) in the mitochondria and this interaction promotes leakage of ROS, mitochondrial dysfunction and cell death (Lustbader et al., 2004).

One possible site of initiation of apoptosis is the synapse. Perturbed synaptic ER Ca^{2+} homeostasis was shown to promote activation of apoptotic cascades (Mattson, 2000). Although $InsP_3Rs$ are mostly found in dendritic shafts and cell bodies, levels of RyRs are highest in axons and dendritic spines (Padua et al., 1992; Padua et al., 1996). Since RyRs are overexpressed and have increased activity in AD brain, these channels may have responsible for synaptic dysfunction and loss that characterize AD (Selkoe, 2002).

Many questions related to the initial mechanistic process that leads to cell death in AD remain to be answered. Nonetheless, there is increasing evidence that ER stress is involved in the process and that both ER and mitochondria (and possibly also other organelles) cooperate in the neuronal cell that occurs in this disease.

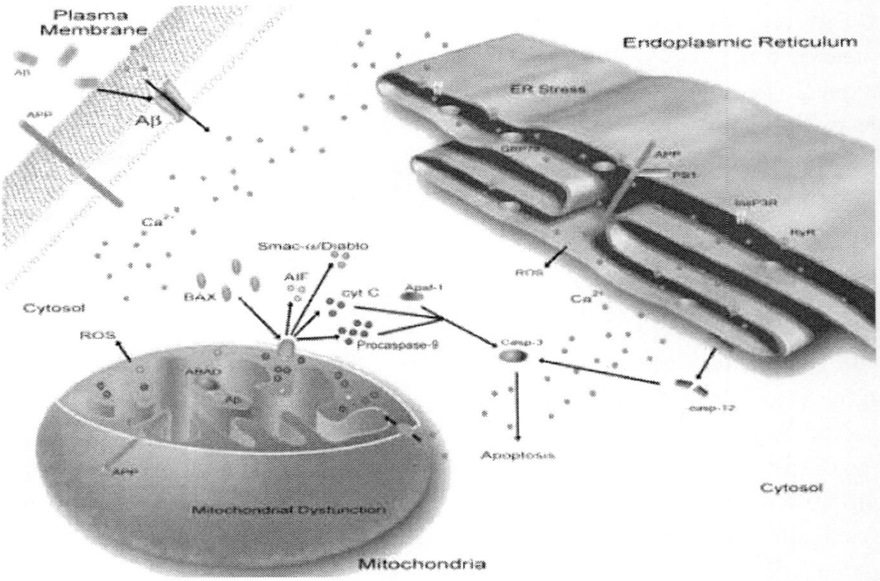

Figure 2. Endoplasmic reticulum-mediated apoptotic pathway in Aβ toxicity. Accumulation of excessive Aβ peptide, due to increase generation through the amyloidogenic processing of APP or decreased clearance, can increase cytosolic Ca^{2+} levels, possibly due to the formation of Aβ oligomers, which assemble into Ca^{2+} channels as they interact with the cell surface, and/or to the increase RyRs levels, increasing Ca^{2+} release from ER. Furthermore, mutated presenilins located at the ER membrane are responsible for the overfilling of the ER leading to larger Ca^{2+} signals. Overexpression of APP also strongly enhances cytosolic Ca^{2+} levels and the transcription factor CHOP. Therefore, Aβ leads to the perturbation of Ca^{2+} homeostasis, can induce UPR, ER stress and activation of caspase-12, which can trigger apoptotic cell death. The ER-mediated apoptotic pathway may then proceed through a mitochondrial process. Due to the close proximity of both organelles, Ca^{2+} released from the ER can perturb Ca^{2+} homeostasis in the mitochondria, initiating the mitochondrial cell death pathway, involving translocation of Bax, release of apoptotic factors (cytochrome-c, AIF) and activation of caspases cascade. Moreover, processed Aβ formed in the mitochondria or imported to this organelle can interact with ABAD protein potentiating mitochondrial dysfunction. Finally, Aβ peptide is also responsible for the formation of ROS, from the ER and/or from the mitochondria, increasing its toxic effects.

3. Potential Therapeutics

Only two classes of medication are currently available for AD patients, confined to symptomatic palliative interventions, namely the acetylcholinesterase inhibitors and the N-methyl-D-aspartate receptor antagonist memantine. Both classes have modest clinical benefits and do not prevent disease progression. Therefore, it is imperative to develop effective therapies that halt or reverse the underlying disease process.

The recent evidences that demonstrate the role of ER dysfunction in AD, support that ER may be a target for therapeutic intervention. Neuroprotective strategies, designed to reduce specifically Ca^{2+} release through RyR with dantrolene (inhibitor of RyRs) or using lithium to slow IP3 turnover (Ferreiro et al., 2006; Devaki et al., 2006), are promising therapeutical targets. In addition, the use of Ca^{2+} chelators, like BAPTA-AM can provide an additional help to reduce intracellular Ca^{2+}. In fact, BAPTA has been shown to block Ca^{2+} levels increase and to inhibit procaspase-12 cleavage (Sokka et al., 2007). Furthermore, manipulation of expression and function of anti-apoptotic Bcl-2 family members, that have been shown to be present in the ER and to regulate ER Ca^{2+} homeostasis, may also bring some improvement to the protection of neurons. Indeed, treatment strategies that are currently being developed include the delivery of anti-apoptotic Bcl-2 family members, small molecule inhibitors that interact directly with Bcl-2 family members, including both endogenously synthetic agents, which increase indirectly the overall expression and/or function of anti-apoptotic Bcl-2 family members (reviewed in Shacka and Roth, 2005). In addition, targeted anti-apoptotic Bcl-2 and Bcl-xL to the ER may also represent a therapeutical tool to regulate ER Ca^{2+} homeostasis.

The accumulation of unfolded proteins has also been a target of the therapeutical strategies. In fact, to diminish the accumulation of unfolded proteins that normally occurs in the ER stress condition, several authors studied the potential use of chemical chaperones (Perlmutter, 2002; Cohen and Kelly, 2003; Romisch, 2004; Welch, 2004). These small organic compounds have chaperone-like activities and have the capacity to assist in the correct folding/assembly of unfolded proteins. The use of drugs that up-regulate the expression of ER stress proteins like GRP78, for example valproate (Wang and Young, 2004), or by stimulating ERAD, which could lead to an improvement in the folding capacity (Mori, 2000). Furthermore, selective chemical inhibitors of eIF2alpha dephosphorylation, like Salubrinal, may be useful to reduce ER stress (Boyce et al., 2005).

Another possibility to reduce the effects of ER stress has been the attempt to reduce oxidative stress that occurs upon this process. Application of exogenous antioxidants, like α-tocopherol (vitamin E) and polyphenols (Floyd, 1999), mobilization of endogenous antioxidant such as glutathione (Butterfield, 2002) or scavengers of oxygen-free radicals (Oyadomari and Mori, 2004) have been tested to reduce intracellular ROS levels. Besides, application of ER-targeted antioxidants may be useful to reduce ROS levels, which are formed in the ER.

The present knowledge of how AD begins and progresses must be enlarged in order to find new therapeutical strategies to delay or prevent the neuronal death that occurs during progression of this devastating disease.

ACKNOWLEDGMENTS

This work was supported by FCT (Portuguese Research Council) project nº POCTI/36101/NSE/2000 and PhD fellowships nº SFRH/BD/14108/2003 and SFRH/BD/28403/2006. We thank João Cardoso (Department of Physics, Faculty of Science and Tecnology, University of Coimbra) for the contribution in the figure design.

REFERENCES

Anandatheerthavarada, H.K., Biswas, G., Robin, M.A., Avadhani, N.G. Mitochondrial targeting and a novel transmembrane arrest of Alzheimer's amyloid precursor protein impairs mitochondrial function in neuronal cells. *J. Cell Biol.* 161: 41-54, 2003.

Ashkenazi, A. Targeting death and decoy receptors of the tumour-necrosis factor superfamily. *Nat. Rev. Cancer.* 2: 420–430, 2002.

Bathia, R., Lin, H., Lal, R. Fresh and globular amyloid β protein induces rapid cellular degeneration. A possible implication for calcium-uptake via AβP-channel, *FASEB J.* 14: 1233-1243, 2000.

Baumann, O., Walz, B. Endoplasmic reticulum of animal cells and its organization into structural and functional domains. *Int. Rev. Cytol. 205:* 149-214, 2001.

Bernardi, P, Petronilli, V, Di Lisa, F, Forte, M. A mitochondrial perspective on cell death. *Trends Biochem. Sci.* 26: 112-117, 2001.

Berridge, M. J. Inositol trisphosphate and calcium signalling. *Nature* 361: 315-325, 1993.

Berridge, M.J. The endoplasmic reticulum: a multifunctional signalling organelle. *Cell Calcium* 32: 235-249, 2002.

Blaustein, M.P. Lederer, W.J. Sodium/calcium exchange: Its physiological implications. *Physiol. Rev.* 79: 763–854, 1999.

Boehning, D., Patterson, R.L., Sedaghat, L., Glebova, N.O., Kurosaki, T., Snyder, S.H. Cytochrome c binds to inositol (1,4,5) trisphosphate receptors, amplifying calcium-dependent apoptosis. *Nat. Cell Biol.* 5: 1051-61, 2003.

Boehning, D, Patterson, RL, Snyder, SH. Apoptosis and calcium: new roles for cytochrome c and inositol 1,4,5-trisphosphate. *Cell Cycle* 3: 252-254(2004).

Bootman, M.D., Lipp, P. Calcium signalling: Ringing changes to the 'bell-shaped curve. *Curr. Biol.* 9: R876–R878, 1999.

Bootman, M.D., Petersen, O.H. , Verkhratsky, A. The endoplasmic reticulum is a focal point for co-ordination of cellular activity. *Cell Calcium* 32: 231-234, 2002.

Boveris, A., Chance, B. The mitochondrial generation of hydrogen peroxide. General properties and effect of hyperbaric oxygen. *Biochem. J.* 134: 707-716, 1973.

Boyce, M., Bryant, K.F., Jousse, C., Long, K., Harding, H.P., Scheuner, D., Kaufman, R.J., Ma, D., Coen, D.M., Ron, D., Yuan J. A selective inhibitor of eIF2alpha dephosphorylation protects cells from ER stress. *Science* 307: 935-939, 2005.

Bubber, P., Haroutunian, V., Fisch, G., Blass, J.P., Gibson, G.E. Mitochondrial abnormalities in Alzheimer brain: mechanistic implications. *Ann. Neurol.* 57: 695-703, 2005.

Budd, S.L., Nicholls, D.G. A reevaluation of the role of mitochondria in neuronal Ca^{2+} homeostasis. *J. Neurochem.* 66: 403- 411, 1996.

Butterfield, D.A. Elevated Glutathione as a Therapeutic Strategy in Alzheimer's Disease. *Drug Develop. Res.* 56: 428–437, 2002.

Cardoso, S.M., Santos, S., Swerdlow, R.H., Oliveira, C.R. Functional mitochondria are required for amyloid beta-mediated neurotoxicity. *FASEB J.* 15: 1439-1341, 2001.

Cardoso, S.M., Swerdlow, R.H., Oliveira, C.R. Induction of cytochrome c-mediated apoptosis by amyloid β 25-35 requires functional mitochondria. *Brain Res.* 931: 117-125, 2002.

Caspersen, C., Wang, N., Yao, J., Sosunov, A., Chen, X., Lustbader, J.W., Xu, H.W., Stern, D., McKhann, G., Yan, S.D. Mitochondrial Abeta: a potential focal point for neuronal metabolic dysfunction in Alzheimer's disease. *FASEB J.* 19: 2040-2041. 2005.

Chan, PH. Role of oxidants in ischemic brain damage. *Stroke* 27: 1124-1129, 1996.

Cheng, T.P., Reese, T.S. Polarized compartmentalization of organelles in growth cones from developing optic tectum. *J. Cell Biol.* 101: 1473–1480, 1985.

Cheung, H.H., Kelly, N.L,, Liston, P., Korneluk, R.G. Involvement of caspase-2 and caspase-9 in endoplasmic reticulum stress-induced apoptosis: a role for the IAPs. *Exp. Cell Res.* 312: 2347-2357, 2006.

Clapham, D.E. Calcium signaling. *Cell* 80: 259-268, 1995.

Cohen, F.E., Kelly, J.W. Therapeutic approaches to protein misfolding diseases. *Nature* 426: 905- 909, 2003.

Copanaki, E., Schurmann, T., Eckert, A., Leuner, K., Muller, W.E., Prehn, J.H., Kogel, D. The amyloid precursor protein potentiates CHOP induction and cell death in response to ER Ca(2+) depletion. Biochim. *Biophys. Acta.* 1773: 157-65, 2007.

Corbett, E.F., Michalak, M. Calcium, a signaling molecule in the endoplasmic reticulum? *Trends Biochem. Sci.* 25: 307–311, 2000.

Dailey, M.E. Bridgman, P.C. Dynamics of the endoplasmic reticulum and other membranous organelles in growth cones of cultured neurons. *J. Neurosci.* 9: 1897-1909, 1989.

Deitch, J.S., Banker, G.A. An electron microscopic analysis of hippocampal neurons developing in culture: early stages in the emergence of polarity. *J. Neurosci.* 13: 4301-4315, 1993.

Devaki, R, Shankar Rao, S., Nadgir, SM. The effect of lithium on the adrenoceptor-mediated second messenger system in the rat brain. *J. Psychiatry Neurosci.* 31: 246-52, 2006.

Dreher, D. Jornot, L., Junod, A.F. Effects of hypoxanthine-xanthine oxidase on Ca^{2+} stores and protein synthesis in human endothelial cells. *Circ. Res.* 76: 388-395. 1995.

Duchen, M.R. Mitochondria and calcium: from cell signalling to cell death. *J. Physiol.* 529: 57-68, 2000.

Ferreiro, E, Oliveira, C.R., Pereira, C. Involvement of endoplasmic reticulum Ca^{2+} release through ryanodine and inositol 1,4,5-triphosphate receptors in the neurotoxic effects induced by the amyloid-beta peptide. *J. Neurosci. Res.* 76: 872-880, 2004.

Ferreiro, E., Resende, R., Costa, R., Oliveira, C.R., Pereira, C.M.F. An endoplasmic-reticulum-specific apoptotic pathway is involved in prion and amyloid-beta peptides neurotoxicity. *Neurobiol. Dis.* 23: 669-678, 2006.

Ferri, K.F., Kroemer, G. Organelle-specific initiation of cell death pathways. *Nat. Cell Biol.* 3: E255-263, 2001.

Fischer, H, Koenig, U, Eckhart, L, Tschachler, E. Human caspase 12 has acquired deleterious mutations. *Biochem Biophys Res Commun.* 293: 722-726, 2002).

Finch E.A., Turner T.J., Goldin S.M. Calcium as a coagonist of inositol 1,4,5-trisphosphate-induced calcium release. *Science* 252: 443-446, 1991.

Floyd, R.A. Antioxidants, oxidative stress, and degenerative neurological disorders. *Proc. Soc. Exp. Biol. Med.* 222: 236-45, 1999.

Foyouzi-Youssefi, R., Arnaudeau, S., Borner, C., Kelley, W.L., Tschopp, J., Lew, D.P., Demaurex, N., Krause, K.H. Bcl-2 decreases the free Ca^{2+} concentration within the endoplasmic reticulum. *Proc. Natl. Acad. Sci. U.S.A.* 97: 5723-5728, 2000.

Frenguelli, B.G., Irving, A.J., Collingridge, G.L. Ca^{2+} stores and hippocampal synaptic plasticity. *Semin. Neurosci.* 8: 301-209, 1996.

Ghribi, O, Herman, MM, Pramoonjago, P, Spaulding, NK, Savory, J. GDNF regulates the A beta-induced endoplasmic reticulum stress response in rabbit hippocampus by inhibiting the activation of gadd 153 and the JNK and ERK kinases. *Neurobiol. Dis.* 16: 417-427, 2004.

Guo, Q., Sopher, B.L., Furukawa, K., Pham, D.G., Robinson, N., Martin, G.M., Mattson, M.P. Alzheimer's presenilin mutation sensitizes neural cells to apoptosis induced by trophic factor withdrawal and amyloid-peptide: involvement of calcium and oxyradicals. *J. Neurosci.* 17: 4212–4222, 1997.

Guo, Q., Sopher, B.L., Furukawa, K., Pham, D.G., Robinson, N., Martin, G.M., Mattson, M.P. Increased vulnerability of hippocampal neurons to excitotoxic necrosis in presenilin-1 mutant knock-in mice. *Nat. Med.* 5: 101-106, 1999a.

Guo, Q., Sebastian, L., Sopher, B.L., Mille,r M.W., Ware, C.B., Martin, G.M., Mattson, M.P. Increased vulnerability of hippocampal neurons from presenilin-1 mutant knock-in mice to amyloid beta peptide toxicity: central roles of superoxide production and caspase activation. *J. Neurochem.* 72: 1019-1029, 1999b.

Hampton, R.Y. IRE1: a role in UPREgulation of ER degradation. Dev. *Cell* 4: 144-146, 2003.

Hardingham, G.E., Arnold, F.J., Bading, H. Nuclear calcium signaling controls CREB-mediated gene expression triggered by synaptic activity. *Nat. Neurosci.* 4: 261-267, 2001.

Harding, H.P., Zhang, Y., Ron, D. Protein translation and folding are coupled by an endoplasmic-reticulum-resident kinase. *Nature* 397: 271-274, 1999.

Harding, H.P., Novoa, I., Zhang, Y., Zeng, H., Wek, R., Schapira, M., Ron, D. Regulated translation initiation controls stress-induced gene expression in mammalian cells. *Mol. Cell* 6: 1099-1108, 2000.

Harding, H.P., Zhang, Y., Zeng, H., Novoa, I., Lu, P.D., Calfon, M., Sadr, N., Yun, C., Popko, B., Paules, R., Stojdl, D.F., Bell, J.C., Hettmann, T., Leiden, J.M., Ron, D. An integrated stress response regulates amino acid metabolism and resistance to oxidative stress. *Mol. Cell* 11: 619-633, 2003.

Haze, K., Yoshida, H., Yanagi, H., Yura, T., Mori, K. Mammalian transcription factor ATF6 is synthesized as a transmembrane protein and activated by proteolysis in response to endoplasmic reticulum stress. *Mol. Biol. Cell* 10: 3787-3799, 1999.

He, Q., Lee, D., Rong, R., Yu, M., Luo, X., Klein, M., El-Deiry, W., Huang, Y., Hussain, A., Sheikh, M., Endoplasmic reticulum calcium pool depletion-induced apoptosis is coupled with activation of the death receptor 5 pathway. *Oncogene* 21: 2623-2633, 2002.

Heinitz, K., Beck, M., Schliebs, R., Perez-Polo, J.R. Toxicity mediated by soluble oligomers of beta-amyloid(1-42) on cholinergic SN56.B5.G4 cells. *J. Neurochem.* 98: 1930-1945, 2006.

Hetz, C., Russelakis-Carneiro, M., Maundrell, K., Castilla, J., Soto, C. Caspase-12 and endoplasmic reticulum stress mediate neurotoxicity of pathological prion protein. *EMBO J.* 22: 5435-5445, 2003.

Hitomi, J, Katayama, T, Eguchi, Y, Kudo, T, Taniguchi, M, Koyama, Y, Manabe, T, Yamagishi, S, Bando, Y, Imaizumi, K, Tsujimoto, Y, Tohyama, M. Involvement of caspase-4 in endoplasmic reticulum stress-induced apoptosis and Abeta-induced cell death. *J. Cell. Biol.* 165: 347-356, 2004.

Hoozemans, J.J.M., Veerhuis, R., Van Haastert, E.S., Rozemuller, J.M., Baas, F., Eikelenboom, P., Scheper. W. The unfolded protein response is activated in Alzheimer's disease. *Acta Neuropathol.* 110: 165-172, 2005.

Huang, H., Zhang, H., Ou, H., Chen, H., Gibson, G.E. α-Keto-β- methyl-n-valeric acid diminishes reactive oxygen species and alters endoplasmic reticulum Ca^{2+} stores. Free. *Radical Biol. Med.* 37: 1779–1789, 2004.

Iino, M. Biphasic Ca^{2+} dependence of inositol 1,4,5-trisphosphate Ca^{2+} release in smooth muscle cells of the guinea pig Taenia caeci. *J. Gen. Physiol.* 95: 1103-1122 1990.

Iwakoshi, N.N., Lee, A.H., Glimcher, L.H. The X-box binding protein-1 transcription factor is required for plasma cell differentiation and the unfolded protein response. *Immunol. Rev.* 194: 29-38, 2003.

Kanaseki, T, Ikeuchi, Y, Tashiro, Y. Rough surfaced smooth endoplasmic reticulum in rat and mouse cerebellar Purkinje cells visualized by quick-freezing techniques. *Cell Struct. Funct.* 23: 373–387, 1998.

Kaufman, R.J. Stress signaling from the lumen of the endoplasmic reticulum: coordination of gene transcriptional and translational controls. *Genes Dev.* 13: 1211-1233, 1999.

Kaufman, R.J. Orchestrating the unfolded protein response in health and disease. *J. Clin. Invest.* 110: 1389-1398, 2002.

Kudo, T., Okumura, M., Imaizumi, K., Araki, W., Morihara, T., Tanimukai, H., Kamagata, E., Tabuchi, N., Kimura, R., Kanayama, D., Fukumori, A., Tagami, S., Okochi, M., Kubo, M., Tanii, H., Tohyama, M., Tabira, T., Takeda, M. Altered localization of amyloid precursor protein under endoplasmic reticulum stress. *Biochem. Biophys. Res. Commun.* 344: 525-530, 2006.

Kuznetsov, G., Brostrom, M.A., Brostrom, C.O. Demonstration of a calcium requirement for secretory protein processing and export. Differential effects of calcium and dithiothreitol. *J. Biol. Chem.* 265: 3932–3939, 1992.

Lee, A.H., Iwakoshi, N.N., Glimcher, L.H. XBP-1 regulates a substet of endoplasmic reticulum resident chaperones genes in the unfolded protein response. *Mol. Cell Biol.* 23: 7448-7459, 2003.

Lei, K., Davis, R.J. JNK phosphorylation of Bim-related members of the Bcl2 family induces Bax-dependent apoptosis. *Proc. Natl. Acad. Sci. USA.* 100: 2432-2437, 2003.

Leissring, M.A., Akbari, Y., Fanger, C.M., Cahalan, M.D., Mattson, M.P., LaFerla, F.M. Capacitative calcium entry deficits and elevated luminal calcium content in mutant presenilin-1 knockin mice. *J. Cell Biol.* 149: 793-797, 2000.

Levesque, L, Annaert, W, Craessaerts, K, Mathews, PM, Seeger, M, Nixon, RA, Van Leuven, F, Gandy, S, Westaway, D, St George-Hyslop, P, De Strooper, B, Fraser, PE. Developmental expression of wild-type and mutant presenilin-1 in hippocampal neurons from transgenic mice: evidence for novel species-specific properties of human presenilin-1. *Mol. Med.* 5: 542–554, 1999.

Liu, C.Y., Schröder, M., Kaufman, R.J. Ligand-independent dimerization activates the stress response kinases IRE1 and PERK in the lumen of the endoplasmic reticulum. *J. Biol. Chem.* 275: 24881-24885, 2000.

Liu, C.Y., Wong, H.N., Schuaerte, J.A., Kaufman, R.J. The protein kinase/endoribonuclease IRE1α that signals the unfolded protein response has a luminal N-terminal ligand-independent dimerization domain. *J. Biol. Chem.* 277: 18346-18356, 2002.

Lodish, H.F., Kong, N., Wikstrom, L. Calcium is required for folding of newly made subunits of the asialoglycoprotein receptor within the endoplasmic reticulum. *J. Biol. Chem.* 267: 12753–12760, 1992.

Lu, P.D., Harding, H.P., Ron, D. Translation reinitiation at alternative open reading frames regulates gene expression in an integrated stress response. *J. Cell. Biol.* 167: 27-33, 2004.

Lustbader, J.W., Cirilli, M., Lin, C., Xu, H.W., Wu, H. et al., ABAD directly links Abeta to mitochondrial toxicity in Alzheimer's disease. *Science* 304: 448-451, 2004.

Marciniak, SJ, Yun, CY, Oyadomari, S, Novoa, I, Zhang, Y, Jungreis, R, Nagata, K, Harding, HP, Ron, D. CHOP induces death by promoting protein synthesis and oxidation in the stressed endoplasmic reticulum. *Genes Dev.* 18: 3066-3077, 2004.

Mattson M.P. Cellular actions of β-amyloid precursor protein and its soluble and fibrillogenic derivatives. *Physiol. Rev.* 77: 1081-1132, 1997.

Mattson, M.P., LaFerla, F.M., Chan, S.L., Leissring, M.A., Shepel, P.N., Geiger, J.D. Calcium signaling in the ER: its role in neuronal plasticity and neurodegenerative disorders. *Trends Neurosci.* 23: 222-229, 2000.

Mattson, M.P. Neurobiology: Ballads of a protein quartet. *Nature* 422: 385-387, 2003.

McCullough, K.D., Martindale, J.L., Klotz, L.O., Aw, T.Y., Holbrook, N.J. Gadd153 sensitizes cells to endoplasmic reticulum stress by down-regulation Bcl-2 and perturbing the cellular redox state. *Mol. Cell. Biol.* 21: 1249-1259. 2001.

Meldolesi, J., Pozzan, T. The endoplasmic reticulum Ca^{2+} store: a view from the lumen. *Trends Biochem. Sci.* 23: 10-14, 1998.

Michalak, M., Parker, J.M.R, Opas, M. Ca^{2+} signaling and calcium binding chaperones of the endoplasmic reticulum. *Cell Calcium* 32: 269-278, 2002.

Mori, K. Tripartite management of unfolded proteins in the endoplasmic reticulum. *Cell* 101: 451-454, 2000.

Nadanaka, S., Okada, T., Yoshida, H., Mori, K A role of disulfide bridges formed in the luminal domain of ATF6 in sensing endoplasmic reticulum stress. *Mol. Cell Biol.* 27: 1027-1043, 2007.

Nakagawa, T., Yuan, J. Cross-talk between two cysteine protease families: activation of caspase-12 by calpain in apoptosis. *J. Cell Biol.* 150: 887-894, 2000.

Nakagawa, T., Zhu, H., Morishima, N., Li, E., Xu, J., Yankner, B.A., Yuan, J. Caspase-12 mediates endoplasmic-reticulum-specific apoptosis and cytotoxicity by amyloid-β. *Nature* 403: 98-103, 2000.

Nakamura, T., Barbara, J.G., Nakamura, K., Ross, W.N. Synergistic release of Ca^{2+} from IP3-sensitive stores evoked by synaptic activation of mGluRs paired with backpropagating action potentials. *Neuron* 24: 727-737, 1999.

Niggli, E. Localized intracellular calcium signaling in muscle: calcium sparks and calcium quarks. *Annu. Rev. Physiol.* 61: 311-335, 1999.

Nishitoh, H., Matsuzawa, A., Tobiume, K., Saegusa, K., Takeda, K., Inoue, K., Hori, S., Kakizuka, A., Ichijo, H. ASK1 is essential for endoplasmic reticulum stress-induced neuronal cell death triggered by expanded polyglutamine repeats. *Genes Dev.* 16: 1345-1355, 2002.

Nutt, L.K., Chandra, J., Pataer, A., Fang, B., Roth, J.A., Swisher, S.G., O'Neil, R.G., McConkey, D.J. Bax-mediated Ca^{2+} mobilization promotes cytochrome c release during apoptosis. *J. Biol. Chem.* 277: 20301- 20308, 2002a .

Nutt, L.K., Pataer, A., Pahler, J., Fang, B., Roth, J., McConkey, D.J., Swisher, S.G. Bax and Bak promote apoptosis by modulating endoplasmic reticular and mitochondrial Ca^{2+} stores. *J. Biol. Chem.* 277: 9219-9225, 2002b.

Obeng, E.A., Boise, L.H. Caspase-12 and caspase-4 are not required for caspase-dependent endoplasmic reticulum stress-induced apoptosis. *J. Biol. Chem.* 280: 29578-29587, 2005.

Okada, T.K., Yoshida, R., Akazawa, M., Negishi, M., Mori, K. Distinct roles of activating transcription factor 6 (ATF6) and double-stranded RNA-activated protein kinase-like endoplasmic reticulum (PERK) in transcription during the mammalian unfolded protein response. *Biochem. J.* 366: 585-594, 2000.

Oyadomari, S., Mori, M. Roles of CHOP/GADD153 in endoplasmic reticulum stress. *Cell Death Differ.* 11: 381-389, 2004.

Padua, R.A. Yamamoto, T., Fyda, D., Sawchuk, M.A., Geiger, J.D., Nagy, J.I. Autoradiographic analysis of $[^3H]$ryanodine binding sites in rat brain. Regional distribution and the effects of lesions on sites in the hippocampus. *J. Chem. Neuroanat.* 5: 63-73, 1992.

Padua, R.A., Nagy, J.L., Geiger, J.D. Subcellular localization of ryanodine receptors in rat brain. *Eur. J. Pharmacol.* 298: 185-189, 1996.

Pashen, W. Dependence of vital cell function on endoplasmic reticulum calcium levels: implications for the mechanisms underlying neuronal cell injury in different pathological states. *Cell Calcium* 29: 1-11, 2001.

Patil, C., Walter, P. Intracellular signaling from the endoplasmic reticulum to the nucleus: the unfolded protein response in yeast and mammals. *Curr. Opin. Cell Biol.* 13: 349-355, 2001.

Perlmutter, D.H. Chemical chaperone: a pharmacological strategy for disorders of protein folding and trafficking. *Ped. Res.* 52: 832- 836, 2002.

Pinton, P., Ferrari, D., Magalhaes, P., Schulze-Osthoff, K., Di Virgilio, F., Pozzan, T., Rizzuto, R. Reduced loading of intracellular Ca(2+) stores and downregulation of capacitative Ca(2+) influx in Bcl-2-overexpressing cells. *J. Cell Biol.* 148: 857-862, 2000.

Pozzan, T., Rizzuto, R., Volpe, P., Meldolesi, J. Molecular and cellular physiology of intracellular calcium stores. *Physiol. Rev.* 74: 595-636, 1994.

Putcha, G.V., Le, S., Frank, S., Besirli, C.G., Clark, K., Chu, B., Alix, S., Youle, R.J., LaMarche, A., Maroney, A.C., Johnson, E.M. Jr. JNK-mediated BIM phosphorylation potentiates BAX-dependent apoptosis. *Neuron* 38: 899–914, 2003.

Rao, R.V., Hermel, E., Castro-Obregon, S., del Rio, G., Ellerby, L.M., Ellerby, H.M., Bredesen, D.E. Coupling endoplasmic reticulum stress to the cell death program. Mechanism of caspase activation. *J. Biol. Chem.* 276: 33869-33874, 2001.

Racay, P., Kaplan, P., Lehotsky, J., Mezesova, V. Rabbit brain endoplasmic reticulum membranes as target for free radicals. Changes in Ca^{2+}-transport and protection by stobadine. Biochem. *Mol. Biol. Int.* 36: 569-577, 1995.

Reddy, P.H. Amyloid precursor protein-mediated free radicals and oxidative damage: implications for the development and progression of Alzheimer's disease. *J. Neurochem.* 96: 1-13, 2006.

Reddy, P.H., Beal, M.F. Are mitochondria critical in the pathogenesis of Alzheimer's disease? *Brain Res. Rev.* 49: 618-632. 2005.

Romisch, K. A cure for traffic jams: small molecule chaperones in the endoplasmic reticulum. *Traffic* 5: 815-820, 2004.

Rose, C.R., Konnerth, A. Stores not just for storage: intracellular calcium release and synaptic plasticity. *Neuron* 31: 519-22, 2001.

Rutkowski, D.T., Kaufman, R. J. A trip to the ER: coping with stress. *Trends in Cell Biol.* 14: 20-28, 2004.

Shacka, JJ, Roth, KA. Regulation of neuronal cell death and neurodegeneration by members of the Bcl-2 family: therapeutic implications. *Curr. Drugs Targets CNS Neurol. Disord.* 4: 25-39, 2005.

Schröder, M., Kaufman, R.J. ER stress and the unfolded protein response. *Mutat. Res.* 569: 29-63, 2005.

Scorrano, L., Oakes, S.A., Opferman, J.T., Cheng, E.H., Sorcinelli, M.D., Pozzan, T., Korsmeyer, S. J. BAX and BAK regulation of endoplasmic reticulum Ca^{2+}: a control Point for apoptosis. *Science* 300: 135-139, 2003.

Selkoe, D.J. Translating cell biology into therapeutic advances in Alzheimer's disease. *Nature* 399: A23–A31, 1999.

Selkoe, D.J. Alzheimer's Disease Is a Synaptic Failure. Science 298: 789-791, 2002.

Shamu, C.E., Walter, P. Oligomerization and phosphorylation of the Ire1p kinase during intracellular signalling from the endoplasmic reticulum to the nucleus. *EMBO J.* 15: 3028-3039, 1996.

Sharp, A.H., McPherson, P.S., Dawson, T.M., Aoki, C., Campbell, K., Snyder, S.H. Differential immunohistochemical localization of inositol 1,4,5 triphosphate- and ryanodine-sensitive Ca^{2+} release channels in rat brain. *J. Neurosci.* 13: 3051–3063. 1993.

Shen, Y., Ballar, P., Apostolou, A, Doong, H., Fang, S. ER stress differentially regulates the stabilities of ERAD ubiquitin ligases and their substrates. *Biochem. Biophys. Res. Commun.* 352: 919-924, 2007.

Shi, Y., Vattem, K.M., Sood, R., An, J., Liang, J., Stramm, L., Wek, R.C. Identification and characterization of pancreatic eukaryotic initiation factor 2 α-subunit kinase, PEK, involved in translational control. *Mol. Cell Biol.* 18: 7499-7509, 1998.

Shi, Y., An, J., Liang, J., Hayes, S.E., Sandusky, G.E., Stramm, L.E., Yang, N.N. Characterization of a mutant pancreatic eIF-2α kinase, PEK, and co-localization with somatostatin in islet delta cells. *J. Biol. Chem.* 274: 5723-5730, 1999.

Simakova, O., Arispe, N.J. Early and late cytotoxic effects of external application of the Alzheimer's Abeta result from the initial formation and function of Abeta ion channels. *Biochemistry* 45: 5907-5915, 2006.

Smith, IF, Hitt, B, Green, KN, Oddo, S, LaFerla, FM. Enhanced caffeine-induced Ca2+ release in the 3xTg-AD mouse model of Alzheimer's disease. *J. Neurochem.* 94: 1711-1718, 2005.

Sokka, A.L., Putkonen, N., Mudo, G., Pryazhnikov, E., Reijonen, S., Khiroug, L., Belluardo, N., Lindholm, D., Korhonen, L. Endoplasmic reticulum stress inhibition protects against excitotoxic neuronal injury in the rat brain. *J. Neurosci.* 27: 901-908, 2007.

Steiner, H., Haass, C. Intramembrane proteolysis by presenilins. *Nat. Rev. Mol. Cell Biol.* 1: 217-224, 2000.

Stutzmann, G.E., LaFerla, F.M., Parker, I. Ca^{2+} signaling in mouse cortical neurons studied by two-photon imaging and photoreleased inositol triphosphate. *J. Neurosci.* 23: 758-765, 2003.

Stutzmann, G.E., Caccamo, A., LaFerla, F.M., Parker, I. Dysregulated IP3 signaling in cortical neurons of knock-in mice expressing an Alzheimer's-linked mutation in presenilin1 results in exaggerated Ca^{2+} signals and altered membrane excitability. *J. Neurosci.* 24: 508-513, 2004.

Stutzmann, G.E., Smith, I., Caccamo, A., Oddo, S., Laferla, F.M., Parker, I. Enhanced ryanodine receptor recruitment contributes to Ca^{2+} disruptions in young, adult, and aged Alzheimer's disease mice. *J. Neurosci.* 26: 5180-5189, 2006.

Supnet, C., Grant, J., Kong, H., Westaway, D., Mayne, M. Amyloid-beta-(1-42) increases ryanodine receptor-3 expression and function in neurons of TgCRND8 mice. *J. Biol. Chem.* 281: 38440-38447, 2006.

Szalai, G., Krishnamurthy, R., Hajnoczky, G. Apoptosis driven by IP(3)-linked mitochondrial calcium signals. *EMBO J.* 18: 6349-6361, 1999.

Tardif, K.D., Waris, G., Siddiqui, A. Hepatitis C virus, ER stress, and oxidative stress. *Trends Microbiol.* 13: 159-163, 2005.

Tirasophon, W., Welihinda, A.A., Kaufman, R.J. A stress response pathway from the endoplasmic reticulum to the nucleus requires a novel bifunctional protein kinase/endoribonuclease (Ire1p) in mammalian cells. *Genes Dev.* 12: 1812-1824, 1998.

Urano, F., Wang, X., Bertolotti, A., Zhang, Y., Chung, P., Harding, H.P., Ron, D. Coupling of stress in the ER to activation of JNK protein kinases by transmembrane protein kinase IRE1. *Science* 287: 664-666, 2000.

Vattem, K.M., Wek, R.C. Reinitiation involving upstream ORFs regulates ATF4 mRNA translation in mammalian cells. *Proc. Natl. Acad. Sci. USA* 101: 11269-11274 2004 .

Verkhratsky, A., Petersen, O.H. The endoplasmic reticulum as an integrating signalling organelle: from neuronal signalling to neuronal death. *Eur. J. Pharmacol.* 447: 141-154, 2002.

Wang, J.F., Young, L.T. Regulation of molecular chaperone GRP78 by mood stabilizing drugs. *Clin. Neurosci. Res.* 4: 281-288, 2004.

Wang, X.Z., Harding, H.P., Zhang, Y., Jolicoeur, E.M., Kuroda, M., Ron, D. Cloning of mammalian Ire1 reveals diversity in the ER stress responses. *EMBO J.* 17: 5708-5717, 1998.

Wang, X.Z., Ron, D. Stress-induced phosphorylation and activation of the transcription factor CHOP (GADD153) by p38 MAP Kinase. *Science* 272: 1347-1349. 1996.

Wang, S., El-Deiry, W.S. Cytochrome c: a crosslink between the mitochondria and the endoplasmic reticulum in calcium-dependent apoptosis. *Cancer Biol. Ther.* 3: 44-46, 2004.

Welch, W.J. Role of quality control pathways in human diseases involving protein misfolding. *Semin. Cell Dev. Biol.* 15: 31-38, 2004.

Wisniewski, T., Ghiso, J., Frangione, B. Biology of Abeta amyloid in Alzheimer's disease. *Neurobiol Aging* 4: 313-328, 1997.

Wrzosek, A. Regulation of Ca^{2+} release from internal stores in cardiac and skeletal muscle. *Acta Biochim. Pol.* 47: 705-723, 2000.

Xu, C., Bailly-Maitre, B, Reed, J.C. Endoplasmic reticulum stress: cell life and death decisions. *Clin. Invest.* 115: 2656-2664, 2005.

Yankner, BA. New clues to Alzheimer's disease: unraveling the roles of amyloid and tau. *Nat. Med.* 2: 850-852, 1996.

Yamamoto, K., Ichijo, H., Korsmeyer, S.J. BCL-2 is phosphorylated and inactivated by an ASK1/Jun N-terminal protein kinase pathway normally activated at G(2)/M. *Mol Cell. Biol.* 19: 8469-8478, 1999.

Ye, J., Rawson, R.B., Komuro, R; Chen, X; Dave, U.P.; Prywes, R.; Brown, M.S.; Goldstein, J.L. ER stress induces cleavage of membrane-bound ATF6 by the same proteases that process SREBPs. *Mol. Cell* 6: 1355-1364, 2000.

Yoneda, T., Imaizumi, K., Oono, K., Yui, D., Gomi, F., Katayama, T., Tohyama, M. Activation of caspase-12, an endoplasmic reticulum (ER) resident caspase, through tumor necrosis factor receptor-associated factor 2-dependent mechanisms in response to ER stress. *J. Biol. Chem.* 276: 13935-13940, 2001.

Yoshida, H.T., Okada, K., Haze, H., Yanagi, H., Yura, T., Mori, K. ATF6 activated by proteolysis directly binds in the presence of NF-Y (CBF) to the cis-acting element responsible for the mammalian unfolded protein response. *Mol. Cell Biol.* 20: 6755-6767, 2000.

Yoshida, H.T., Okada, K., Haze, H., Yanagi, H., Yura, T., Negishi, M., Mori, K. Endoplasmic reticulum-stress-induced formation of transcription factor complex ERSF including NF-Y (CBF) and activating transcription factors 6α and 6β that activates the mammalian unfolded protein response. *Mol. Cell Biol.* 21: 1239-1248, 2001.

Yoshida, H.T., Matsui, T, Hosokawa, N., Kaufman, R.J., Nagata, K., Mori, K A time-dependent phase shift in the mammalian unfolded protein response. *Dev. Cell* 4: 265-271, 2003.

Zhang, M., Yamazaki, T., Yazawa, M., Treves, S., Nishi, M., Murai, M., Shibata, E., Zorzato, F., Takeshima, H. Calumin, a novel Ca^{2+}-binding transmembrane protein on the endoplasmic reticulum. Cell Calcium: doi:10.1016/j.ceca.2006.11.009, 2007

Zhao, L., Ackerman, S.L. Endoplasmic reticulum stress in health and disease. Curr. Opin. Cell Biol. 18:444-452, 2006.

In: Cognitive Science Compendium, Volume 1
Editor: Miao-Kun Sun

ISBN: 978-1-60456-597-3

Chapter 15

THE P300 COMPONENT OF THE EVENT-RELATED BRAIN POTENTIAL AND BAYES' THEOREM

Bruno Kopp*
Research Institute of Cognitive Neurology,
Klinikum Braunschweig and
University of Technology Carolo-Wilhelmina at
Braunschweig, Germany

ABSTRACT

The impact of the methodology of electroencephalography-based brain imaging on cognitive neuroscience has not been fully acknowledged yet. In particular, event-related brain potentials (ERPs) provide unique information to the study of the neural basis of human cognition. The article describes a Bayesian theory of the P300, perhaps the most studied ERP component. The theory was derived from several sources of knowledge. Firstly, empirical knowledge about factors that determine P300 amplitude and latency is considered. Here, the well recognized probability effect on the P300 proved to be the crucial factor. Secondly, Bayesian decision theory is shortly introduced. Recently, the conceptual framework of Bayesian statistics has been successfully applied to the sensorimotor system, and neuroscientific research on the Bayesian brain has just now been ignited. Thirdly, recent empirical knowledge about the neural basis of decision making is considered. A decision making model is shortly introduced because of its elegant simplicity, and due to the fact that it can be interpreted as a Bayesian decision maker. A Bayesian theory of the P300 arises from the integration of these different traces of knowledge, according to which P300 amplitude varies as function of Bayesian belief revision and P300 latency varies as a function of the duration of this revision process.

* Correspondence and requests for reprints: Research Institute of Cognitive Neurology, Klinikum Braunschweig and University of Technology Carolo-Wilhelmina at Braunschweig, Salzdahlumer Str. 90, 38126 Braunschweig, Germany E-mail: b.kopp@klinikum-braunschweig.de

The theory posits that the human brain codes and computes with probabilities, in a Bayesian manner, and that the P300 offers a window on the Bayesian brain.

Keywords: P300, electroencephaolography (EEG), event-related potentials (ERPs), Bayes' theorem, probability, decision making, human brain.

Electroencephalography (EEG) is widely adopted for the non-invasive assessment of cortical processing, mainly due to the high temporal resolution of EEG-based brain imaging. Developments in EEG acquisition technology and data processing capability have allowed for the identification and characterization of specific deflections comprising activities associated with given experimental stimuli or responses. These time-locked waveforms are identified as Event-Related Potentials (ERPs). The P300, first described by Sutton et al., (1965), is perhaps the most-studied ERP component, partly due to its relatively large amplitude and facile elicitation in experimental contexts. The P300 deflection emerges as a positivity, typically appearing approximately 300 to 500 ms following stimulus presentation. Timing of this component may range widely, however, from 250 ms and extending to 700 ms, with amplitude varying from +5 μV to +20 μV for auditory and visual ERPs.

Elicitation of the P300 may be achieved through an experimental *oddball* paradigm, in which subjects are exposed to a continuous succession of two types of stimuli, one presented regularly and the other displayed sporadically (Ritter and Vaughan, 1969). The P300 is elicited if an active classification task is added (Picton et al., 1995) in which each event belongs to one of two categories. For example, an experimental oddball paradigm may ask subjects to press a button when detecting male names (*target stimuli*) embedded in a sequence containing male and female names (Kutas et al., 1977). The cornerstone of P300 research is that its amplitude is inversely proportional to the probability of the eliciting stimulus category (*probability effect*) and that the P300 is larger to target than non-target stimuli at the same probability level (*target effect*; Duncan-Johnson and Donchin, 1977). Thus, a markedly increased P300 amplitude is usually observed in response to rare target stimuli of an oddball experimental paradigm. Other factors documented to alter P300 amplitude include stimulus sequence (Squires et al., 1976; Duncan-Johnson and Donchin, 1982), stimulus quality, attention, task difficulty and task relevance of the eliciting stimulus (for reviews see Linden, 2005; Luck, 2005; Nieuwenhuis et al., 2005; Picton 1992; Polich and Kok,1995; Soltani and Knight, 2000).

McCarthy and Donchin (1981) proposed that P300 latency varies as a function of factors governing stimulus evaluation time. As outlined above, the P300 amplitude is larger to infrequent stimuli, particularly when those stimuli are targets. Thus, the P300 must be generated after the stimulus has been categorized according to the rules of the task. This, in turn, implies that manipulations that postpone stimulus categorization (e.g., those that increase the time required for perceptual processing) must increase P300 latency, whereas its latency should not be dependent on post-categorization processes, such as response selection. However, the empirical issue whether or not the P300 is involved in processes that are related to response selection is still a matter of controversy (see Coles et al., 1995; Leuthold and Sommer, 1998).

Most well-characterized is the *P3b*, or "classical P3" (the term P300 refers to this P3b sub-component), with parietal topography. In oddball experimental paradigms, infrequent stimuli that are irrelevant to the task (distractors) but more salient than the targets, evoke the *novelty P3* or *P3a* (Courchesne et al., 1975; Squires et al., 1975), which peak earlier than the P3b and have a more frontocentral scalp distribution than the P3b. ERP waveform factor analyses indicate that the novelty P3 and P3a are identical, but that they are in fact dissociable from the P3b (Spencer et al., 2001). The P3a was conceived as a correlate of attentional orienting to salient distractors (Friedman et al., 2001), reflecting involuntary switching of attention to deviant events (Escera et al., 1998). However, evidence is accumulating which suggests that task irrelevance of a deviant stimulus is not an antecedent condition for eliciting the P3a (Debener et al., 2005; Kopp et al., 2006).

The P300 amplitude and Bayes' theorem were never linked together in a formal way, although the original article by Sutton et al., . (1965) concluded that stimuli that "resolve the subject's uncertainty" elicit the P300, and despite the fact that the term *subjective probability* was continuously used to denote the functional significance of the P300 (e.g., Duncan-Johnson and Donchin, 1977). De Swart et al., (1981) cite some early, yet never persecuted Bayesian notions in the context of research on the P300. Friston (2005) repeatedly mentions the P300 in his Bayesian theory of cortical processing, but his article does not provide an explicit derivation of the Bayesian qualities of the P300.

Instead of that, the available evidence was summarized by theorists that the P300 is a cortical process supporting the formulation of an internal environmental model in which a stimulus needs to be evaluated. According to this "context-updating" theory (Donchin, 1981), the P300 is invoked in the service of *strategic* rather than *tactical* information processing. Whereas the tactical stream selects specific responses at any instance, the strategic stream is involved in the planning and control of behavior. Strategic processes comprise the long-term setting of priorities, the setting of biases, the deployment of attention, or the mapping of probabilities on the environment. In short, this stream is composed of a class of metacontrol processes which is concerned with maintaining a proper representation of the environment. According to an authoritative portrayal of the context-updating theory (Donchin and Coles, 1988), it "... asserts nothing more than that the P300 is elicited by processes associated with the maintenance of our model of the context of the environment" (p. 370). Alternatively, the "context-closure" theory (Verleger, 1988) emerged as an antagonist to context-updating theory, reflecting the concept that the P300 reflects activity of a memory trace remodelling post-detection of a target stimulus (Desmedt, 1980). However, these proposals were not followed by convincing lines of experiments providing support for them, probably due to the fact that the broad concepts of "context-updating" and "context-closure" do not specify how they are related to particular experimental manipulations (Luck, 2005).

Johnson (1984, 1986) proposed a triarchic theory of the P300 amplitude, according to which

$$\text{P300 amplitude} = f\{ T \times [(1/P) + M] \} \quad (1)$$

with P = subjective probability (experimental variables affecting global stimulus probability, sequential expectancies), M = stimulus meaning (experimental variables affecting task complexity, stimulus complexity, stimulus value), T = transmitted stimulus information proportion of the total information received by the subject (with values between 0 and 1). Without going into the details of the triarchic theory here, it reminds us that the P300 amplitude is influenced by a variety of determinants. The purpose of this article is to provide a closer look at the *P*-factor of the triarchic theory of the P300 and to suggest that it possesses Bayesian qualities.

I start my line of reasoning with the statement that the interpretation, but not the mathematical properties, of probability is controversial. According to the "frequentist" interpretation, probabilities are merely limiting relative frequencies. However, in cognitive science applications, probabilities refer to "degrees of belief" or *subjective* probabilities (Gigerenzer and Murray, 1987). The subjective interpretation of probability generally aims to evaluate conditional probabilities, $P\ (h_j/d)$, that is (so-called inverse) probabilities of alternative hypotheses h_j about states of reality, given certain data, d, i.e. input available to the senses. Application of elementary rules of probability theory leads to Bayes' theorem (1763),

$$P\ (h_j/d) = [P\ (d/h_j)\ P\ (h_j)]\ /\ P\ (d) \qquad (2)$$

In words, the components of Bayesian inference are $P\ (h_j)$ that reflects the prior belief in a particular state before the sensory information is received and $P\ (d/h_j)$ that equals the *likelihood* of the sensory input given the hypothesized state. According to Bayes' theorem, the probability of the state given the sensory input, i.e. the *posterior* $P\ (h_j/d)$, is obtained by multiplying the prior by the likelihood and by normalizing the product (by scaling it so that the sum of the probabilities over all possible states sums to one). The posterior belief can then be used as the new prior belief for another perception, so that a belief can be updated within a series of perceptions sequentially through time.

Probability has only recently become a major focus of attention in the cognitive sciences. Indeed, classic work by Kahneman, Tversky and their colleagues (Kahneman et al., 1982; Kahneman and Tversky, 2000), but also earlier work (Phillips and Edwards, 1966), suggested that human cognition might be non-probabilistic in fundamental ways. Many of these studies may be best described as textbook probability reasoning experiments, and people are often poor at solving this kind of problems or "… fall victim to a range of probabilistic fallacies" (Chater et al., 2006, p. 288). But the fact that people seem to be poor probabilists is not in conflict with the hypothesis that aspects of cortical processing can be modeled in probabilistic terms. To use Chater et al.,'s (2006) perspicuous example, the fact that Fourier analysis is hard to understand does not imply that it is not fundamental to perception.

The issue in probabilistic analyses of human cognition is to understand what is believed, or what can be inferred, about objects in the environment (Kersten et al., 2004), about regularities linking environmental objects (Tenenbaum et al., 2006), or about states of the motor system (Körding and Wolpert, 2006). Human visual perception is a subfield of cognitive science where Bayesian decision theory has been applied successfully (Geisler and Kersten, 2002; Kersten and Yuille, 2003; Stocker and Simoncelli, 2006). Von Helmholtz (1867) long ago regarded perceptions as being similar to predictive hypotheses and described visual perceptions as unconscious inferences from sensory data and knowledge derived from the past. Thus, the Bayesian decision theory mainly adds the probabilistic nature of the unconscious inferences to von Helmholtz' theory. The Bayesian theory of perception posits that perception is an unconscious process of probabilistic decision making. Perceptions are conceptualized as beliefs in hypotheses about states of reality, based on sensory evidence modifying subjective probabilities from previous experience, i.e. as evidence-based revision of subjective probability. Ultimately, these considerations lead to a Bayesian theory of cortical processing according to which the Bayesian brain represents information probabilistically, by coding and computing with probabilities (Friston, 2005; Knill and Pouget, 2004; Rao, 2005). A major challenge for cognitive neuroscientists is to test these ideas experimentally.

The neural mechanisms of perceptual decision making are only beginning to be understood although decision making plays a pivotal role in translating perception into action (Gold and Shadlen, 2001; Platt, 2002; Schall, 2001; Smith and Ratcliff, 2004). In perceptual decision making studies, single-unit recordings were obtained from monkeys that were trained to discriminate noisy visual stimuli and to report their perceptual decisions using eye movements. Neural substrates of perceptual decisions have been found in the frontal eye fields (FEF), the lateral intraparietal area (LIP), and the dorsolateral prefrontal cortex (DLPFC). The decision-related signals are usually considered to implement the decision transformation in these tasks, converting sensory representations into decision variables that evolve over time and that are capable of guiding behavior. The activity in these cells corresponds well with that of the behavioral eye movement decisions (Smith and Ratcliff, 2004). Thus, the temporal dynamics of the firing rates of neurons in the parietal and prefrontal cortex and of behavioral decisions have been described in parallel and they evidently converge.

In a general sense, neurons in sensorimotor structures accumulate evidence that leads to alternative categorizations (stimulus encoding), while other groups of neurons prepare and initiate overt responses (response preparation). Sequential sampling models (SSMs) are quantitative decision models that build on this fundamental distinction. According to SSMs, decisions are made by mechanisms that accumulate information to a response criterion (see Smith and Ratcliff, 2004, for a taxonomy of SSMs). SSMs have been successful in explaining behavioral data obtained from speeded decisions in perceptual and memory tasks (Ratcliff and Smith, 2004). As an example of a particularly economical SSM, Figure 1 illustrates the *LATER* model (*L*inear *A*pproach to *T*hreshold with *E*rgodic *R*ate; Carpenter and Williams, 1995; Sinha et al., 2006). According to the model, a decision signal S rises linearly at a rate r from a starting level S_0 until it reaches a threshold level S_T when a response is initiated. On different trials, r varies randomly in a Gaussian manner, resulting in a skewed distribution of response latencies.

Manipulations that lead to increases of the rate of rise r shorten response times, whereas any decrease of r prolongs response times. This empirically derived model is strengthened by its theoretical interpretation as a Bayesian decision maker. The decision signal S is identified with the perceived probability of the existence of a situation demanding a response. S_0 then corresponds to prior probability, r with the rate of arrival of sensory information, and S_T with a criterion posterior probability (Carpenter and Williams, 1995; Sinha et al., 2006).

The rest of this article introduces a Bayesian theory of the P300. Two hypotheses constitute the core of this theory. Firstly, the Bayesian theory of the P300 posits that

$$\text{P300 amplitude} = f\{ P(h_j/d) - P(h_j) \} \qquad (3).$$

In words, P300 amplitude varies as a function of the difference between posterior probability and prior probability, i.e. as a function of the magnitude of evidence-based probability revision. Several basic empirical facts support the validity of the amplitude hypothesis of the Bayesian theory of the P300. To begin with, rare stimuli in experimental

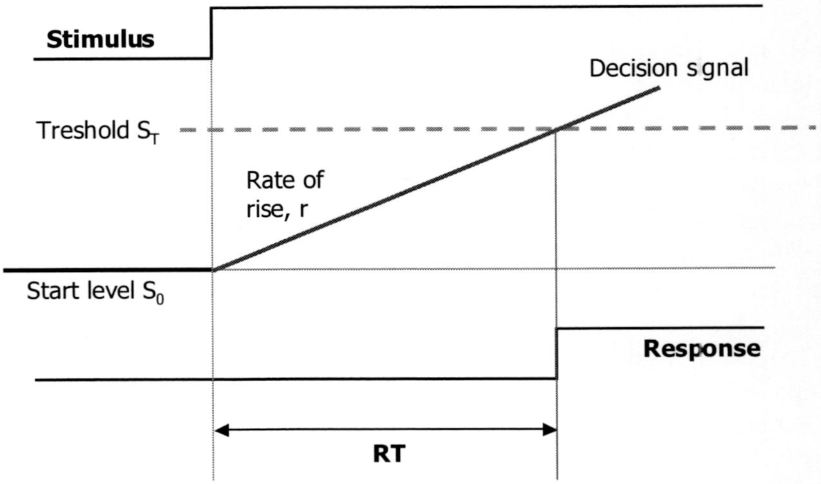

Figure 1. Illustration of the LATER model. A decision signal, initially S_0, starts to rise in response to a target stimulus of an experimental oddball paradigm at a constant rate *r* until it crosses a fixed threshold level, S_T. Response time (RT) is a function of the time at which the decision signal crosses S_T.

oddball paradigms elicit the P300 if and only if the processing of these stimuli involves task-relevant decision making (Picton et al., 1995). Most importantly, the probability effect on the P300 amplitude that was reported by Duncan-Johnson and Donchin (1977) and in an immense number of other oddball studies is also compatible with the amplitude hypothesis of the Bayesian theory of the P300. This holds true if one assumes (1) that subjective prior probability matches the relative frequency of stimuli and (2) that posterior probability is constant, i.e. that it does not vary as a function of the probability of stimuli. Figure 2 illustrates both assumptions. Inspection of Figure 2 reveals that at a given rate of rise *r*, P300 amplitude is inversely proportional to the probability of the eliciting stimulus (the probability effect). In contrast, even powerful manipulations of the rate of rise *r* should not affect P300 amplitude.

Secondly, the Bayesian theory of the P300 posits that

$$\text{P300 latency} = f\{\, t(S_T) - t(S_0) \,\} \tag{4}$$

(t = time). In words, P300 latency varies as a function of the time interval between the starting level S_0 and the threshold level S_T. Inspection of Figure 2 reveals that at a given rate of rise *r*, P300 latency, akin to the P300 amplitude, is inversely proportional to the probability of the eliciting stimulus. In sharp contrast to what was predicted for the P300 amplitude, however, manipulations of the rate of rise *r* should strongly affect P300 latency.

It is important to notice that this Bayesian theory of the P300 is not primarily a theory about the P300. As outlined above, it is well recognized that P300 amplitude is controlled, in a multiplicative fashion, directly by the task relevance of the eliciting stimulus and inversely by the relative frequency of the stimulus (Johnson, 1984, 1986). Yet, the Bayesian theory of the P300 does not capture the basic interaction of the two factors that largely determine P300 amplitude (Picton et al., 1995) nor does it account for the target effect on P300 amplitude. Instead of that, the Bayesian theory of the P300 is an attempt to model the probability effect on P300 amplitude and latency, other things being equal. The theory adds some specific assumptions – that were derived from Bayes' theorem (Bayes, 1763) and from the LATER model (Carpenter and Williams, 1995; Sinha et al., 2006) – to the directly observable fact of the P300 probability effect. Its formulation thus reflects a cognitive theory about the kind of information processing performed by the human brain (Rösler, 1983). In particular, the Bayesian theory of the P300 posits that the human brain codes and computes with probabilities, in a Bayesian manner (Friston, 2005; Knill and Pouget, 2004; Rao, 2005). According to the theory, adequate research on the P300 offers a window on the Bayesian brain: Does the human cortex actually act like a Bayesian decision maker? The answer to this question is of great theoretical importance since it would help to shape valid theories about perception and cortical processing.

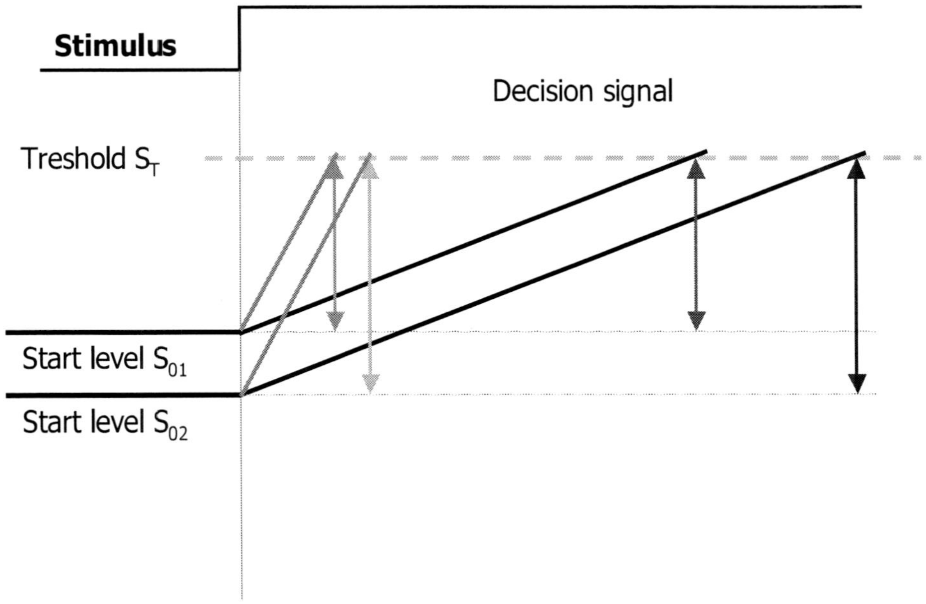

Figure 2. Illustration of the Bayesian theory of the P300. Target stimuli of an experimental oddball paradigm are presented at two relative frequencies, so that $S_{01} > S_{02}$. The probability effect on the amplitude of the P300 is identifiable as the length difference between the blue (turquoise) and the red (orange) arrows. Manipulations of the rate of rise r do not affect P300 amplitude, but they affect P300 latency.

The major challenge for cognitive neuroscientists is to design the adequate experiment to test these ideas. Here, the fundamental problem is to comply with the ceteris paribus constraint. For example, manipulations of target probability in usual experimental oddball paradigms also change the absolute time between targets, i.e. the target-to-target time interval. Gonsalvez and colleagues have argued that target-to-target interval may better explain target-P300 amplitude modulation than target probability (e.g., Gonsalvez and Polich, 2002). Thus, the experiment designed to test the Bayesian theory of the P300 must dissolve the confound between target probability and target-to-target interval. These methodological subtleties notwithstanding, future P300 research will likely provide a window on the nature of the neural basis of decision making in the human brain.

Acknowledgments

The research reported in this article was supported by the ZNS – Hannelore Kohl Stiftung.

References

Bayes T. An essay towards solving a problem in the doctrine of chance. *Philos. Trans R. Soc* 53: 370-418, 1763.

Carpenter RHS, Williams MLL. Neural computation of log likelihood in the control of saccadic eye movements. *Nature* 377: 59-62, 1995.

Chater N, Tenenbaum JB, Yuille A. Probabilistic models of cognition: conceptual foundations. *Trends Cogn. Sci.* 10: 287-291, 2006.

Coles MGH, Smid HGOM, Scheffers MK, Otten LJ. Mental chronometry and the study of human information processing. In MD Rugg, MGH Coles (Eds.), *Electrophysiology of mind and brain* (pp. 86-131). Oxford, Oxford University Press. 1995.

Courchesne E, Hillyard SA, Galambos R. Stimulus novelty, task relevance and the visual evoked potential in man. *Electroencephalogr. Clin. Neurophysiol.* 39: 131-143, 1975.

Debener S, Makeig S, Delorme A, Engel AK. What is novel in the novelty oddball paradigm? Functional significance of the novelty P3 event-related potential as revealed by independent component analysis. *Brain Res. Cogn. Brain Res.* 22: 309–321, 2005.

Desmedt JE. P300 in serial tasks: an essential post-decision closure mechanism. *Prog. Brain Res.* 54: 682-686,1980.

De Swart JH, Kok A, Das-Smaal EA. P300 and uncertainty reduction in a concept-identification task. *Psychophysiology* 18: 619–629, 1981.

Donchin E. Surprise!...Surprise? *Psychophysiology* 18: 493–513, 1981.

Donchin E, Coles MGH. Is the P300 component a manifestation of context updating? *Behav. Brain Sci.* 11: 357-374,1988.

Duncan-Johnson CC, Donchin E. On quantifying surprise: the variation of eventrelated potentials with subjective probability. *Psychophysiology* 16: 53–55, 1977.

Duncan-Johnson CC, Donchin E. The P300 component of the event-related brain potential as an index of information processing. *Biol. Psychol.* 14: 1-52, 1982.

Escera C, Alho K, Winkler I, Näätänen R. Neural mechanisms of involuntary attention to acoustic novelty and change. *J. Cogn. Neurosci.* 10: 590-604, 1998.

Friedman D, Cycowicz YM, Gaeta H. The novelty P3: an event-related brain potential (ERP) sign of the brain's evaluation of novelty. *Neurosci. Biobehav. Rev.* 25: 355-373, 2001.

Friston K. A theory of cortical responses. *Philos. Trans. R. Soc. Lond. B. Biol. Sci.* 360: 815-836, 2005.

Geisler WS, Kersten D. Illusions, perception and Bayes. *Nat. Neurosci.* 5: 508-510, 2002.

Gigerenzer G, Murray DJ. *Cognition as intuitive statistics.* Hillsdale, NJ, Erlbaum. 1987.

Gold JI, Shadlen MN. Neural computations that underlie decisions about sensory stimuli. *Trends Cogn. Sci.* 5: 10-16, 2001.

Gonsalvez CL, Polich, J. P300 amplitude is determined by target-to-target interval. *Psychophysiology* 39: 388-396, 2002.

Johnson R Jr. P300: a model of the variables controlling its amplitude. *Ann. NY Acad. Sci.* 425: 223-229, 1984.

Johnson R Jr. A triarchic model of P300 amplitude. *Psychophysiology* 23: 367- 384, 1986.

Kahneman D, Slovic P, Tversky A. *Judgment under uncertainty: heuristics and biases*. New York, Cambridge University Press. 1982.

Kahneman D, Tversky A. *Choices, values, and frames*. New York, Cambridge University Press, 2000.

Kersten D, Mamassian P, Yuille A. Object perception as Bayesian inference. *Annu. Rev. Psychol.* 55: 271–304, 2004.

Kersten D, Yuille A. Bayesian models of object perception. *Curr. Opin. Neurobiol.* 13: 150-158, 2003.

Knill DC, Pouget A. The Bayesian brain: the role of uncertainty in neural coding and computation. *Trends Neurosci.* 27: 712-719, 2004.

Körding KP, Wolpert DM. Bayesian decision theory in sensorimotor control. *Trends Cogn. Sci.* 10: 319-326, 2006.

Kopp B, Tabeling S, Moschner C, Wessel K. Fractionating the neural mechanisms of cognitive control. *J. Cogn. Neurosci.* 18: 949-965, 2006.

Kutas M, McCarthy G, Donchin E. Augmenting mental chronometry: the P300 as a measure of stimulus evaluation time. *Science* 197: 792-795, 1977.

Leuthold H, Sommer W. Postperceptual effects and P300 latency. *Psychophysiology* 35: 34-46, 1998.

Linden DE. The P300: where in the brain is it produced and what does it tell us? *Neuroscientist* 11: 563-576, 2005.

Luck SJ. *An Introduction To The Event-Related Potential Technique*. Cambridge, MA, MIT Press. 2005.

McCarthy G, Donchin E. A metric for thought: a comparison of P300 latency and reaction time. *Science* 211: 77-80, 1981.

Nieuwenhuis S, Aston-Jones G, Cohen JD. Decision making, the P3, and the locus coeruleus-norepinephrine system. *Psychol. Bull.* 131: 510-532, 2005.

Phillips LD, Edwards W. Conservatism in a simple probability inference task. *J Exp. Psychol.* 72: 346-354, 1966.

Picton TW. The P300 wave of the human event-related potential. *J. Clin. Neurophysiol.* 9: 456-79,1992.

Picton TW, Lins OG, Scherg M. The recording and analysis of event-related potentials. In F Boller, J Grafman (Eds.), *Handbook of neuropsychology*, Vol. 10 (pp. 3-73). Amsterdam, Elsevier.1995.

Platt ML. Neural correlates of decisions. *Curr. Opin. Neurobiol.* 12: 141-148, 2002.

Polich J, Kok A. Cognitive and biological determinants of P300: an integrative review. *Biol. Psychol.* 41: 103-46, 1995.

Rao RP. Bayesian inference and attentional modulation in the visual cortex. *Neuroreport* 16: 1843-1848, 2005.

Ratcliff R, Smith PL. A comparison of sequential sampling models for two-choice reaction time. *Psychol. Rev.* 111: 333-367, 2004.

Ritter W, Vaughan HG Jr. Averaged evoked responses in vigilance and discrimination: a reassessment. *Science* 164: 326-328, 1969.

Rösler F. Endogenous ERPs and cognition: probes, prospects, and pitfalls in matching pieces of the mind-body puzzle. In AWK Gaillard, W Ritter (Eds.), *Tutorials in ERP-research: endogenous components* (pp. 9-36). Amsterdam, North-Holland. 1983.

Schall JD. Neural basis of deciding, choosing and acting. *Nat Rev Neurosci* 2: 33- 42, 2001.

Sinha N, Brown JTG, Carpenter RHS. Task switching as a two-stage decision process. *J. Neurophysiol.* 95: 3146-3153, 2006.

Smith PL, Ratcliff R. Psychology and neurobiology of simple decisions. *Trends Neurosci.* 27: 161-168, 2004.

Soltani M, Knight RT. Neural origins of the P300. *Crit. Rev. Neurobiol.* 14: 199-24, 2000.

Spencer KM, Dien J, Donchin E. Spatiotemporal analysis of the late ERP responses to deviant stimuli. *Psychophysiology* 38: 343-358, 2001.

Squires NK, Squires KC, Hillyard SA. Two varieties of long-latency positive waves evoked by unpredictable auditory stimuli in man. *Electroencephalogr. Clin. Neurophysiol.* 38: 387-401, 1975.

Squires KC, Wickens C, Squires NK, Donchin E. The effect of stimulus sequence on the waveform of the cortical event-related potential. *Science* 193: 1142- 1146, 1976.

Stocker AA, Simoncelli EP. Noise characteristics and prior expectations in human visual speed perception. *Nat. Neurosci.* 9: 578-585, 2006.

Sutton S, Braren M, Zubin J, John ER. Evoked-potential correlates of stimulus uncertainty. *Science* 150: 1187-1188,1965.

Tenenbaum J, Griffiths TL, Kemp C. Theory-based Bayesian models of inductive learning and reasoning. *Trends Cogn. Sci.* 10: 309-318, 2006.

Verleger R. Event-related potentials and cognition: A critique of the context updating hypothesis and an alternative interpretation of P3. *Behav. Brain Sci.* 11: 343-356, 1988.

von Helmholtz H. *Handbuch der Physiologischen Optik. Hamburg*, Voss. (1867).

In: Cognitive Science Compendium, Volume 1
Editor: Miao-Kun Sun
ISBN: 978-1-60456-597-3

Chapter 16

MEDIAL PREFRONTAL CORTEX AND PAVLOVIAN CONDITIONING

D. A. Powell*

Shirley L. Buchanan Neuroscience Laboratory
Dorn VA Medical Center, Columbia, SC 29209
Department of Psychology, University of South Carolina
Columbia, SC 29208
Department of Neuropsychiatry and Behavioral Science
University of South Carolina School of Medicine
Columbia, SC 29208

ABSTRACT

The prefrontal cortex plays a central role in complex behaviors that require working memory, long term planning, appropriate response selection, etc. It has been known for some time that processing of emotional memories involves the medial prefrontal cortex (mPFC). Such emotional memories can be studied in animal models by using Pavlovian (classical) conditioning of autonomic adjustments. In the present review, papers reporting use of these techniques are first examined. It has now become apparent that, not only do learned autonomic responses involve mPFC control, but somatomotor Pavlovian conditioning also involves the mPFC. However, the conditions under which mPFC exerts control of somatomotor Pavlovian conditioning involve circumstances in which the conditions for acquisition of the response are not optimal, suggesting that prefrontal mechanisms are accessed only when higher level processing is required. Studies evaluating somatomotor conditioning, as a function of prefrontal involvement, are also reviewed, and a strategy for further understanding prefrontal control of these processes is outlined.

* Correspondence and requests for reprints: D.A. Powell, Ph.D., Shirley L. Buchanan Neuroscience Laboratory (151A), Wm. Jennings Bryan Dorn VA Medical Center, 6439 Garners Ferry Road, Columbia, SC 29209-1639, Phone: (803) 695-6821, FAX: (803) 695-7942, EMAIL donnie.powell@med.va.gov

ABBREVIATIONS

Aβ	amyloid-beta peptide;
ABAD	Abeta-binding alcohol dehydrogenase;
AD	Alzheimer's disease;
AIF	apoptosis-inducing factor;
APP	amyloid precursor protein;
Ask1	Apoptosis-signal-regulating-kinase-1;
ATF4	activating transcription factor 4;
ATF6	activating transcription factor 6;
CICR	Ca^{2+}-induced Ca^{2+} release;
eIF2α	eukaryotic initiation factor-2 alpha;
ER	endoplasmic reticulum;
ERAD	ER-associated protein degradation;
GRP78	glucose-regulated protein 78;
InsP3R	inositol 1,4,5-triphosphate receptor;
IRE1	inositol requiring transmembrane kinase/endonuclease;
JNK	c-Jun N-terminal kinase;
PERK	RNA-dependent protein Kinase (PKR)-like ER membrane-localized kinase;
PS1	presenilin-1;
PS2	presenilin-2;
ROS	reactive oxygen species;
RyR	ryanodine receptor;
SERCA	sarco-endoplasmic reticulum ATPase;
TRAF2	TNF-associated factor 2;
UPR	Unfolded Protein Response.

1. INTRODUCTION

The prefrontal cortex (PFC) plays a central role in complex behaviors that require working memory, long-term planning, appropriate response selection, etc. (Damasio, 1994; Fuster, 1997). Consequently, it is also involved in wide range of neuropsychiatric disorders ranging from the degenerative diseases of old age (e.g., frontotemporal dementias, Alzheimer's Disease, Parkinson's Disease, etc.) to schizophrenia and posttraumatic stress disorders (see Goldman-Rakic and Selemon, 1997; Rauch and Shin, 1999). Prefrontal function has been studied in a variety of ways, using computational models (O'Reilly, 2006), animal models (Fuster, 1997; Powell, McLaughlin and Chachich, 2000), clinical material (Damasio, 1994), and more recently various brain scanning techniques (e.g., Buckner and Petersen, 1996). The primary objective of the present paper is to elucidate the role that medial prefrontal cortex (mPFC) plays in Pavlovian (classical) conditioning of autonomic and somatomotor responses. These Pavlovian conditioning models will first be briefly described. The next section will review research on the role of mPFC in conditioning of non-specific autonomic and somatomotor responses. A third section will review data implicating the medial prefrontal cortex in Pavlovian conditioning of specific somatomotor responses. A fourth section examines the role of the mPFC in extinction, and a final section summarizes the relevance of this basic research and its application to a specific model of mPFC function.

2. EXPERIMENTAL ANALYSIS OF PAVOLVIAN CONDITIONING

2.1. Specific and Non-Specific Responses

During Pavlovian, or classical conditioning, a signal, or conditioned stimulus (CS) consistently precedes an unconditioned stimulus (US). The US elicits an unconditioned reflex (UR) but as a result of pairings of the CS and US a conditioned response (CR) eventually occurs during the signal (CS), that in many cases resembles the original UR. The CR is thus a new or learned response, since it did not originally occur to the CS and is a response that is elicited by the CS/US contingency. This procedure is schematically illustrated in Figure 1.

In the rabbit eyeblink (EB) and nictitating membrane (NM) conditioning model, developed by Gormezano (1966), eyelid closure and nictitating membrane extension occurs as CRs to neutral conditioned stimuli (CSs), such as a tones, which systematically precede an airpuff or a brief periorbital electric shock train that serves as the US (see Figure 1). Note that the occurrence of the learned EB or NM response during the CS has no effect on the later occurrence of corneal airpuff or eyeshock reinforcer, as would be the case in operant conditioning. In the latter paradigm, occurrence of the CR would either prevent US occurrence or attenuate its magnitude (Gormezano and Coleman, 1973). This means that acquisition of the CR depends solely on the association between the CS and US, and not on any environmental effects CR occurrence may have. This type of learning is thus often referred to as "associative learning".

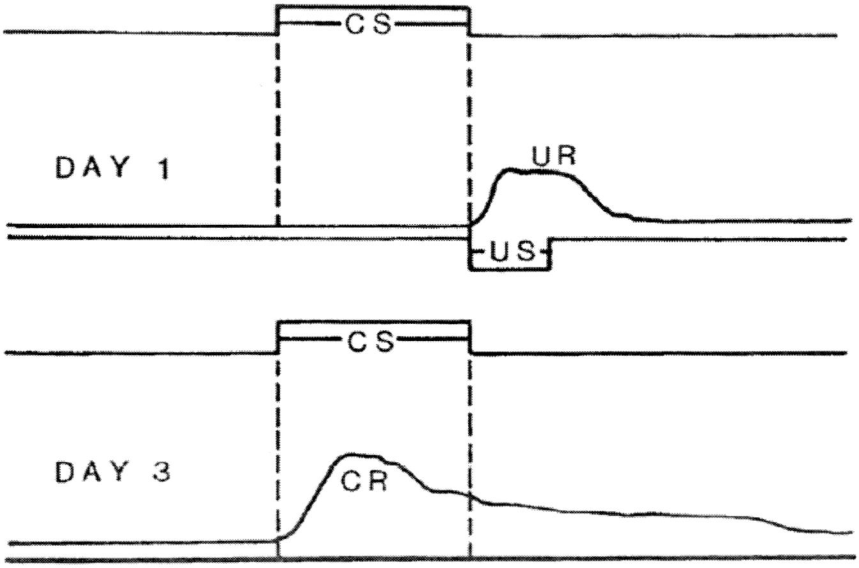

Figure 1. Polygraph tracing showing the topography of the unconditioned and conditioned eyeblink response as assessed during the first session of training (UR) and during the third session of training (CR). A .5 sec tone was the CS and the US was a 100 msec corneal airpuff.

As has been noted by others (e.g., Schneiderman, 1972), when animals are exposed to classical conditioning contingencies that elicit a specific somatomotor response (e.g., EB or NM responses) a number of non-specific responses not usually assessed are concomitantly elicited. Although non-specific conditioned responses may also consist of somatomotor behaviors (Dykman et al., 1965) learned autonomic or visceral changes have been most often studied, including heart rate (HR), systemic blood pressure, skin conductance changes, etc. (see Obrist, 1981; Cohen and Randall, 1984; Smith and Devito, 1984; Weinberger and Diamond, 1987). It should be emphasized that the parametric features of Pavlovian conditioned somatomotor responses, which have been focus of most research on classical conditioning, and the concomitantly occurring non-specific visceral responses are very different (Powell et al., 2000). For example, the temporal parameters that are optimal for somatomotor and visceral conditioning may be quite dissimilar (Powell, Lipkin and Milligan, 1974; Schneiderman, 1972). Autonomic CRs are typically greatest in magnitude when the interval between the onset of the CS and US or interstimulus interval (ISI) is fairly long, e.g., from 4-8 secs (Powell and Kazis, 1976). However, little or no somatomotor conditioning occurs at these relatively long CS/US intervals. Rather the CS/US interval that produces the most rapid EB or NM conditioning is .5 sec or less (Schneiderman and Gormezano, 1964; Gormezano, 1972). A second striking difference between learned somatomotor and visceral responses is that the latter are acquired within just a few trials, usually ten trials or less, whereas many trials, up to 100, may be required before the first somatomotor CR occurs (see Lennartz and Weinberger, 1992, 1994; Kehoe and Mccrae, 1994 for a discussion of this issue when extremely long intersession intervals are employed).

Learned autonomic changes have been referred to as nonspecific responses, since they occur regardless of the nature of the conditioning contingencies (e.g., Powell, Buchanan and Gibbs, 1990; Prokasy, 1984; Weinberger and Diamond, 1987). Thus, signaled electric shock USs, delivered to either the orbital region or the foot pad of animals, results in a host of such non-specific CRs to the signal that are quite similar regardless of the site of application of the US. The learned somatomotor response in each case, however, is specific to the US, consisting of an EB CR in the former case and leg flexion CR in the latter case. Consequently, such responses are usually referred to as specific conditioned responses.

Figure 2 illustrates the dramatic differences observed in these two response systems in rabbits exposed to classical conditioning contingencies. It depicts the results of an experiment in which a 1216 Hz, 1 sec, 75 db tone was employed as the CS and a 250 msec, 3 mA, AC periorbital electric shock train was the US. The top panel of Figure 2 shows percent EB CRs over ten acquisition sessions consisting of 60 CS/US presentations each (conditioning group) and a single extinction session. A nonassociative control group received a random sequence of unpaired CS/US presentations (explicitly unpaired CS/US group). EB CRs increased from a low near-zero rate during session 1 to greater than 90% by session 10 in the conditioning group; the explicitly unpaired group showed virtually no EB responses.

The bottom panel of Figure 2 shows mean change in duration from pre-CS baseline of the third interbeat interval of the electrocardiogram following tone onset, as a function of acquisition sessions in the same two groups of animals. The third interbeat interval was chosen because it was the last one that was available for analysis in all animals before CS offset and the occurrence of the US. This interbeat interval usually represents the largest change from pre-CS baseline (Powell and Levine-Bryce, 1988). The duration of this interbeat interval, referred to as heart period, is the reciprocal of HR; as can be seen, it was lengthened by 5-10 msec as a result of training. Thus, HR CRs consisted of decelerations from pre-CS baseline. However, the heart period changes evoked by the tone in the unpaired group were much smaller and more variable, suggesting that the tone-evoked HR decelerations in the paired groups were associative. Note, however, that no acquisition function is apparent in the conditioning group, as was the case for the EB data above. The reason for this can be seen in Figure 3, which shows mean change in the third interbeat interval over trials during the initial session. In Figure 3, the acquisition function for heart period is clearly apparent. Change in heart period decreased in both groups across the first 3 to 5 trials, representing habituation of the decelerative cardiac component of the orienting reflex. The orienting reflex is the initial response to novel stimulation and normally also consists of bradycardia (Powell and Kazis, 1976). After habituation of the orienting reflex, the conditioning group demonstrated a second bradycardiac response of greater than 10-20 msec, whereas the unpaired group continued to show small and variable responses of 1-5 msec. Thus, the decelerative HR CR appeared by trial 5 and had reached its maximum magnitude by trial 10, well before EB CRs began to occur in any animal.

Plotting the HR change associated with the largest post-CS interbeat interval may, however, mask important temporal changes in the pattern of the HR CR. Thus, when beat x beat changes in CS-evoked HR change are plotted as a function of CS duration, a characteristic pattern of changes is observed. This pattern is shown in Figure 4, which shows the HR change from pre-CS baseline for 2 consecutive daily acquisition sessions, averaged over trials, as a function of post-CS interbeat intervals. The data are from eight rabbits that received differential HR conditioning using a more optimal 4 sec ISI (see Buchanan and Powell, 1989). There is an initial abrupt, but relatively small HR deceleration during the first, second, and sometimes third interbeat interval in response to both CS+ (reinforced) and CS- (not reinforced) presentations.

This HR deceleration becomes larger during subsequent interbeat intervals, increasing almost monotonically and reaching a maximum at tone termination when the US would be expected to occur. This regular pattern of the autonomic CR also occurs during human HR and skin conductance conditioning (Ohman, 1988) and has been recognized for some time. The initial small and short-duration component has been referred to as the "registration" or "orienting" component of the CR and the later-occurring and larger magnitude response has been referred to as the "contingency" component (Maxwell, Powell and Buchanan, 1994).

This characteristic shape of the HR CR not only has functional implications for the role of conditioned bradycardia with regard to behavior in general but also may be important for understanding the central nervous system (CNS) control of learned autonomic adjustments. For example, separate groups of neurons in the prefrontal cortex show differential CS-evoked changes during the contingency and registration components of the heart rate CR (Maxwell et al., 1994). A similar classification of neurons was also discovered in the mediodorsal nucleus of the thalamus, which is the thalamic projection nucleus to the prefrontal cortex (Chachich, Buchanan and Powell, 1997). These conditioned bradycardiac changes have also been shown to occur in a wide variety of Pavlovian conditioning situations, including the use of both visual and tactile CSs and in which either an air puff or sweetened water was used as the US (Powell, Gibbs, Maxwell and Levine-Bryce, 1993). The airpuff also elicits EB CRs, but the water is an appetitive CS which evokes jaw movements (JM) as skeletal CRs (see below).

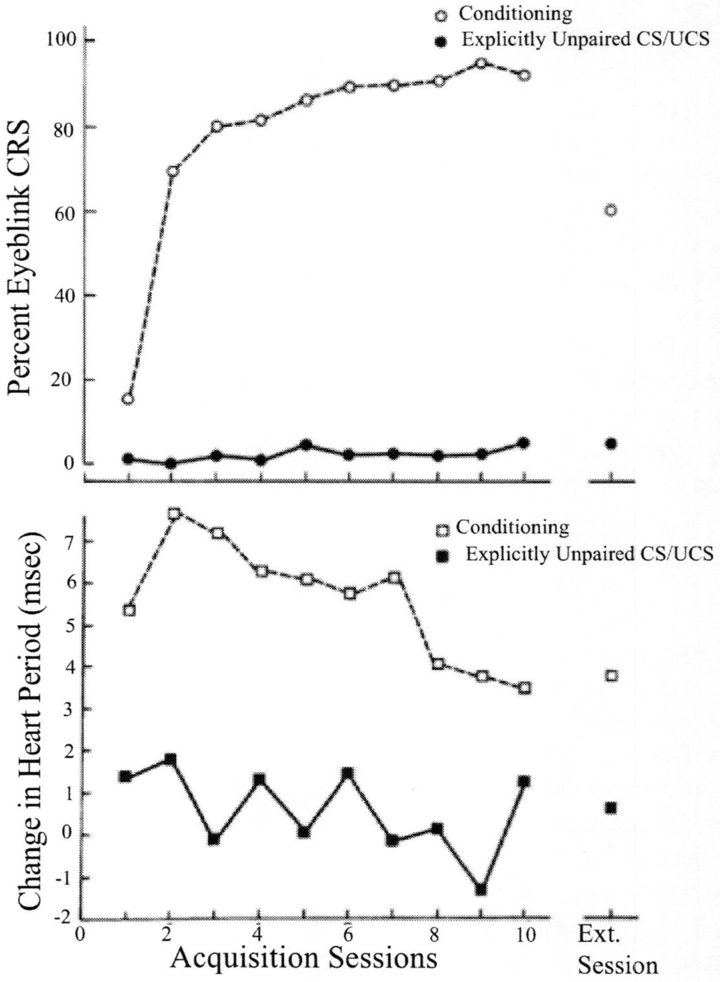

Figure 2. Top: Conditioned eyeblink responses of rabbits that received CS/US presentations (conditioning group) and rabbits that received explicitly unpaired CS/US presentations (explicitly unpaired group) as a function of ten 60-trial conditioning sessions. The CS was a 1 sec, 75 db, 1216 Hz tone that was followed by 250 msec, 3 mA, periorbital shock train. Bottom: Mean change in heart period from pre-CS baseline of the third post-CS interbeat interval of the same rabbits that received conditioning or explicitly unpaired CS/US presentations. The third interbeat interval is shown because it was the last one available for analysis in all animals prior to CS termination and is usually associated with the largest heart period change from pre-CS baseline. (From Powell et al., 1989).

Leg flexion conditioning in the rabbit is also accompanied by bradycardia (Powell and Lipkin, 1975). Learned HR decelerations thus appear to be a generalized phenomena of Pavlovian conditioning, regardless of the type of somatomotor CR. There is also some evidence that interference with the occurrence of this decelerative HR CR may adversely affect acquisition of specific EB CRs (Albiniak and Powell, 1980; Joseph and Powell, 1980; Kazis, Milligan and Powell, 1973; Powell, 1979).

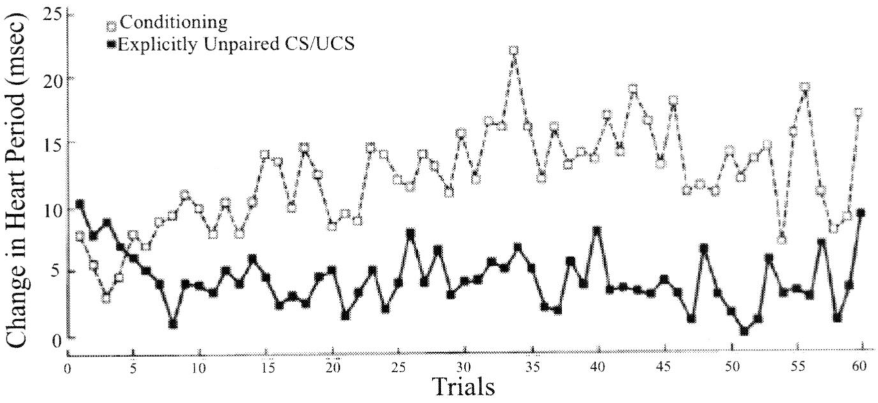

Figure 3. Change from pre-CS baseline in heart period of the third interbeat interval of rabbits that received Pavlovian eyeblink conditioning as indicated in Figure 2. The data are shown for conditioning and explicitly unpaired CS/US groups as a function of blocks of trials during the first session of training. (From Powell et al., 1989).

2.2. Theoretical Implications

The above-described results suggest that if Pavlovian HR and EB conditioning are both valid animal models of associative learning, they must represent different aspects of learning, since the parametric features of the stimuli required to elicit them and their rate of acquisition are different. Single process models of learning and memory thus may be inadequate to explain even simple associative learning phenomena, such as classical conditioning. This point has also been made by others (e.g., Lennartz and Weinberger, 1992). Indeed, it is well accepted that different aspects of learning and memory are related to underlying demands of the task being studied and the type of cognitive resources required (Kilstrom, 1987; Olton, 1979; Schacter, 1992; Squire, 1992; 1994; Tulving, 1985). These theoretical approaches to learning and memory have resulted in a typology of memory in which different kinds of learning and memory are used to accomplish different sets of goals. All of these models assume that different kinds of knowledge exist and therefore separate cognitive systems have evolved to deal with them, and moreover that different CNS mechanisms underlie the acquisition, storage and retrieval of these different kinds of knowledge (Squire, 1992).

Although memory systems have been characterized in a variety of ways by different theorists, considerable data support a model that includes at least two separate kinds of cognitive functioning. Squire (1992) has characterized these two kinds of processing as declarative and non-declarative. Declarative processing refers to information that is consciously retrieved, is dependent on intact hippocampal functioning, and, in many cases, is associated with a specific prior event (episode). This kind of learning thus refers to acquisition of facts regarding objects and relationships in one's environment as well as memory of the specific episodes during which these events occurred. By contrast, non-declarative memory involves acquisition of a) habits and skills, b) priming effects, c) learning of dispositions (including classical and operant conditioning), and d) non-associative effects such as habituation and sensitization (see Squire, 1992). Such learning may not necessarily occur at a conscious level and thus may not be associated with specific episodes in the past and, by implication, do not require hippocampal processing.

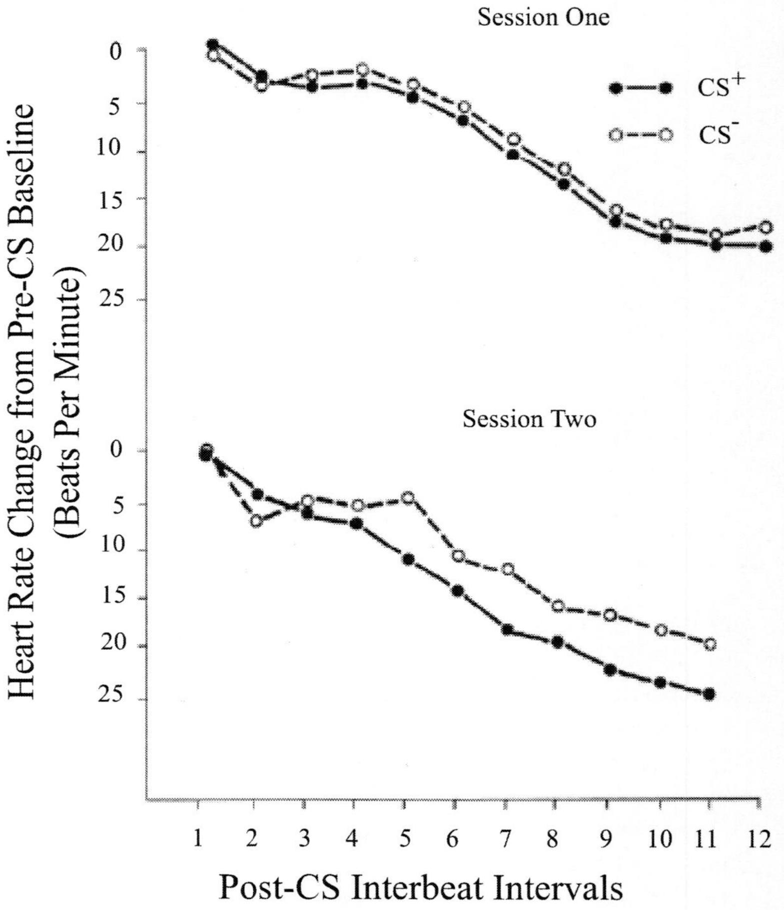

Figure 4. Heart rate change from pre-CS baseline in a group of 8 rabbits that received two sessions of differential conditioning in which either a 1216 Hz or 404 Hz, 75 db, 4.0 sec tone served as either CS+ (paired with US) or CS- (not paired with US). The data are shown in beats per minute, as a function of 12 interbeat intervals after the onset of the CS. (Adapted from Buchanan & Powell, 1989).

The separate acquisition functions and different stimulus parameters required for acquisition of the EB CR and other specific responses, such as the leg flexion or JM responses and non-specific visceral responses, such as the HR or skin conductance CRs, also tap different processes and therefore are dependent upon different CNS mechanisms. A six stage empirical model of somatomotor classical conditioning, depicted in Figure 5, reflects this idea.

It indicates a) the hypothetical constructs, b) behavioral correlates, and c) possible neuroanatomical structures associated with each stage. The six behavioral stages represented in this model are meant to describe the sequential events that take place when any mammalian organism is exposed to signals (i.e., CSs) that predict significant events (USs). However, the fact that the behavioral components of this model occur sequentially does not mean that the brain mechanisms or changes underlying these behaviors also occur in a sequential fashion. Indeed, it is highly likely that, depending on the type of paradigm being studied, parallel processing in different brain structures occurs simultaneously. Thus, the major point such a model is designed to make is that association-formation, as a result of exposure to classical conditioning contingencies, produces multiple memory traces involving both excitatory and inhibitory processes even at its simplest level. Such an analysis thus implies that multiple brain structures must be involved. It is now clear that this is the case. The brain structures listed in Figure 5, associated with these different stages of learning during classical conditioning, have varying degrees of experimental support, based on current evidence, as is discussed in more detail below.

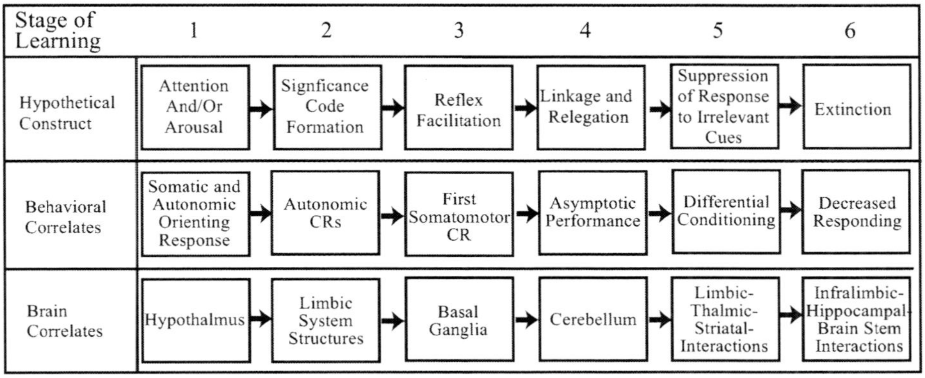

Figure 5. Diagram of a 6-stage model of classical conditioning, showing the hypothetical constructs, behavioral indices, and postulated neuroanatomical structures/mechanisms involved in each stage. The degree of experimental support for the latter varies depending upon the stage involved, but substantial evidence exists implicating the indicated structures in the behaviors listed in each case. Although the model suggests sequential processing in a step wise order, as noted in the text, some types of processing are almost certainly simultaneous and take place in a parallel distributed fashion.

2.3. Appetitive Classical Conditioning

The EB response system discussed thus far is a defensive reaction in response to an aversive US. Appetitive classical conditioning has also been studied in the rabbit using the JM response that occurs in response to presentation of water as the unconditioned stimulus. Appetitive Pavlovian conditioning training thus involves the repeated pairings of a tone (CS) and an intraoral injection of a 1 cc pulse of water (US) to rabbits that have limited access to water. A polygraph tracing illustrating the occurrence of EB and JM CRs and URs is illustrated in Figure 6.

The conditioned HR changes that accompany JM conditioning are illustrated in Figure 7. As this figure shows, the initial HR response during JM conditioning consists of bradycardia, but with continued training learned tachycardia begins to occur, as shown in Figure 8 which shows CS-elicited HR changes during JM conditioning over five consecutive daily sessions. As this figure shows, although CS-evoked HR slowing occurred during session 1, this response quickly began to become accelerative during sessions 2 and 3.

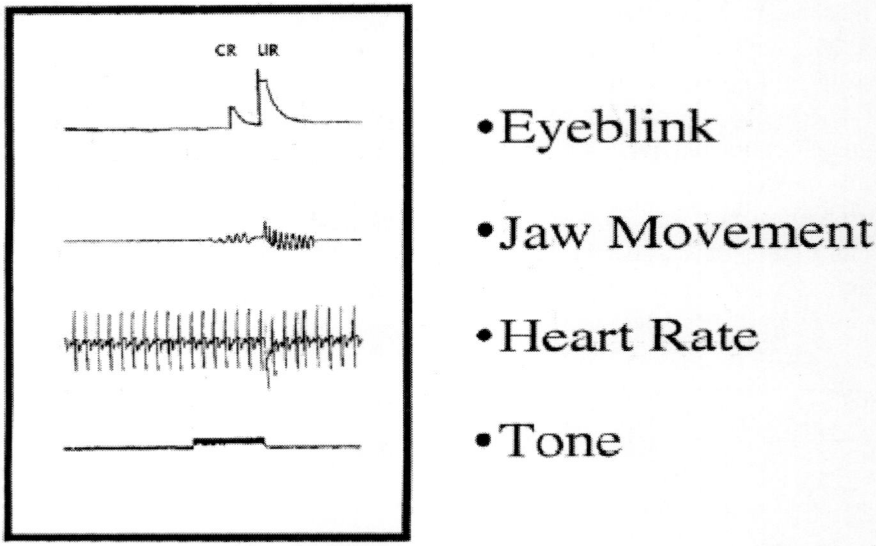

Figure 6. Polygraph tracing showing eyeblink and jaw movement conditioned responses (CR) and unconditioned responses (UR) to a tone conditioned stimulus. The accompanying ECG is illustrated for the JM response. Note the characteristic sinusoidal nature of the JM CR and UR as opposed to the discrete on/off character of the EB responses.

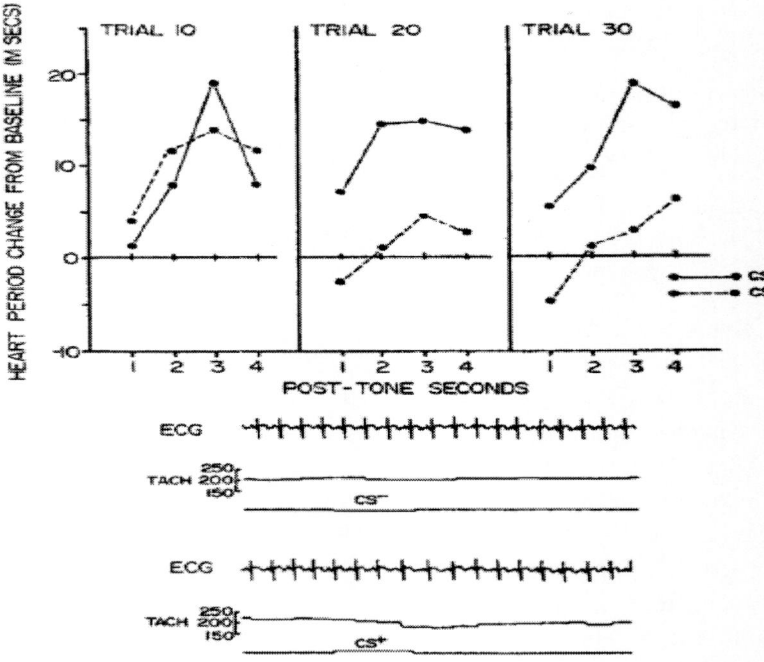

Figure 7. Top: Mean heart period (HP) change from pre-CS baseline during three nonreinforced test trials of rabbits that received 1-sec, 304- or 1216-Hz tones as CSs and a 1-ml pulse of water as a US. The US was systematically paired with the CS+ but never paired with the CS-. Data are shown as a function of four post-CS sec. Bottom: Polygraph tracing of typical cardiac response to appetitive CS+ and CS-. (ECG = electrocardiogram, Tach = tachometric equivalents of HR in beats per minute). (From Powell et al., 1993).

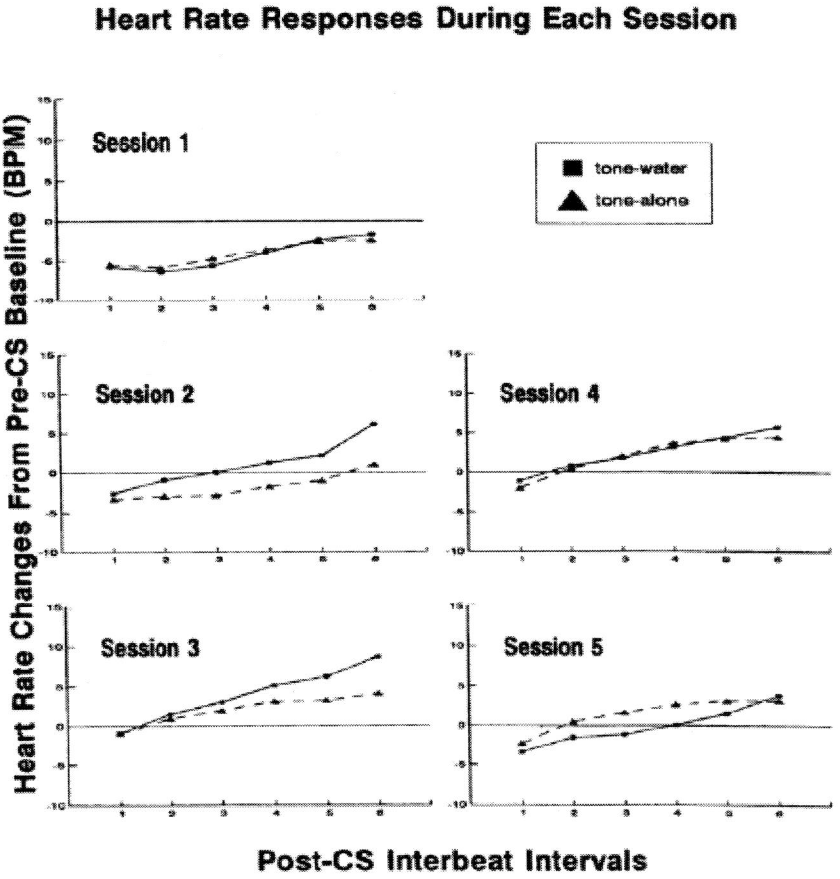

Figure 8. CS-evoked HR change in beats per minute (BPM) from pre-CS baseline, that occurred in response to a tone paired with water (tone-water) and a tone not paired with water (tone-alone), over five training sessions, as a function of six post-tone interbeat intervals (IBIs).

Of some interest in this regard is the question of why the cardiac changes accompanying appetitive conditioning consist of HR accelerations, while those accompanying defensive conditioning consist of cardiac decelerations. We know from earlier experiments that the latter response is primarily vagally controlled, although there is a small contribution produced by sympathetic inhibition (e.g., Kazis, Duncan and Powell, 1973). In order to determine the peripheral control of the learned tachycardia accompanying appetitive conditioning an experiment was conducted in which atenolol (a beta 1 receptor antagonist) and methylscopolamine (a peripheral vagal antagonist) were administered to rabbits undergoing JM conditioning (Powell, McLaughlin, Churchwell, Elgarico and Parker, 2002). The results of this study strongly suggest that the HR accelerations accompanying JM conditioning are produced by vagal inhibition. Thus, methylscopolamine abolished the conditioned tachycardia accompanying JM conditioning. However, atenolol, rather than attenuating this response, actually enhanced HR CR magnitude, which is also compatible with the suggestion that this response is totally controlled by vagal inhibition. These data are illustrated in Figure 9.

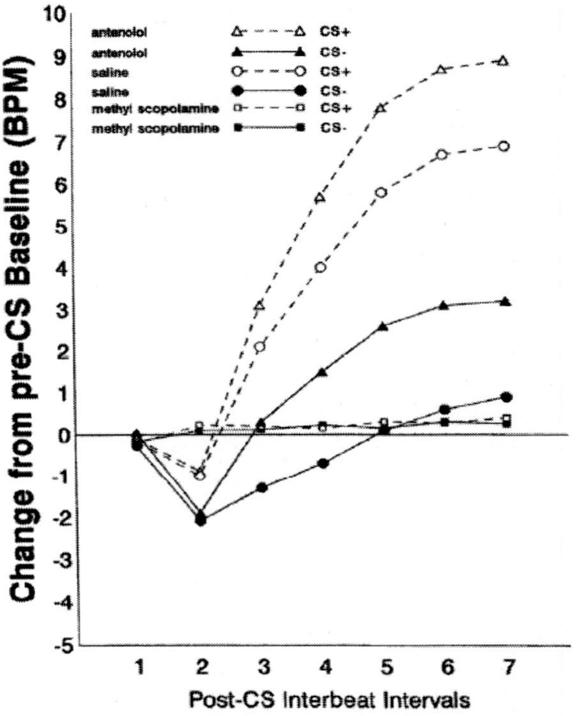

Figure 9. Mean heart rate change in beats per minute (BPM) from pre-CS (conditioned stimulus) baseline for groups of rabbits (n = 8 each) that received the peripheral autonomic antagonists as indicated. Data are presented as a function of 7 interbeat intervals that occurred during CS presentation. One acoustic stimulus served as CS+ and was always followed by water while another acoustic stimulus had no programmed consequences (CS-). (From Powell et al., 2002).

Interestingly, whereas vagal blockade with methylscopolamine decreased JM conditioning, atenolol actually facilitated it. These data thus suggest that although conditioned HR changes during appetitive and defensive conditioning are in opposite directions, they are both vagally controlled.

2.4. Fear Conditioning

A considerable amount of research has also been done, primarily in rodents, on what has come to be called "fear conditioning". In most of these experiments rats are exposed to a tone or light as a conditioned stimulus, which is paired with mild electric foot shock as a US. Under these conditions a variety of responses, which are usually referred to as "fear responses", are exhibited upon subsequent presentations of the CS. Typical fear indices have included a) freezing (or movement arrest, e.g. see LeDoux et al., 1986); b) enhancement of muscular reflexes (e.g., potentiated startle, e.g., see Davis, 1997); c) analgesia, or decreased pain sensitivity (e.g., see Fanselow, 1986); d) 22 Khz ultrasonic vocalizations (e.g., see Blanchard et al., 1991; Lee and Kim, 2004); and e) alterations in autonomic nervous system activities, as described in detail above. A critical review of Pavlovian fear conditioning has recently been published (Kim and Jung, 2006). Most of the evidence, as reviewed by Kim and

Jung (2006), suggests that fear conditioning is mediated by the amygdala, although the mPFC also plays a role, as is described below.

3. Medial Prefrontal Cortex and Pavlovian Conditioning of Non-Specific Responses

The studies of Lofving (1961) in the cat and Kaada (1951) in primates, dogs and cats were the first to suggest that an anterior midline cortical region mediates a cardioinhibitory mechanism. Similar mechanisms exist not only in anterior cingulate cortex of the rabbit, but throughout the entire midline prefrontal region (Buchanan and Powell, 1982aandb; Buchanan, Valentine and Powell, 1985; Powell, Watson and Maxwell, 1994). This area is shown on diagrams of the rabbit's brain in Figure 10. The mPFC includes Brodmann's areas 8, 24, 25 and 32 in the rabbit. These areas, along with the more lateral insular PFC (PFCi; see Figure 10a), constitute the portions of the PFC from which bradycardia and depressor responses can be elicited by electrical stimulation (Buchanan, Thompson, Maxwell and Powell, 1994; Buchanan and Thompson, 1990; Powell et al., 1994). Moreover, lesions of mPFC greatly attenuate the bradycardia associated with classical conditioning, but damage to more lateral isocortex and insular PFC (PFCi) does not (Buchanan and Powell, 1982aandb; Chachich and Powell, 1999; Powell et al., 1994). Unconditioned cardiac responding to neither the CS nor US is affected by mPFC lesions (Buchanan and Powell, 1982a). Neuronal activity in the mPFC is also highly correlated with conditioned bradycardia, whereas that of the more lateral insular PFC is not (Gibbs and Powell, 1988, 1991; Gibbs, Prescott and Powell, 1992; Powell, Watson and Maxwell, 1994; also see below). mPFC lesions also eliminate the initial deceleration component of the HR CR during JM conditioning (see above), but have no effect on either the JM CR itself, or on the later-occurring conditioned tachycardia that accompanies JM conditioning. Thus, the mPFC appears to mediate the associative characteristics of inhibitory cardiovascular adjustments, a hypothesis supported by findings showing that functional anatomical pathways between the mPFC and brain stem cardiac control centers have been demonstrated in several species, including the rabbit (Buchanan, Thompson, Maxwell and Powell, 1994).

It has also been shown that the amygdala central nucleus (ACN) is involved in cardioinhibitory changes (Kapp, Frysinger, Gallagher and Hasleton, 1979; Kapp, Gallagher, Underwood and McNall and Whitehorn, 1982; Schwaber, Kapp, Higgins and Rapp, 1982). A host of other kinds of data also implicate the amygdala in learned emotional behaviors (see Adolphs, 1999; Bard and Mountcastle, 1948). The mPFC and ACN thus appear to exert parallel control over inhibitory cardiovascular adjustments, although reciprocal connections between the baso-lateral amygdala (ABL) and the mPFC have also been demonstrated (Buchanan and Powell, 1980; McDonald, 1991). In addition, considerable evidence also suggests that the plasticity associated with classical conditioning of autonomic changes may originate in structures in the primary sensory systems (LeDoux, 2000). Thus, autonomic CRs evoked by acoustic CSs are mediated by a pathway from the medial division of the medial geniculate nucleus (MGN) to the lateral nucleus of the amygdala (ALN) (LeDoux, 2000). Since the ALN also receives somatosensory input (LeDoux, Romanski and Xagoraris, 1989), it may be the site of the initial plasticity underlying "emotional" learning. The ALN projects to both the ACN and ABL nuclei, which projects in turn to the mPFC; thus the ALN could very well provide the plasticity associated with both mPFC and ACN control of conditioned bradycardia.

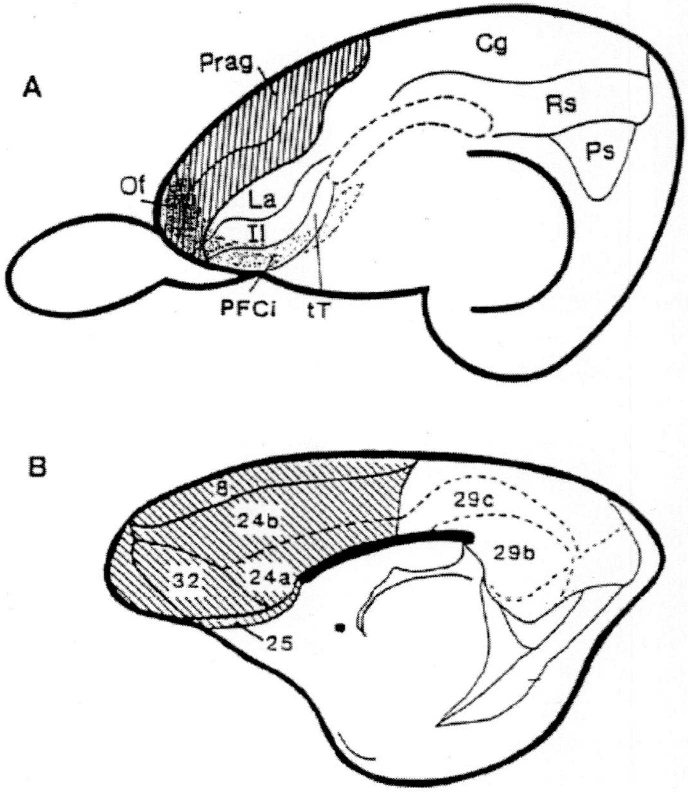

Figure 10. A: Schematic diagram of the rabbit brain, illustrating the structures in the prefrontal cortex from which cardioinhibition is elicited by electrical stimulation. The striped area indicates the portions of the midline PFC that have been demonstrated to be cardioactive, including all of the precentral agranula area (Prag) and the more dorsal portions of the anterior limbic area (La). Both of these relatively agranular (or dysgranular) prefrontal areas receive projections from the lateral mediodorsal nucleus of the thalamus (MDN), but this projection is much stronger to the more ventral anterior limbic area. The stippled area indicates a three dimensional representation of the lateral orbital and agranular insular areas (PFCi), which also receive projections from the thalamic mediodorsal nucleus, but these projections arise primarily from the medial MDN. Other abbreviations: Cg - posterior cingulate cortex; Re - retrosplenial cortex; Ps - post subicular area. This nomenclature is adapted from Rose and Woolsey [1948]. Note that it is somewhat different from that normally used to refer to the architectonic areas of the prefrontal cortex in the rat. B: Diagram of the medial aspect of the rabbit's brain showing the Brodmann numbers assigned to the medial prefrontal cortex based on an anatomical analysis by Vogt, Sikes, Swadlow and Weyland (1986). The striped area is that part of the prefrontal area which receives the projections of the mediodorsal nucleus of the thalamus and thus comprises the medial part of the agranular prefrontal cortex.

Another structure that participates in cardiac conditioning is the anterior cerebellar vermis. Cerebellar vermal lesions greatly attenuate HR conditioning in both rabbits and rats (Supple and Kapp, 1993; Supple and Leaton, 1990). It is not clear why removing only a portion of the autonomic brainstem inputs (i.e., prefrontal and amygdala inputs were still intact in these animals) should virtually abolish the response. Presumably, these different inputs control autonomic responding under varying stimulus conditions, but the exact nature of these conditions has not yet been determined.

The effects of medial prefrontal lesions have also been studied on fear conditioning. Morgan and LeDoux (1995) in rats and Voiuimba et al., (2000) in mice found that lesions of the dorsomedial mPFC (area 32 or 24) enhanced fear reactivity by increasing freezing. However, most of the other work on the mPFC and fear conditioning has focused on the role of the mPFC in fear extinction. Morgan, Romanski and LeDoux (1993) first reported that mPFC lesions delayed fear extinction. However, these lesions were in the ventromedial PFC. i.e., infralimbic (area 25) or ventral prelimbic (area 32) cortex. Such lesions, however, had no effect on initial acquisition of freezing, indicating that the ventromedial PFC is specifically involved in extinction of the CS-induced fear response. However, Gewirtz, Falls and Davis (1997), using both pre- and post-conditioning lesions, found no difference between control and ventral mPFC lesioned rats in extinction, using both the freezing response and startle reflex as indices of the fear CR. Quirk, Russo, Barron and Lebron (2000) also reported that ventral mPFC lesions did not affect acquisition or within trial extinction of CS-induced freezing, but impaired recall of extinction on the next day. In summary, compatible with the data reviewed above, these studies suggest that the more dorsal aspects of the mPFC (i.e., areas 24 and 32) mediate acquisition of non-specific responses, while the ventral portions of area 32 and area 25 are involved in extinction. Thus, it seems clear that different mPFC areas are not only architectonically different, but have separate behavioral functions.

If the mPFC is a necessary CNS structure for the elaboration of conditioned non-specific adjustments, neuronal changes in this area of the brain should be evoked by classical conditioning contingencies. This appears to be the case, as described in several recent publications (Gibbs and Powell, 1988; Gibbs aand Powell, 1991; Gibbs, Prescott and Powell, 1992; Maxwell, Powell and Buchanan, 1994; McLaughlin, Powell and White, 2002; Powell and Ginsberg, 2005). An early study compared mPFC neuronal changes during EB and HR conditioning, using eyeshock as the US in one group of subjects, and airpuff as the US, in a second group (Powell, Maxwell and Penney, 1996). Although EB and NM conditioning occurred in both groups, HR conditioning was not significantly different from a pseudoconditioning control group when the airpuff was the US. Moreover, CS-evoked multiple unit activity (MUA) was considerably greater in the eyeshock than airpuff group, suggesting that the mPFC neuronal discharge associated with classical conditioning may indeed be in support of an emotional component of behavior, assuming that eyeshock is a more aversive US than the airpuff. In addition, recent findings that mPFC lesions attenuate the facilitation of the EB UR produced by Pavlovian conditioning, but have no effect on prepulse inhibition of the EB UR support this contention (McLaughlin, Flaten, Chachich and Powell, 2001), since conditioned EB reflex modification has an affective component, while prepulse inhibition does not (e.g., see Ison and Hoffman, 1983).

Later studies have focused on single unit activity in the mPFC during HR conditioning (Gibbs and Powell, 1988, 1991; Gibbs et al., 1992; Maxwell et al., 1994; McLaughlin et al.,, 2002; Powell and Ginsberg, 2005). These studies all show that both CS-evoked excitatory and inhibitory discharge occurs in the neurons in the mPFC, although some biphasic cells were observed as well. Of some concern in these electrophysiological studies has been the question of whether these CS-evoked changes originate in the mPFC or reflect the action of peripheral end-organ activity carried to forebrain areas by autonomic afferents. It has been known for some time that vagal, as well as sympathetic afferents, reach both areas 24 and 32 of the mPFC (Newman, 1974). A recently completed series of studies (Powell and Ginsberg, 2005) examined this question by studying HR conditioning and associated mPFC single unit activity in animals that were administered atenolol (a peripheral adrenergic beta blockade) or methylscopolamine (peripheral vagal blockade). Unit activity under these conditions was compared to the discharge of the same cells during vehicle sessions. Neither spike amplitude nor CS-evoked activity of over 84 cells was affected by peripheral autonomic blockade, suggesting that this activity is centrally initiated in the mPFC and not the result of peripheral afferent feedback.

Another important aspect of this study was the finding that the discharge of a subset of these cells showed trial-related changes in discharge that was related to acquisition of the HR discrimination between a reinforced CS+ that was paired with the US and non-reinforced CS- that was not. It was not possible in previous studies to draw this conclusion, since recordings in these structures were made during extinction after training was complete. In these studies recordings were made from a stiff fixed electrode that was acutely lowered into the brain. Data collection during extinction was necessary, because it is difficult to record from a single cell over successive trials during acquisition using this procedure due to head movements produced by the presentation of the eyeshock US. However, in a more recent study (i.e., Powell and Ginsberg, 2005), chronic recording of multiunit activity was accomplished via fine wires resting in the brain. Off line separation of single units from such activity can then be obtained over trials so that single cell activity can be associated with the acquisition of the behavioral CR (i.e., the CS-evoked HR changes). This study showed that a subset of mPFC cells show trial-related changes in discharge that were related to the magnitude of the HR CR.

Neurophysiological recording studies also point toward the involvement of the mPFC in fear conditioning and extinction. Milad and Quirk (2002), for example, showed that neurons in the infralimbic cortex were related to extinction memory. Although these cells were not responsive to the CS during conditioning or on the first day of extinction, they showed significant increases in response to CS presentations on day 2 of extinction, suggesting that these cells signal memory for extinction. Moreover, Quirk et al., (2000) found that infralimbic stimulation paired with the CS reduced freezing and accelerated extinction learning. Consequently, it can be concluded from these studies that the dorsomedial mPFC, including areas 32 and 24, are involved in acquisition of fear conditioning, but more ventral prelimbic and infralimbic cortex appear to be involved in extinction. These electrophysiological results are thus compatible with the lesion results discussed previously.

4. Medial Prefrontal Cortex and Pavlovian Conditioning of Specific Somatomotor Responses

4.1. Delay Versus Trace Conditioning

An essential neural substrate for somatomotor, i.e., EB, NM, and leg flexion, conditioning appears to exist in extrapyramidal cerebellar and pontine structures (Thompson, 2005). Lesions of the deep nuclei of the cerebellum and/or perhaps the cerebellar cortex completely prevent EB or NM conditioning from occurring in the ipsilateral eye, without affecting the UR (Thompson, 2005). Damage to the substantia nigra and caudate nucleus also retards, but does not abolish, acquisition of the EB CR (Kao and Powell, 1988; Powell, Mankowski and Buchanan, 1978). However, neither cerebellar nor nigrostriatal lesions affect HR conditioning (Kao and Powell, 1988; Lavond, Lincoln, McCormick and Thompson, 1984; Powell et al., 1978). A variety of other kinds of evidence (e.g., electrophysiological changes) strongly support the contention that an essential neural substrate for somatomotor learning exists in extrapyramidal structures (Thompson, 2005).

It is clear then that separate substrates mediate autonomic and somatomotor conditioning; i.e., much evidence, as reviewed above, suggests that corticolimbic structures mediate learned autonomic adjustments but cerebellar and neostriatal structures mediate somatomotor learning. However, there are interconnections between these two systems, especially in the PFC, providing for possible functional interactions between them (e.g., see Buchanan, Thompson, Maxwell and Powell, 1994; Powell et al., 1990; Powell, McLaughlin and Chachich, 2000). Cortico-limbic structures, including the amygdala, hippocampus and mPFC may also at least partially determine somatomotor response acquisition, even though the essential neural substrate for acquisition and expression of this response is in the cerebellum. Much of the evidence for this model of classical conditioning comes from studies showing impairments of EB conditioning by hippocampectomy in the trace, but not simple, delay conditioning paradigms (Maxwell, Powell and Buchanan, 1994; Reiman, Lane, Ahern, Schwartz and Davidson and Nadel, 2000).

The difference between delay and trace conditioning is schematically illustrated in Figure 11. In the trace paradigm (bottom) there is a brief temporal gap between the termination of the CS and onset of the US. This procedure can be contrasted with the normal delay procedure (top), in which the CS and US either overlap and co-terminate, or US onset occurs simultaneously with CS termination. Acquisition is normally slower using the trace paradigm, and hippocampal lesions differentially impair acquisition during trace conditioning (Weiss and Disterhoft, 1996), but have no effect on or may actually facilitate acquisition in the delay paradigm (see Thompson, 2005 for a recent review). Although the mechanisms for these effects are not clearly understood, it is assumed by most that under ideal conditions, i.e., when the conditioning parameters are optimal for acquisition of the response (i.e., during delay conditioning with relatively short ISIs), limbic system input from the hippocampus and other structures (e.g., mPFC) is not necessary for the memory trace to be carried through the interstimulus interval. However, during trace procedures, in which the CS is not physically present to activate appropriate memory mechanisms, cortico-limbic input to extrapyramidal structures is necessary for the memory to persist through the trace period (Weiss and Disterhoft, 1996). It is possible that the neural substrate for this kind of interaction between extrapyramidal and cortico-limbic structures is through the efferent projections from the PFC to the striatum and possibly pons (e.g., Groenewegen, Berendse and Haber, 1993; Mogensen, Ciriello, Garland and Wu, 1987). Cerebellar inputs to the PFC via the thalamus have also been demonstrated (Middleton and Strick, 1994). A cerebellar-striatal-thalamic-PFC module that controls associative learning is thus a possibility, as has been suggested by others (Houk and Wise, 1995; Weiss and Disterhoft, 1996). Such models suggest that classical conditioning involves multiple distributed brain structures, as discussed in detail above.

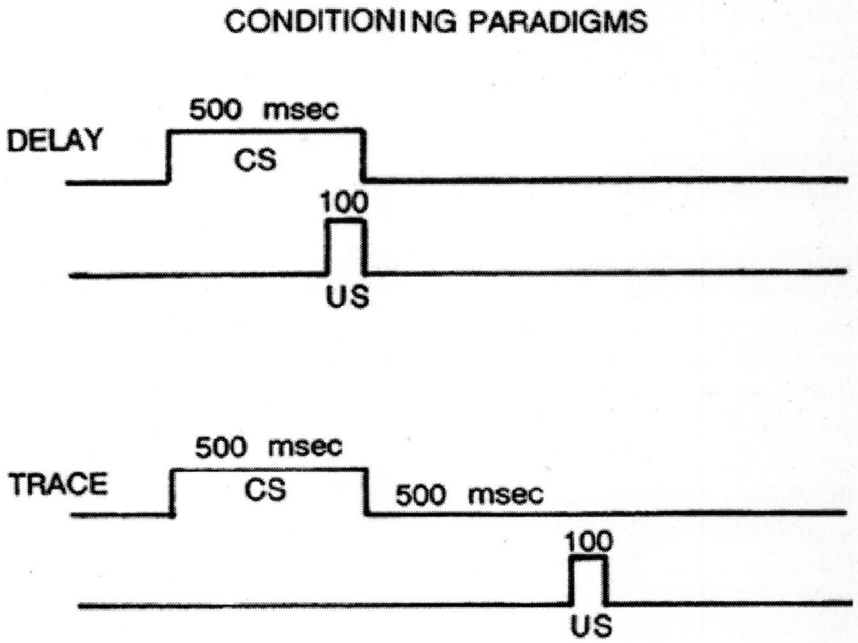

Figure 11. Schematic diagram of delay and trace conditioning. The top two traces illustrate the presentation of a 500 msec conditioned stimulus (CS) which overlaps and coterminates with a 100 msec unconditioned stimulus (US). The bottom two tracings show a 500 msec CS separated from the 100 msec unconditioned stimulus by a 500 msec trace period.

Also supporting such a model are findings showing that EB CR acquisition using other less-than-optimal paradigms is affected by damage to cortico-limbic structures. For example, hippocampal lesions also impair EB performance in discrimination/reversal paradigms (Buchanan and Powell, 1980; Powell and Buchanan, 1980). In this paradigm one CS (CS+) is followed by the US (e.g., an airpuff or periorbital shock) and another CS (CS-) is not; the reversal procedure involves switching the CS+ and CS- after a discrimination between them is attained so that the original CS+ becomes the CS- and vice versa. Typically, in animals with hippocampal lesions, mild impairments in the original differentiation between the CS+ and CS- are obtained, but severe reversal impairments are observed (Buchanan and Powell, 1980). Impairments in a discrimination/reversal paradigm have also been demonstrated subsequent to damage to the mediodorsal nucleus (MDN) of the thalamus, which provides the specific thalamic projections to the mPFC, (Buchanan, 1991). Similar impairments as a result of mPFC damage have also recently been reported (Chachich and Powell, 2004) It was also recently found that damage to the MDN produces a dramatic effect on EB conditioning when other parameters are less than optimal for acquisition, such as when reinforcement schedules are extremely lean (Buchanan, Penney, Tebbutt and Powell, 1997) or when the ISI is longer than is optimal. However, only modest lesion effects are obtained when optimal parameters are employed.

These kinds of findings suggest that although separate substrates mediate the autonomic and somatomotor components of associative learning, there are interactions between them, when conditions are not optimal for acquisition. A series of studies of EB conditioning in human subjects, reported by Squire and colleagues (Clark and Squire, 1998; Clark, Manns and Squire, 2001; Manns, Clark and Squire, 2000) reinforces this conclusion. In a first study, four human subjects with amnesia due to midline diencephalic (viz. hippocampal and/or thalamic) damage, were compared to normal controls on both delay and trace differential conditioning. The amnesic patients showed little evidence of EB conditioning on a trace task with a 250 ms CS and a 1000 ms trace period, but showed normal conditioning on a delay task with a 1000 ms interstimulus interval. Even more interesting, a post-experimental true-false test, designed to determine the extent of the subjects' "conscious awareness" of the experimental contingencies, showed that none of the 4 brain damaged subjects were "aware" of these contingencies. Moreover, several normal subjects also failed to show trace conditioning and were also "unaware" of the contingencies, whereas normal subjects that acquired the trace EB CR were all able to verbalize the contingences, based on the post-experimental test. Both "aware" and "non-aware" normal subjects, however, acquired the delay EB CR. These findings led the authors to conclude that perhaps EB trace conditioning should be classified as a declarative task, since it depends both on "conscious awareness" of the experimental contingencies and intact hippocampal functioning, whereas delay conditioning is non-declarative (viz. procedural), since it was normally acquired by all subjects whether "aware" or "unaware". More importantly, brain damaged subjects acquired delay conditioning normally.

Further, Daum and colleagues have shown that patients with temporal lobe lesions are impaired on a conditioned discrimination task in which a visual stimulus served as a signal for the occurrence of an auditory stimulus that signaled a corneal airpuff (Daum, Channon, Polkey and Gray, 1991), but showed normal acquisition, compared to non-brain damaged controls, on an EB simple discrimination task (Daum, Channon and Gray, 1992). Gabrieli, McGlinchey-Berroth, Carrilo, Gluck, Cermak and Disterhoft (1995) also reported intact delay EB conditioning in amnesic patients, whereas this same group reported impaired trace EB conditioning in temporal lobe damaged amnesic patients (McGlinchey-Berroth, Carrillo, Gabrieli, Braun and Disterhoft, 1997). Several reports, however, have indicated that human subjects with cerebellar damage fail to show normal delay EB conditioning (Daum, Schugens, Ackerman, Lutzenberger, Dichgans and Birbaumer, 1993; Solomon, Stowe and Pendleberry, 1989; Topka, Valls-Sole, Massaquoi and Hallett, 1993), reinforcing the conclusion based on animal studies, described above, that cerebellar structures are necessary for delay EB conditioning.

All human studies reported so far thus support conclusions drawn from non-human animals; viz. that cortico-limbic damage has no effect on EB conditioning using optimal delay conditioning parameters, but severely impairs conditioning when these parameters are not optimal, as during a conditioned discrimination paradigm or during trace conditioning. On the other hand, cerebellar damage impairs both delay and trace conditioning, even under optimal circumstances. These findings thus clearly support the conclusion that simple delay conditioning comprises a non-declarative task, whereas other kinds of EB conditioning may be classified as declarative, based on Squire's definition of these two types of learning and memory (Squire, 2002).

4.2. Studies of Trace Eyeblink Conditioning in Rabbits

As described above, several studies have shown that damage to the hippocampus impairs trace, but not delay EB conditioning in the rabbit (James, Hardiman and Yeo, 1987; Moyer, Deyo and Disterhoft, 1990; Port et al., 1986). Moreover, Chachich, Penney and Powell (1996) showed retrograde labeling of efferents to the mPFC in the dorsal but not ventral subiculum or CA1 or CA2 cells of the hippocampus proper in the rabbit. This experiment thus showed that the PFC of the rabbit, like the rat, cat and primate, receives information based on hippocampal processing. These findings lead to the supposition that prefrontal processing, as a result of hippocampal input, might also be related to somatomotor classical conditioning under non-optimal conditions. Indeed, Disterhoft and colleagues have shown that damage to more posterior mPFC results in deficits in the trace but not delay EB conditioned response in the rabbit (Kronforst-Collins and Disterhoft, 1998; Weible, McEchron and Disterhoft, 2000). However, attempts to replicate these findings have not met with total success. Using similar CS durations to that of Kronforst-Collins and Disterhoft (1998), who used a 100 msec CS, and Weible et al., (2000), who used a 250 msec CS, McLaughlin, Skaggs, Churchwell and Powell (2002) were not able to demonstrate deficits in trace EB conditioning as a result of mPFC damage. However, extending the CS duration to 500 msec resulted in trace but not delay conditioning deficits.

Later experiments have shown that a critical variable in studies of mPFC control of trace EB conditioning is the type of US employed (i.e., eyeshock or airpuff). For example, Oswald, Knuckley, Mahan, Sanders and Powell (2006) compared groups of rabbits with mPFC lesions or sham lesions, half of which were subjected to a periorbital shock US and half a corneal airpuff US. Using a short CS duration (300 msec), similar to that employed in the original Disterhoft studies, these investigators found that, although conditioning deficits were obtained in the animals that received the corneal airpuff US, these deficits were not observed in animals that were administered the periorbital shock US (Oswald et al., 2006). As noted, however, McLaughlin et al., (2002) obtained mPFC-produced deficits in lesioned animals in which eyeshock was the US using a longer 500 msec CS duration. A later experiment thus compared periorbital shock and corneal airpuff in groups of rabbits that received sham or mPFC lesions, but in which both the CS and trace duration was 500 msec. Only thirty acquisition trials were employed, with a 1 min average intertrial interval (ITI), since longer ITIs result in faster learning (e.g., Prokasy, Grant and Myers, 1958). Even when utilizing these extremely optimal learning parameters, however, large differences in EB conditioning performance between animals given periorbital shock and airpuff USs occurred; i.e., lesion induced acquisition reductions were much greater using the corneal airpuff than the periorbital shock US, although the differences between the sham and lesioned groups were statistically significant in both instances. Differences between the shock and airpuff groups were also highly significant ($p<.0001$). In this experiment a pseudoconditioning control group was also employed in which random unpaired presentations of tone and shock occurred. Both lesion and sham groups that received periorbital shock as the US showed significantly higher percent EB CRs than the unpaired groups. Although EB CRs of the sham-airpuff group were also greater than that of the unpaired control group, percent EB CRs of the lesion-airpuff group were not. This finding thus suggests that associative learning failed to occur in the mPFC lesioned animals that received the relative unaversive airpuff US. Consequently, it is clear that mPFC damage retards trace EB conditioning, but its effects are more severe when the relatively unaversive airpuff serves as the US.

It has been reported that, although original differential conditioning is not impaired, reversal conditioning is severely impaired by mPFC lesions (Chachich and Powell, 1999), again an effect similar to hippocampal lesions (Buchanan and Powell, 1980). In a recent experiment the effects of continuous (100%) versus partial (25%) reinforcement on Pavlovian delay and trace EB conditioning was studied in rabbits with either mPFC lesions or sham lesions (Powell, Churchwell and Burriss, 2005). mPFC lesions impaired trace conditioning in both reinforcement conditions. Partial reinforcement retarded EB conditioning in both the trace and delay paradigms, but this impairment was significantly greater during trace conditioning in rabbits with mPFC lesions. These experiments thus clearly demonstrate that damage to the mPFC produces differential impairments in EB conditioning when non-optimal conditioning parameters are employed. These conditions include reversal training, trace conditioning, and partial reinforcement.

4.3. Studies of Post-Training mPFC Lesions on Trace Eyeblink Conditioning

Post-training lesions of the mPFC also produce a dramatic effect on the subsequent performance of the EB response (Powell, Skaggs, Churchwell and McLaughlin, 2001), suggesting that the effects of mPFC damage on EB conditioning involve memory rather than encoding processes. Although a small effect is obtained during delay conditioning, a more severe deficit is produced when the trace conditioning paradigm is employed. Accompanying HR changes, however, are unaffected. This aspect of mPFC control of EB conditioning has been studied in some detail in both rabbits and rats. In a recent experiment (Simon, Knuckley, Churchwell and Powell, 2005) rabbits were trained on trace eyeblink conditioning until they reached a criterion of ten consecutive EB conditioned responses. Electrolytic lesions were then made in the mPFC centered on Brodmann's area 32 at five different time intervals after training. These included immediately, 24 hours, 1 week, 2 weeks, and 1 month after training. Separate groups of animals received sham lesions at the same time intervals after training. After a two week post-operative recovery period, all animals were retested for three days on trace conditioning using the same parameters that were employed during pre-operative training. Substantial EB performance deficits occurred in the animals with mPFC lesions, compared to sham lesioned animals, on the first day of retesting in all five groups. However, by the second or third day of retesting the rabbits with lesions were performing at a level that was comparable to that of sham animals. Animals that received more posterolateral lesions of neocortex, however, did not show post-operative conditioning deficits, suggesting that damage to area 32 is required to produce these deficits. Moreover, a comparison of percentage EB CRs of animals with post-operative damage with that of animals that received mPFC lesions before training, suggests that mPFC post-training lesions produced damage to a retrieval process and not to storage or an acquisition process.

A recently completed study (Oswald, Maddox and Powell, 2006) further supports these conclusions. In this experiment rabbits were trained on trace eyeblink conditioning to criterion of 10 consecutive EB CRs. One week following acquisition, ibotenic acid lesions or sham surgeries were made in the mPFC centered on area 32. Following a one week post-operative recovery period all animals were retrained for 4 consecutive days using the same parameters, given one week off and then retrained for another 4 days. Mean EB conditioning performance deficits in the lesioned group occurred on the first and second days of both 1-week and 2-week post-training periods. However, by the third and fourth day of retraining, like the animals in the earlier study, lesioned animals were performing at a level comparable to that of the sham animals. These findings strongly indicate that the mPFC is involved in retrieval rather than consolidation or storage since lesioned animals were impaired at both retesting times, but were able to relearn the task. Also the fact that ibotenic acid was employed, which destroys only the soma, and not fibers of passage, suggests that intrinsic mPFC cells are involved in this retrieval process.

Similar studies have been reported by Takehara and colleagues in rodents (Takehara, Kawahara, Takatsuki and Kirino, 2002; Takehara, Kawahara and Kiriono, 2003; Takahara-Nishiuchi, Kawahara and Kirino, 2005). In a first experiment, rats received aspiration lesions of the hippocampus, prefrontal cortex, or cerebellum after they had learned a trace EB conditioning task to a criterion. As expected, hippocampal lesions impaired performance only early during post-operative testing. However, lesions of the mPFC produced post-operative performance deficits at one and two weeks or one month following training. Unlike the study in rabbits reported above, however, mPFC lesions in rats appeared to produce increasingly greater performance deficits as a function of length of time after training that the lesion was made. It is difficult to understand why lesions that interfered with more remote memory had the greatest effect in this study, whereas in the study by Simon et al., (2005) this was not the case. However, like the rabbit study, the study by Takehara et al., (2003) in rats showed that mPFC lesions produced their greatest performance deficits on first post-operative testing day, but performance approached that of sham animals by day 3.

Takehara et al., (2003), as well as Christian and Thompson (2003) suggest that mPFC processing of information results in a reorganization of memory processes from temporary storage in the hippocampus to more permanent storage in the mPFC. However, the results of the papers by Simon et al., (2005) and Oswald et al., (2006) suggest that mPFC is important in retrieval and not storage of information. A subsequent study by Takehara-Nishiuchi, Nakao, Kawahara, Matsuki and Kirino (2006) was designed to further investigate this possibility. In this experiment, NMDA receptor blockade in the mPFC was applied during the first or two weeks after initial training, which resulted in marked impairments in memory retention measured 6 weeks after training. However, the same treatment had no effect if performed during the third or fourth weeks following training. The authors suggest that their results indicate that successful establishment of remotely acquired memory requires activation of NMDA receptors in the mPFC at least during the initial week of the post-learning period. The present author suggests, based on a wide range of available data as described above, that permanent storage of the EB CR in both trace and delay conditioning is in the deep nuclei of the cerebellum.

4.4. Studies of Appetitive Conditioning

Recent experiments have shown that damage to the mPFC also impairs JM conditioning (McLaughlin and Powell, 1999), but the HR changes in this case are unaffected. Similar experiments, in which post-training lesions were studied on JM conditioning, also revealed lesion-induced deficits in JM performance, compared to rabbits with sham lesions

(McLaughlin, Powell and White, 2002), and, like the EB experiments, the accompanying CS-evoked HR changes were unaffected by the lesion. These findings are extremely important since they suggest that the mPFC is involved in retrieval of memories necessary for evoking learned somatomotor behaviors, as noted previously (McLaughlin and Powell, 1999; Powell et al., 2002).

Electrophysiological recording studies have also focused on mPFC participation in JM conditioning. In a recent experiment, single unit discharge was assessed in a differential aversive/appetitive conditioning paradigm, in which one CS signaled water, another eyeshock, and a third no US (McLaughlin, Powell and White, 2002). The signal associated with water produced pronounced reliable JM conditioning, while the signal associated with eyeshock produced EB conditioning; somatomotor responses to the third signal were nonassociative. Accompanying CS-evoked single unit activity in area 32 showed a variety of CS-related changes; interestingly some cells responded to the appetitive signal, while a different population of cells responded to the aversive signal. However, several cells responded to both CSs. This experiment thus suggests a heterogeneous population of units in the mPFC that respond to classical conditioning contingencies of both a defensive as well as appetitive nature.

Another interesting model of appetitive conditioning in rodents has been described by Holland (1977, 1984). This investigator determined that there were two types of classically conditioned behavior in rats in appetitive food situations. First, a conditioning-dependent enhancement of orienting responses elicited by the CS prior to conditioning occurs, and second, behaviors which were originally evoked by the food US, but not the CS, also occur. With regard to the latter, both visual and auditory CSs come to elicit behaviors that resemble those elicited by food delivery itself; "rats stand relatively motionless with their heads inserted in the recessed area where food is delivered and/or make short, rapid movements of the head toward the food cup". These behaviors are described by Holland as "US generated" behaviors. However, in addition to these US-elicited behaviors, both auditory and visual CSs also acquire CS-related cues that are specifically oriented toward the CS. Food reinforced visual cues come to elicit rearing of the hind legs and orientation towards a light source, whereas auditory cues produce a conditioned startle response.

Gallagher, Graham and Holland (1990) determined that lesions of the central nucleus of the amygdala abolished CS-generated CRs to both visual and auditory CSs, although the original orienting responses and habituation to CSs were comparable in the lesion and control groups. However, rats with central nucleus lesions readily acquired the US generated CRs. These conditioned behaviors have not been studied as a function of medial prefrontal damage, but Gallagher, McMahan and Schoenbaum (1999) found that orbitofrontal cortex lesions did not abolish the initial acquisition of CS-related behaviors, nor did the lesions abolish taste aversion learning during a second training phase. However, in a test for devaluation of the reinforcer lesioned rats exhibited no change in conditioned responding to the CS, which contrasted with control rats in which devaluation of the US produced a significant decrease in approach to the food cup during presentation of the CS. Although preliminary, the results using this rodent model of appetitive classical conditioning is compatible with the work that has been done previously in the rabbit, as described above.

5. Role of mPFC in Extinction

Also as described above, there has been increased recent interest in the neural mechanisms of extinction of both autonomic and somatomotor conditioned responses. Extinction is defined as a decrease in CR frequency and/or amplitude that results from the removal of the unconditioned stimulus from the contingency; i.e., the CS is presented alone unpaired with the US. Although Pavlov (1927) discovered and used the term extinction in his now famous classical conditioning experiments in dogs, recent research has revealed that extinction procedures are much more complex than early investigators originally thought. It is now accepted that extinction is not simply a decline in responses due to a temporal or "forgetting" process, but is most likely due to acquisition of an active inhibitory process that is itself associative in nature. This idea is very similar to Pavlov's original concept of conditioned inhibition (Pavlov, 1927). Robleto, Poulos and Thompson (2004) recently reviewed possible brain mechanisms involved in extinction of the classically conditioned EB response, and Myers and Davis (2002) have published a similar review of the behavioral and neural analysis of extinction of fear conditioning. It is important to note that the recent interest in extinction has an applied basis, since the experimental analysis of fear conditioning has been viewed as an animal model of the anxiety neuroses (Myers and Davis, 2002). Thus, the neural mechanisms by which fear is inhibited is important information for clinical intervention of patients suffering from fear disregulation.

Myers and Davis (2002) concluded that, on the whole, the search for an inhibitory brain structure has not been very fruitful, as no one structure has emerged that appears to be essential for extinction. A good deal of early work focused on the hippocampus and its septal connections as a possible mechanism underlying extinction (Schmajuk, 1984; Thomas and Gash, 1988). Other studies suggest that the amygdala mediates extinction, as well as acquisition of fear conditioning, as recently reviewed by Kim and Jung (2006). More recent work has focused upon the sensory cortex (Quirk, Armony and LeDoux, 1997). However, again data on the role of the sensory cortex in extinction have not been convincing. More recent research, which shows some promise, however, are studies involving the prefrontal cortex. Quirk et al., (1997) and LeDoux et al., (1989) in their studies of the sensory cortical control of extinction both felt that the sensory cortex likely feeds extinction-related information forward to other cortical areas, such as the PFC. Morgan, Romanski and LeDoux (1993) reported that mPFC lesions had no effect on initial acquisition of fear conditioning, but retarded subsequent extinction to a tone CS. However, as described above, similar work conducted by Gerwitz et al., (1997) and Quirk et al., (2000) found no differences between lesion and control groups on extinction. More recently, however, Quirk et al., (2000) performed an experiment that appears to resolve these inconsistencies. These investigators examined the effects of PFC lesions on acquisition, extinction and recovery of conditioned fear responses over two days. On the first day PFC lesions had no effect on acquisition or extinction of conditioned freezing or suppression of bar pressing. However, on the following day sham rats had recovered only 27% of their acquired freezing, whereas lesioned rats recovered 86%, which was not significantly different from a control group that never received extinction. These data thus suggest that the PFC is involved in consolidation of extinction learning. Even more importantly, Milad and Quirk (2002) found that neurons in the ventral mPFC (i.e., the infralimbic cortex; Brodmann's area 25) showed increased activity, not during acquisition of a bar suppression and freezing response, but during its extinction, thus implicating area 25 in extinction of fear conditioning.

With regard to extinction of the EB CR, Krupa and Thompson (2003) showed that blocking expression of the CR can block extinction. These investigators inactivated the motor nuclei necessary for expression of the EB CR during CS alone extinction training, which resulted in complete prevention of extinction over three to six days. The interpostius nucleus of the cerebellum is also widely regarded to be necessary for acquisition and expression of the EB CR (see above). Hardiman, Ramnani and Yeo (1996) have reported that reversible inactivation of the interpositus nuclear region with muscimol during CS alone extinction training also completely prevented extinction and expression of the CR. These results were replicated by Robelto and Thompson (2003). These data thus seem to implicate the interpositus nucleus in extinction, as well as acquisition of the EB CR, similar to the role that the amygdala plays with regard to fear conditioning, as described above. Another brain stem structure possibly involved in EB extinction is the inferior olive (IO) (Medina et al., 2002; Nelson et al., 1989; Thompson, 2005). Like the study of fear extinction, the hippocampus was shown in earlier studies to be involved in somatomotor extinction, but the hippocampus does not appear to be essential for extinction of the delay CR (e.g., Moyer et al., 1990). There have been no studies of the role of the prefrontal cortex in extinction of somatomotor learning.

6. Pavlovian Conditioning and mPFC Function

Neuronal activity in the PFC is necessary for maintaining information related to ongoing sensory stimulation. O'Reilly (2006) in a recent review points out that mPFC shows sustained neural firing that is robust in the face of potentially distracting information, which allows a focus on the task at hand to the exclusion of all other distracting information, suggesting that the mPFC may be characterized as inhibiting task-irrelevant information. However, an equally important characteristic of mPFC function is that this maintenance of context must be rapidly updated by ongoing changes in stimulation. Such changes are critical for behavioral flexibility, viz. the ability to quickly adapt to the changing demands of the environment. The rapid updating of information in the mPFC is presumed to occur as a result of dynamic gating mechanisms which modulate the stability of mPFC active maintenance (O'Reilly, 2006). O'Reilly suggests that two basic mechanisms exist for carrying out this activity, which involve opening a gate to ongoing mPFC function as a result of a) actions of the neuromodulator dopamine and b) actions of the neostriatum and thalamus, which provide either a "go" or "no go" status to the mPFC.

A major question then is how the effects of mPFC manipulations on Pavlovian conditioning can be integrated with these models. Clearly, occurrence of a CS provides information that requires updating the sensory status of the mPFC. The present findings suggest that this is a multi-stage process, and as noted above, a multi-stage model of Pavlovian conditioning is illustrated in Figure 5. At it simplest and most empirical level, however, three hierarchically arranged and sequentially identifiable stages of information processing are involved in any task that involves associative learning. These include (a) non-specific trace registration associated with attentional processes, such as those underlying orienting and non-associative forms of learning (Stage 1 in Figure 5); (b) the initial registration of associative stimulus significance, as reflected, for example, by the development of non-specific (i.e., visceral) CRs (Stage 2 in Figure 5); and (c) the elaboration of situationally appropriate (i.e., adaptive) somatomotor CRs (stages 3 and 4 of Figure 5. The latter would include for example EB or NM CRs. According to this model changes in the limbic forebrain, including the mPFC, are an integral basis of stage 2 memory traces (e.g., see Buchanan and Powell, 1982a), whereas extrapyramidal (especially cerebellar) plasticity mediates stage 4 memories. Other evidence, as described in Powell and Levine-Bryce (1989), suggest that hypothalamic mechanisms may be sufficient for producing the autonomic changes associated with stage 1 ORs and their habituation.

Information processing theory may provide an heuristic model for understanding mPFC's participation in this process. According to information processing models external stimuli are first processed in a sensory register and those stimuli that are significant are further processed by an attention mechanism and stored in short term or working memory. Working memory has access to least two long term memory registers: i.e., declarative and procedural registers. Declarative memory refers to memory for facts of semantic or episodic nature, and is discussed at some length above. It is thus memory that is available to consciousness, and which therefore can be declared. Procedural memory on the other hand is rule-based memory and contains the habits and skills necessary for dealing with objects both physical and symbolic in one's environment.

Based on such a model, it is suggested that stimuli are selected from the typically defined sensory registers to be placed in working or short term memory if they are (a) novel, or (b) serve as signals for reinforcers (i.e., USs). For this process to function adequately with regard to the second of these principles (i.e., reinforcement) it must sample or in some way come into contact with the current motivational state of the organism. Consequently, there must be another kind of register (or cognitive process) that is sensitive to the current motivational state of the organism and which compares the events in sensory memory with previously occurring events presumably in working memory, to determine if any are signals for reinforcers that are germane to this motivational state. For this reason the present author has suggested that a "reinforcement cue register" is a necessary part of mPFC function (Powell et al., 1990; see Figure 12). The events contained in such a register must come to be there as a consequence of learning. Such learning would consist of classical conditioning, since the events to be compared are signals. Hence, the rehearsal process is an important mechanism for determining not only the contents of long term memory, but also the contents of the reinforcement cue register. In fact the contents of this register might be expected to come directly from working memory based upon the number of previous pairings of the events, i.e., the CS and reinforcer, as well as the current motivational status of the organism. The operation of this register is thus based on an mPFC comparison of events in the sensory register with those in the reinforcement cue register. If a match occurs this event determines the procedure that ultimately produces the desired behavior (also see Gray, 1982).

At a neurophysiological level the occurrence of a specific need must allow the reinforcement cue register selection process access to those neuronal cells or cell systems that have previously signaled the occurrence of reinforcers that met these particular needs. In essence then, the reinforcement cue register may be conceived as providing momentary access to the comparator and working memory in the mPFC to determine whether the current contents of the sensory register match any stimuli that have been highlighted in short term memory by current motivational status as being important. The reinforcement cue register thus would not consist of reinforcing stimuli per se, but rather signals (viz. cues) that previously predicted reinforcers associated with the current motivational status. It would therefore consist of the representations of conditional reinforcers. More importantly, for present purposes it is further suggested that the mPFC plays a major role in the establishment of the secondary reinforcing properties of the cues associated with reinforcers by providing these signals with affective qualities.

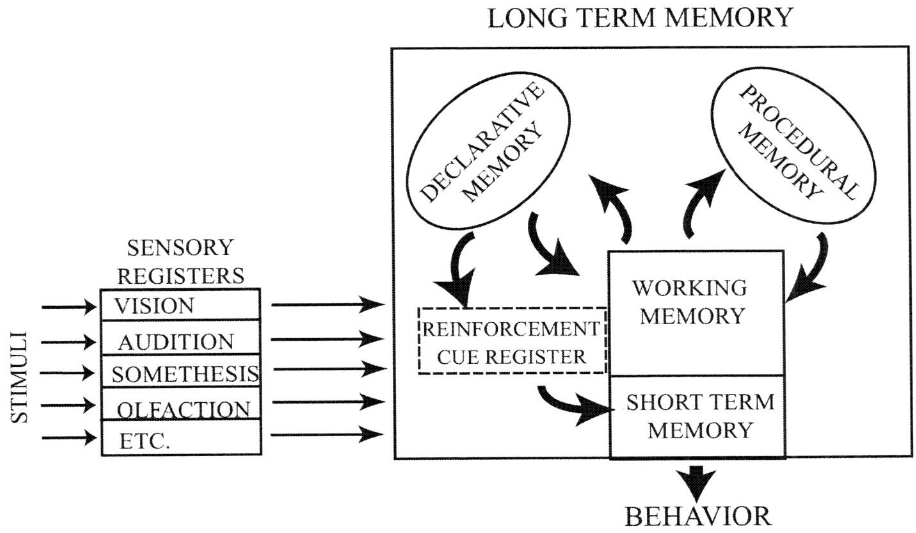

Figure 12. Diagram of brain processing model illustrating the relationship between a postulated "reinforcement cue register" which compares incoming information from sensory systems with elements of working memory, which have been highlighted by current motivational status. A possible role for the mPFC is to associate the signals in such a register with their affective significance. (From Powell, 1989).

It is clear that the affective characteristics of stimuli are not represented physically, (i.e., neurally) in the mPFC, but rather that the mPFC is involved in establishing the importance of such stimuli via a comparison process such as that described above. This is the familiar question of whether acquisition, encoding, storage of information, or retrieval of memory is involved. Considerable evidence suggests that, although such a process might involve mPFC function, no permanent neural representations are stored in the mPFC. Rather the mPFC is involved only in increasing the momentary activation of a subset of neural elements that are simultaneously represented in the sensory registers and in the reinforcement cue register of working memory. The findings of Orona and Gabriel (1983), for example, that MUA in the mPFC reaches a maximum during differential avoidance conditioning long before the behavioral discrimination is made supports this hypothesis. Similarly, CS-evoked multiple unit activity of the mPFC increases initially during acquisition of the cardiac changes associated with Pavlovian conditioning but declines during later trials (Gibbs and Powell, 1988). Also supporting this hypothesis is data described above, which indicates that the essential aspect of this mechanism is associated with retrieval of the CS/US engram from permanent storage in the cerebellum (Simon et al., 2005). Such retrieval would obviously only occur when there is a match of stimuli in the reinforcement cue register with relevant cues in the sensory register.

7. Summary

The prefrontal cortex is a complex heterogeneous structure, which provides for analysis and processing of information related to decision making, based on a variety of information, including the affective properties of stimuli. The present review has focused on this latter aspect of prefrontal functioning and has been limited to the medial prefrontal cortex, since it has been demonstrated that this area of the PFC participates in Pavlovian conditioning of visceral responses. A major point has been made that Pavlovian conditioning consists of a multi-process or set of procedures involving on the one hand non-specific, primarily visceral or autonomic responses, and on the other associative learning underlying conditioning of specific somatomotor responses. An attempt has been made to place the experimental paradigms of various classical conditioning procedures in the context of overall mPFC function. While the emphasis has been on Pavlovian conditioning procedures, it is clear that this is only one aspect of mPFC processing. It is, nevertheless, an important aspect of mPFC functioning, since associative learning is one of the major ways in which new information is acquired and processed so as to produce adaptive somatomotor behaviors. Although much is known about this aspect of mPFC function, obviously considerable questions remain to be answered. One of the purposes of the present review is not only to provide information regarding mPFC control of Pavlovian conditioning, but to raise questions which still need to be answered by future research.

Acknowledgments

Preparation of this paper was supported by DVA Institutional Medical Research funds awarded to the William Jennings Bryan Dorn VA Medical Center. The author thanks Elizabeth Hamel for assistance with manuscript preparation and Andrew Pringle, Jr. for preparation of the illustrations. I am also grateful to Geoff Collier, Barbara Oswald, Jim Appel, and Jack Ginsberg, all of whom read and made insightful comments on an earlier version of this manuscript.

References

Adolphs, R. (1999). The human amygdala and emotion. *Neuroscientist*, 5, 125-137.

Albiniak, B. A., and Powell, D. A. (1980). Peripheral autonomic mechanisms and Pavlovian conditioning in the rabbit (oryctolagus cuniculus). *Journal of Comparative and Physiological Psychology*, 94, 1101-1113.

Bard, P., and Mountcastle, V. B. (1948). Some forebrain mechanisms involved in expression of rage with special reference to suppression of angry behavior. *Research Publications of the Association of Nervous and Mental Diseases*, 27, 362-404.

Blanchard, R. J., Blanchard, D. C., Agullana, R., and Weiss, S. M. (1991). Twenty-two khz alarm cries to presentation of a predator, by laboratory rats living in visible burrow systems. *Physiology and Behavior*, 50, 967-972.

Buchanan, S. L. (1991). Differential and reversal Pavlovian conditioning in rabbits with mediodorsal thalamic lesions: Assessment of heart rate and eyeblink responses. *Experimental Brain Research*, 86, 174-181.

Buchanan, S. L., and Powell, D. A. (1980). Divergencies in Pavlovian conditioned heart rate and eyeblink responses produced by hippocampectomy in the rabbit. *Behavioral and Neural Biology*, 30, 20-38.

Buchanan, S. L., and Powell, D. A. (1982a). Cingulate cortex: Its role in Pavlovian conditioning. *Journal of Comparative and Physiological Psychology*, 96, 755-774.

Buchanan, S. L., and Powell, D. A. (1982b). Cingulate damage attentuates conditioned bradycardia. *Neuroscience Letters*, 29, 261-268.

Buchanan, S. L., and Powell, D. A. (1989). Parasagittal thalamic knife cuts and cardiac changes. *Behavioural Brain Research*, 32, 241-253.

Buchanan, S. L., and Thompson, R. H. (1990). Mediodorsal thalamic lesions and Pavlovian conditioning of heart rate and eyeblink responses in the rabbit. *Behavioral Neuroscience*, 104, 912-918.

Buchanan, S. L., Penney, J., Tebbutt, D., and Powell, D. A. (1997). Lesions of the mediodorsal nucleus of the thalamus and classical eyeblink conditioning under less than optimal stimulus conditions: Role of partial reinforcement and interstimulus interval. *Behavioral Neuroscience*, 111, 1075-1085.

Buchanan, S. L., Thompson, R. H., Maxwell, B. L., and Powell, D. A. (1994). Efferent connections of the medial prefrontal cortex in the rabbit. *Experimental Brain Research*, 100, 469-483.

Buchanan, S. L., Valentine, J. D., and Powell, D. A. (1985). Autonomic responses are elicited by electrical stimulation of the medial but not lateral frontal cortex in rabbits. *Behavioural Brain Research*, 18, 51-62.

Buckner, R. L., and Petersen, S. E. (1996). What does neuroimaging tell us about the role of prefrontal cortex in memory retrieval? *Seminars in the Neurosciences,* 8, 47-55.

Chachich, M. E., and Powell, D. A. (2004). The role of claustrum in Pavlovian heart rate conditioning in the rabbit (oryctolagus cuniculus): Anatomical, electrophysiological, and lesion studies. *Behavioral Neuroscience*, 118, 514-525.

Chachich, M., and Powell, D. A. (1999). Both medial prefrontal and amygdala central nucleus lesions abolish heart rate classical conditioning, but only prefrontal lesions impair reversal of eyeblink differential conditioning. *Neuroscience Letters*, 257, 151-154.

Chachich, M., Buchanan, S., and Powell, D. A. (1997). *Characterization of single-unit activity in the mediodorsal nucleus of the thalamus during expression of differential heart rate conditioning in the rabbit.* 129-141.

Chachich, M., Penney, J., and Powell, D. A. (1996). Subicular lesions disrupt but do not abolish classically conditioned bradycardia in rabbits. *Behavioral Neuroscience*, 110, 707-717.

Christian, K. M., and Thompson, R. F. (2003). Neural substrates of eyeblink conditioning: Acquisition and retention. *Learning and Memory*, 11, 427-455.

Clark, R. E., and Squire, L. R. (1998). Classical conditioning and brain systems: The role of awareness. *Science*, 280, 77-81.

Clark, R. E., Manns, J. R., and Squire, L. R. (2001). Trace and delay eyeblink conditioning: Contrasting phenomena of declarative and nondeclarative memory. *Psychological Science*, 12, 304-308.

Cohen, D. H., and Randall, D. C. (1984). Classical conditioning of cardiovascular responses. *Annual Review of Physiology*, 46, 187-197.

Damasio, A. R. (1994). *Descartes error - emotion, reason and the human brain*. New York, NY: GP Putnam.

Daum, I., Channon, S., and Gray, J. A. (1992). Classical conditioning after temporal lobe lesions in man: Sparing of simple discrimination and extinction. *Behavioural Brain Research*, 52, 159-165.

Daum, I., Channon, S., Polkey, C. E., and Gray, J. A. (1991). Classical conditioning after temporal lobe lesions in man: Impairment in conditional discrimination. *Behavioral Neuroscience*, 105, 396-408.

Daum, I., Schugens, M. M., Ackermann, H., Lutzenberger, W., Dichgans, J., and Birbaumer, N. (1993). Classical conditioning after cerebellar lesions in humans. *Behavioral Neuroscience*, 107, 748-756.

Daum, I., Schugens, M. M., Breitenstein, C., Topka, H., and Spieker, S. (1996). Classical eyeblink conditioning in parkinson's disease. *Movement Disorders*, 11, 639-646.

Davis, M. (1997). Neurobiology of fear responses: The role of the amygdala. *Journal of Neuropsychiatry and Clinical Neurosciences*, 9, 382-402.

Dykman, R. A., Mack, R. L., and Ackerman, P. T. (1965). The evolution of autonomic and motor components of the non-avoidance conditioned response in the dog. *Psychophysiology*, 1, 209-230.

Fanselow, M. S. (1986). Conditioned fear-induced opiate analgesia: A competing motivational state theory of stress analgesia. *Annals of the New York Academy of Sciences*, 476, 40-54.

Fuster, J. M. (1997). *The prefrontal cortex: Anatomy, physiology, and neuropsychology of the frontal lobe*. Philadelphia, PA: Lippincott-Raven.

Gabrieli, J. D. E., McGlinchey-Berroth, R., Carrillo, M. C., Gluck, M. A., Cermak, L. S., and Disterhoft, J. F. (1995). Intact delay-eyeblink classical conditioning in amnesia. *Behavioral Neuroscience*, 109, 819-827.

Gallagher, M., Graham, P. W., and Holland, P. C. (1990). The amygdala central nucleus and appetitive Pavlovian conditioning: Lesions impair one class of conditioned behavior. *The Journal of Neuroscience*, 10, 1906-1911.

Gallagher, M., McMahan, R. W., and Schoenbaum, G. (1999). Orbitofrontal cortex and representation of incentive value in associative learning. *The Journal of Neuroscience*, 19, 6610-6614.

Gewirtz, J. C., Falls, W. A., and Davis, M. (1997). Normal conditioned inhibition and extinction of freezing and fear-potentiated startle following electrolytic lesions of medial prefrontal cortex in rats. *Behavioral Neuroscience*, 111, 712-726.

Gibbs, C. M., and Powell, D. A. (1988). Neuronal correlates of classically conditioned bradycardia in the rabbit: Studies of the medial prefrontal cortex. *Brain Research*, 442, 86-96.

Gibbs, C. M., and Powell, D. A. (1991). Single-unit activity in the dorsomedial prefrontal cortex during the expression of discriminative bradycardia in rabbits. *Behavioural Brain Research*, 43, 79-92.

Gibbs, C. M., Prescott, L. B., and Powell, D. A. (1992). A comparison of multiple-unit activity in the medial prefrontal and agranular insular cortices during Pavlovian heart rate conditioning in rabbits. *Experimental Brain Research*, 89, 599-610.

Goldman-Rakic, P. S., and Selemon, L. D. (1997). Functional and anatomical aspects of prefrontal pathology in schizophrenia. *Schizophrenia Bulletin*, 23, 437-458.

Gormezano, I. (1966). Classical conditioning. In J. B. Sidowski (Ed.), *Experimental methods and instrumentation in psychology* (pp. 385-420). New York NY: McGraw Hill.

Gormezano, I. (1972). *Investigations of defense and reward conditioning in the rabbit.* In A. H. Black and W. F. Prokasy (Eds.), . York, NY: Appleton-Century-Crofts.

Gormezano, I., and Coleman, S. R. (1973). The law of effect and CR contingent modification of the ucs. *Conditional Reflex*, 8, 41-56.

Gray, J. A. (1982). *The neuropsychology of anxiety: An inquiry into the functions of the septo-hippocampal system*. New York NY: Oxford University Press.

Groenewegen, H. J., Berendse, H. W., and Haber, S. N. (1993). Organization of the output of the ventral striatopallidal system in the rat: Ventral pallidal efferents. *Neuroscience*, 57, 113-142.

Hardiman, M. J., Ramnani, N., and Yeo, C. H. (1996). Reversible inactivations of the cerebellum with muscimol prevent the acquisition and extinction of conditioned nictitating membrane responses in the rabbit. *Experimental Brain Research*, 110, 235-247.

Holland, P. C. (1977). Conditioned stimulus as a determinant of the form of the Pavlovian conditioned response. *Journal of Experimental Psychology*, 3, 77-104.

Holland, P. C. (1984). Origins of behavior in Pavlovian conditioning. *Psychology, Learning and Motivation*, 18, 129-174.

Houk, J. C., and Wise, S. P. (1995). Distributed modular architectures linking basal ganglia, cerebellum, and cerebral cortex: Their role in planning and controlling action. *Cerebral Cortex*, 5, 95-110.

Ison, J. R., and Hoffman, H. S. (1983). Reflex modification in the domain of startle: Ii. The anomalous history of a robust and ubiquitous phenomenon. *Psychological Bulletin*, 94, 3-17.

James, G. O., Hardiman, M. J., and Yeo, C. H. (1987). Hippocampal lesions and trace conditioning in the rabbit. *Behavioural Brain Research*, 23, 109-116.

Joseph, J. A., and Powell, D. A. (1980). Peripheral 6-hydroxydopamine administration in the rabbit (oryctolagus cuniculus): Effects on Pavlovian conditioning. *Journal of Comparative and Physiological Psychology*, 94, 1114-1125.

Kaada, B. R. (1951). Somato-motor, autonomic and electrocorticographic responses to electrical stimulation of rhinencephalic and other structures in primates, cat and dog. *Acta Physiologica Scandinavica*, 23, 1-285.

Kao, K. T., and Powell, D. A. (1988). Lesions of the substantia nigra retard Pavlovian eyeblink but not heart rate conditioning in the rabbit. *Behavioral Neuroscience*, 102, 515-525.

Kapp, B. S., Frysinger, R. C., Gallagher, M., and Haselton, J. R. (1979). Amygdala central nucleus lesions: Effect on heart rate conditioning in the rabbit. *Physiology and Behavior*, 23, 1109-1117.

Kapp, B. S., Gallagher, M., Underwood, M. D., McNall, C. L., and Whitehorn, D. (1982). Cardiovascular responses elicited by electrical stimulation of the amygdala central nucleus in the rabbit. *Brain Research*, 234, 251-262.

Kazis, E., Duncan, S., and Powell, D. A. (1973). *Cholinergic and adrenergic control of heart rate changes in the rabbit*. Bulletin of the Psychonomic Society, 3, 41-43.

Kazis, E., Milligan, W. L., and Powell, D. A. (1973). Autonomic-somatic relationships: Blockade of heart rate and corneoretinal potential. *Journal of Comparative Physiological Psychology*, 84, 98-110.

Kehoe, E. J., and Macrae, M. (1994). Classical conditioning of the rabbit nictitating membrane response can be fast or slow: Implications for lennartz and weinberger's (1992) two-factor theory. *Psychobiology*, 22, 1-4.

Kihlstrom, J. F. (1987). The cognitive unconscious. Science, 237, 1445-1452.

Kim, J. J., and Jung, M. W. (2006). Neural circuits and mechanisms involved in Pavlovian fear conditioning: A critical review. *Neuroscience and Biobehavioral Reviews*, 30, 188-202.

Kronforst-Collins, M. A., and Disterhoft, J. F. (1998). Lesions of the caudal area of rabbit medial prefrontal cortex impair trace eyeblink conditioning. *Neurobiology of Learning and Memory*, 69, 147-162.

Krupa, D. J., and Thompson, R. F. (2003). Inhibiting the expression of a classically conditioned behavior prevents its extinction. *Journal of Neuroscience*, 23, 10577-10584.

Lavond, D. G., Lincoln, J. S., McCormick, D. A., and Thompson, R. F. (1984). Effect of bilateral lesions of the dentate and interpositus cerebellar nuclei on conditioning of heart-rate and nictitating membrane/eyelid responses in the rabbit. *Brain Research*, 305, 323-330.

LeDoux, J. E. (2000). Emotion circuits in the brain. *Annual Review of Neuroscience*, 23, 155-184.

LeDoux, J. E., Romanski, L., and Xagoraris, A. (1989). Indelibility of subcortical emotional memories. *Journal of Cognitive Neuroscience*, 1, 238-243.

LeDoux, J. E., Sakaguchi, A., Iwata, J., and Reis, D. J. (1986). Interruption of projections from the medial geniculate body to an arch-neostriatal field disrupts the classical conditioning of emotional responses to acoustic stimuli. *Neuroscience*, 17, 615-627.

Lee, T., and Kim, J. J. (2004). Differential effects of cerebellar, amygdalar, and hippocampal lesions on classical eyeblink conditioning in rats. *The Journal of Neuroscience*, 24, 3242-3250.

Lennartz, R. C., and Weinberger, N. M. (1992). Analysis of response systems in Pavlovian conditioning reveals rapidly versus slowly acquired conditioned responses: Support for two factors, implications for behavior and neurobiology. *Psychobiology*, 20, 93-119.

Lennartz, R. C., and Weinberger, N. M. (1994). A comparison of nonspecific and nictitating membrane conditioned responses: Additional support for two-factor theories. *Psychobiology*, 22, 5-15.

Lofving, B. (1961). Cardiovascular adjustments induced from the rostral cingulate gyrus, with special reference to sympatho-inhibitory mechanisms. *Acta Physiologica Scandinavica*, 53 (Suppl. 184), 1-82.

Manns, J. R., Clark, R. E., and Squire, L. R. (2000). Awareness predicts the magnitude of single-cue trace eyeblink conditioning. *Hippocampus*, 10, 181-186.

Maxwell, B., Powell, D. A., and Buchanan, S. L. (1994). Multiple and single unit activity in area 32 (prelimbic region) of the medial prefrontal cortex during Pavlovian heart rate conditioning in the rabbit. *Cerebral Cortex*, 4, 230-246.

McDonald, A. J. (1991). Organization of amygdaloid projections to the prefrontal cortex and associated striatum in the rat. *Neuroscience*, 44, 1-14.

McGlinchey-Berroth, R., Carrillo, M. C., Gabrieli, J. D. E., Brawn, C. M., and Disterhoft, J. F. (1997). Impaired trace eyeblink conditioning in bilateral, medial-temporal lobe amnesia. *Behavioral Neuroscience*, 111, 873-882.

McLaughlin, J., and Powell, D. A. (1999). Pavlovian heart rate and jaw movement conditioning in the rabbit: Effects of medial prefrontal lesions. *Neurobiology of Learning and Memory*, 71, 150-166.

McLaughlin, J., Flaten, M. A., Chachich, M., and Powell, D. A. (2001). Medial prefrontal lesions attenuate conditioned reflex facilitation but do not affect prepulse modification of the eyeblink reflex in rabbits. *Experimental Brain Research,* 140, 318-325.

McLaughlin, J., Powell, D. A., and White, J. D. (2002). Characterization of the neuronal changes in the medial prefrontal cortex during jaw movement and eyeblink Pavlovian conditioning in the rabbit. *Behavioural Brain Research*, 132, 117-133.

McLaughlin, J., Skaggs, H., Churchwell, J., and Powell, D. A. (2002). Medial prefrontal cortex and Pavlovian conditioning: Trace versus delay conditioning. *Behavioral Neuroscience*, 116, 37-47.

Medina, J. F., Repa, J. C., Mauk, M. D., and LeDoux, J. E. (2002). Parallels between cerebellum- and amygdala-dependent conditioning. *Nature Reviews Neuroscience*, 3, 122-131.

Middleton, F. A., and Strick, P. L. (1994). Anatomical evidence for cerebellar and basal ganglia involvement in higher cognitive function. *Science*, 266, 458-461.

Milad, M. R., and Quirk, G. J. (2002). Neurons in medial prefrontal cortex signal memory for fear extinction. *Nature*, 420, 70-74.

Mogensen, G. J., Ciriello, J., Garland, J., and Wu, M. (1987). Ventral pallidum projections to mediodorsal nucleus of the thalamus: An anatomical and electrophysiological investigation in the rat. *Brain Research*, 404, 221-230.

Morgan, M. A., and LeDoux, J. E. (1995). Differential contribution of dorsal and ventral medial prefrontal cortex to acquisition and extinction of conditioned fear in rats. *Behavioral Neuroscience*, 109, 681-688.

Morgan, M. A., Romanski, L. M., and LeDoux, J. E. (1993). Extinction of emotional learning: Contribution of medial prefrontal cortex. *Neuroscience Letters*, 163, 109-113.

Moyer, J. R., Jr., Deyo, R. A., and Disterhoft, J. F. (1990). Hippocampectomy disrupts trace eye-blink conditioning in rabbits. *Behavioral Neuroscience*, 104, 243-252.

Myers, K. M., and Davis, M. (2002). Behavioral and neural analysis of extinction. *Neuron*, 36, 567-584.

Nelson, B. J., Adams, J. C., Barmack, N. H., and Mugnaini, E. (1989). Comparative study of glutamate decarboxylase immunoreactive boutons in the mammalian inferior olive. *Journal of Comparative Neurology*, 286, 514-539.

Obrist, P. A. (1981). *Cardiovascular psychophysiology: A perspective*. New York, New York: Plenum Press.

Ohman, A. (1988). Nonconscious control of autonomic responses: A role for Pavlovian conditioning? *Biological Psychology*, 27, 113-135.

Olton, D. S. (1979). Mazes, maps, and memory. *American Psychologist*, 34, 583-596.

O'Reilly, R. C. (2006). Biologically based computational models of high-level cognition. *Science*, 314, 91-94.

Orona, E., and Gabriel, M. (1983). Multiple-unit activity of the prefrontal cortex and mediodorsal thalamic nucleus during acquisition of discriminative avoidance behavior in rabbits. *Brain Research*, 263, 295-312.

Oswald, B., Knuckley, B., Mahan, K., Sanders, C., and Powell, D. A. (2006). Prefrontal control of trace versus delay eyeblink conditioning: Role of the unconditioned stimulus. *Behavioral Neuroscience*, in press.

Oswald, B. B., Maddox, S., and Powell, D. A. (2006). Post-training ibotenic acid lesions to the medial prefrontal cortex interfere with retrieval of the Pavlovian conditioned eyeblink response. *Neurobiology of Learning and Memory*, submitted.

Pavlov, I. (1927). *Conditioned reflexes: An investigation of the physiological activity of the cerebral cortex*. New York, NY: GV Anrep (Trans) Oxford University.

Port, R. L., Romano, A. G., and Patterson, M. M. (1986). Stimulus duration discrimination in the rabbit: Effects of hippocampectomy on discrimination and reversal learning. *Physiological Psychology*, 14, 124-129.

Powell, D. A. (1979). Peripheral and central muscarinic cholinergic blockade: Effects on Pavlovian conditioning. *Bulletin of Psychonomic Society*, 14, 161-164.

Powell, D. A., and Buchanan, S. L. (1980). Autonomic-somatic relationships in the rabbit (oryctolagus cuniculus): Effects of hippocampal lesions. *Physiological Psychology*, 8, 455-462.

Powell, D. A., and Churchwell, J. (2002). Mediodorsal thalamic lesions impair trace eyeblink conditioning in the rabbit. *Learning and Memory*, 9, 10-17.

Powell, D. A., and Ginsberg, J. P. (2005). Single unit activity in the medial prefrontal cortex during Pavlovian heart rate conditioning: Effects of peripheral autonomic blockade. *Neurobiology of Learning and Memory*, 84, 200-213.

Powell, D. A., and Kazis, E. (1976). Blood pressure and heart rate changes accompanying classical eyeblink conditioning in the rabbit (oryctolagus cuniculus). *Psychophysiology*, 13, 441-447.

Powell, D. A., and Levine-Bryce, D. (1988). A comparison of two model systems of associative learning: Heart rate and eyeblink conditioning in the rabbit. *Psychophysiology*, 25, 672-682.

Powell, D. A., and Levine-Bryce, D. (1989). Conditioned bradycardia in the rabbit: Effects of knife cuts and ibotenic acid lesions in the lateral hypothalamus. *Experimental Brain Research*, 76, 103-121.

Powell, D. A., and Lipkin, M. (1975). Heart rate changes accompanying differential classical conditioning of somatic responses in the rabbit. *Bulletin of the Psychonomic Society*, 5, 28-30.

Powell, D. A., Buchanan, S. L., and Gibbs, C. M. (1990). Role of the prefrontal-thalamic axis in classical conditioning. In H. B. M. Uylings, C. G. V. Eden, J. P. C. D. Bruin, M. A. Corner and M. G. P. Feenstra (Eds.), *The prefrontal cortex: Its structure function and pathology progress in brain research* (Vol. 85, pp. 433-466). Amsterdam The Netherlands: Elsevier Science Publishers BV.

Powell, D. A., Churchwell, J., and Burriss, L. (2005). Medial prefrontal lesions and Pavlovian eyeblink and heart rate conditioning: Effects of partial reinforcement on delay and trace conditioning in rabbits (oryctolagus cuniculus). *Behavioral Neuroscience*, 119, 180-189.

Powell, D. A., Gibbs, C. M., Maxwell, B., and Levine-Bryce, D. (1993). On the generality of conditioned bradycardia in rabbits: Assessment of CS and US modality. *Animal Learning and Behavior*, 21, 303-313.

Powell, D. A., Lipkin, M., and Milligan, M. L. (1974). Concomitant changes in classically conditioned heart rate and corneoretinal potential discrimination in the rabbit (oryctolagus cuniculus). *Learning and Motivation*, 5, 532-547.

Powell, D. A., Mankowski, D., and Buchanan, S. L. (1978). Concomitant heart rate and corneoretinal potential conditioning in the rabbit (oryctolagus cuniculus): Effects of caudate lesions. *Physiology and Behavior*, 20, 143-150.

Powell, D. A., Maxwell, B. L., and Penney, J. A. (1996). Neuronal activity in the medial prefrontal cortex during Pavlovian eyeblink and nictitating membrane conditioning. *The Journal of Neuroscience*, 16, 6296-6306.

Powell, D. A., McLaughlin, J., and Chachich, M. (2000). Classical conditioning of autonomic and somatomotor responses and their central nervous system substrates. In J. E. Steinmetz and D. S. Woodruff-Pak (Eds.), *Eyeblink classical conditioning: Volume 2 animal models* (pp. 257-286). Boston, MA: Kluwer Academic Publishers.

Powell, D. A., McLaughlin, J., Churchwell, J., Elgarico, T., and Parker, A. (2002). Heart rate changes accompanying jaw movement Pavlovian conditioning in rabbbits: Concomitant blood pressure adjustments and effects of peripheral autonomic blockade. *Integrative Physiological and Behavioral Science*, 37, 215-227.

Powell, D. A., Skaggs, H., Churchwell, J., and McLaughlin, J. (2001). Post-training lesions of the medial prefrontal cortex impair performance of Pavlovian eyeblink conditioning but have no effect on concomitant heart rate changes. *Behavioral Neuroscience*, 115, 1029-1038.

Powell, D. A., Watson, K., and Maxwell, B. (1994). Involvement of subdivisions of the medial prefrontal cortex in learned cardiac adjustments. *Behavioral Neuroscience*, 108, 294-307.

Prokasy, W. F. (1984). Acquisition of skeletal conditioned responses in Pavlovian conditioning. *Psychophysiology*, 21, 1-13.

Prokasy, W. F., Grant, D. A., and Meyers, N. A. (1958). Eyelid conditioning as a function of unconditioned stimulus intensity and intertrial interval. *Journal of Experimental Psychology*, 55, 242-246.

Quirk, G. J., Armony, J. L., and LeDoux, J. E. (1997). Fear conditioning enhances different temporal components of tone-evoked spike trains in auditory cortex and lateral amygdala. *Neuron*, 19, 613-624.

Quirk, G. J., Russo, G. K., Barron, J. L., and Lebron, K. (2000). The role of ventromedial prefrontal cortex in the recovery of extinguished fear. *The Journal of Neuroscience*, 20, 6225-6231.

Rauch, S. L., and Shin, L. M. (1999). Functional neuroimaging studies in posttraumatic stress disorder. *Annals New York Academy of Sciences*, 821, 83-98.

Reiman, E. M., Lane, R. D., Ahern, G. L., Schwartz, G. E., Davidson, R. J., and Nadel, L. (2000). In: *Positron emission tomography in the study of emotion, anxiety, and anxiety disorders* (pp. 389-406). New York, NY: Oxford University Press.

Robleto, K., Poulos, A. M., and Thompson, R. F. (2004). Brain mechanisms of extinction of the classically conditioned eyeblink response. *Learning and Memory,* 11, 517-524.

Robleto, K., and Thompson, R. F. (2003). Effects of muscimol infusions on the extinction of the classically conditioned nictitating membrane response in the rabbit. *2003 Abstract Viewer/Itinerary Planner*, Program No. 87.6, Online.

Rose, J. E., and Woolsey, C. N. (1948). *The orbitofrontal cortex and its connections with the mediodorsal nucleus in rabbits, sheep and cat.* Association for Research in Nervous and Mental Diseases, 27, 210-232.

Schacter, D. L. (1992). Understanding implicit memory - a cognitive neuroscience approach. *American Psychologist*, 47, 559-569.

Schmajuk, N. A. (1984). Psychological theories of hippocampal function. *Physiological Psychology*, 12, 166-183.

Schneiderman, N. (1972). Response system divergencies in aversive classical conditioning. In A. H. Black and W. F. Prokasy (Eds.), *Classical conditioning ii: Current theory and research* (pp. 341-376). New York NY: Appleton-Centry-Crofts.

Schneiderman, N., and Gormezano, I. (1964). Conditioning of the nictitating membrane of the rabbit as a function of cs-us interval. *Journal of Comparative and Physiological Psychology*, 57, 188-195.

Schwaber, J. S., Kapp, B. S., Higgins, G. A., and Rapp, P. R. (1982). Amygdaloid and basal forebrain direct connections with the nucleus of the solitary tract and the dorsal motor nucleus. *Journal of Neuroscience*, 2, 1424-1438.

Simon, B., Knuckley, B., Churchwell, J., and Powell, D. A. (2005). Post-training lesions of the medial prefrontal cortex interfere with subsequent performance of trace eyeblink conditioning. *The Journal of Neuroscience*, 25, 10740-10746.

Smith, C. N., Clark, R. E., Manns, J. R., and Squire, L. R. (2005). Acquisition of differential delay eyeblink classical conditioning is independent of awareness. *Behavioral Neuroscience*, 119, 78-86.

Smith, O. A., and DeVito, J. L. (1984). Central neural integration for the control of autonomic responses associated with emotion. *Annual Review of Neuroscience,* 7, 43-65.

Solomon, P. R., Stowe, G. T., and Pendlbeury, W. W. (1989). Disrupted eyelid conditioning in a patient with damage to cerebellar afferents. *Behavioral Neuroscience,* 103, 898-902.

Squire, L. R. (1992). Memory and the hippocampus: A synthesis from findings with rats, monkeys, and humans. *Psychological Review*, 99, 195-231.

Squire, L. R., Sporns, O., and Tononi, G. E. (1994). In: *Memory and forgetting: Long-term and gradual changes in memory storage* (pp. 243-269). New York, NY: Academic Press.

Supple, W. F., and Leaton, R. N. (1990). Cerebellar vermis: Essential for classically conditioned bradycardia in the rat. *Brain Research*, 509, 17-23.

Supple, W. F., Jr., and Kapp, B. S. (1993). The anterior cerebellar vermis: Essential involvement in classically conditioned bradycardia in the rabbit. *The Journal of Neuroscience*, 13, 3705-3711.

Takehara, K., Kawahara, S., and Kirino, Y. (2003). Time-dependent reorganization of the brain components underlying memory retention in trace eyeblink conditioning. *The Journal of Neuroscience*, 23, 9897-9905.

Takehara, K., Kawahara, S., Takatsuki, K., and Kirino, Y. (2002). Time-limited role of the hippocampus in the memory for trace eyeblink conditioning in mice. *Brain Research*, 951, 183-190.

Takehara-Nishiuchi, K., Kawahara, S., and Kirino, Y. (2005). NMDA receptor-dependent processes in the medial prefrontal cortex are important for acquisition and the early stage of consolidation during trace, but not delay eyeblink conditioning. *Learning and Memory*, 12, 606-614.

Takehara-Nishiuchi, K., Nakao, K., Kawahara, S., Matsuki, N., and Kirino, Y. (2006). Systems consolidation requires postlearning activation of NMDA receptors in the medial prefrontal cortex in trace eyeblink conditioning. *The Journal of Neuroscience*, 26, 5049-5058

Thomas, G. J., and Gash, D. M. (1988). Differential effects of hippocampal ablations on dispositional and representational memory in the rat. *Behavioral Neuroscience*, 102, 635-642.

Thompson, R. F. (2005). In search of memory traces. *Annual Review of Psychology,* 56, 1-23.

Topka, H., Valls-Sole, J., Massaquoi, S. G., and Hallett, M. (1993). Deficit in classical conditioning in patients with cerebellar degeneration. *Brain*, 116, 961-969.

Tulving, A. (1985). How many memory systems are there? *American Psychologist*, 40, 385.

Vogt, B. A., Sikes, R. W., Swadlow, H. A., and Weyand, T. G. (1986). Rabbit cingulate cortex: Cytoarchitecture, physiological border with visual cortex, and afferent cortical connections of visual, motor, postsubicular, and intracingulate origin. *The Journal of Comparative Neurology*, 248, 74-94.

Vouimba, R. M., Garcia, R., Baudry, M., and Thompson, R. F. (2000). Potentiation of conditioned freezing following dorsomedial prefrontal cortex lesions does not interfere with fear reduction in mice. *Behavioral Neuroscience,* 114, 720-724.

Weible, A. P., McEchron, M. D., and Disterhoft, J. F. (2000). Cortical involvement in acquisition and extinction of trace eyeblink conditioning. *Behavioral Neuroscience*, 114, 1058-1067.

Weinberger, N. M., and Diamond, D. M. (1987). Physiological plasticity in auditory cortex: Rapid induction by learning. *Progress in Neurobiology*, 29, 1-55.

Weiss, C., and Disterhoft, J. F. (1996). Eyeblink conditioning, motor control, and the analysis of limbic-cerebellar interactions. *Behavioral and Brain Sciences*, 19, 479-527.

INDEX

A

D

E

F

G

H

I

J

M

N

Q

R

S

U

V

W

Y

Z